Lecture Notes in Physics

The Lecture Notes in Physics

The series Lecture Notes in Physics (LNP), founded in 1969, reports new developments in physics research and teaching – quickly and informally, but with a high quality and the explicit aim to summarize and communicate current knowledge in an accessible way. Books published in this series are conceived as bridging material between advanced graduate textbooks and the forefront of research and to serve three purposes:

- to be a compact and modern up-to-date source of reference on a well-defined topic

- to serve as an accessible introduction to the field to postgraduate students and nonspecialist researchers from related areas

- to be a source of advanced teaching material for specialized seminars, courses and schools

Both monographs and multi-author volumes will be considered for publication. Edited volumes should, however, consist of a very limited number of contributions only. Proceedings will not be considered for LNP.

Volumes published in LNP are disseminated both in print and in electronic formats, the electronic archive being available at springerlink.com. The series content is indexed, abstracted and referenced by many abstracting and information services, bibliographic networks, subscription agencies, library networks, and consortia.

Proposals should be sent to a member of the Editorial Board, or directly to the managing editor at Springer:

Christian Caron
Springer Heidelberg
Physics Editorial Department I
Tiergartenstrasse 17
69121 Heidelberg / Germany
christian.caron@springer.com

T. Belloni (Ed.)

The Jet Paradigm

From Microquasars to Quasars

 Springer

Editor
Tomaso Belloni
INAF
Osservatorio Astronomico di
Brera
Via E. Bianchi, 46
23807 Merate
Italy
tomaso.belloni@brera.inaf.it

Belloni, T. (Ed.), *The Jet Paradigm: From Microquasars to Quasars*, Lect. Notes Phys. 794 (Springer, Berlin Heidelberg 2010), DOI 10.1007/978-3-540-76937-8

Lecture Notes in Physics ISSN 0075-8450   e-ISSN 1616-6361
ISBN 978-3-540-76936-1   e-ISBN 978-3-540-76937-8
DOI 10.1007/978-3-540-76937-8
Springer Heidelberg Dordrecht London New York

Library of Congress Control Number: 2009939070

Cover design: Integra Software Services Pvt. Ltd., Pondicherry

Printed on acid-free paper

Springer is part of Springer Science+Business Media (www.springer.com)

To my father: a glimpse of what lies behind the sky

Preface

The discovery of the first case of superluminal radio jets in our galaxy in 1994 from the bright and peculiar X-ray source GRS 1915+105 has opened the way to a major shift in the direction of studies of stellar-mass accreting binaries. The past decade has seen an impressive increase in multi-wavelength studies. It is now known that all black hole binaries in our galaxy are radio sources and most likely their radio emission originates from a powerful jet. In addition to the spectacular events related to the ejection of superluminal jets, steady jets are known from many systems. Compared with their supermassive cousins, the nuclei of active galaxies, stellar-mass X-ray binaries have the advantage of varying on time scales accessible within a human life (sometimes even much shorter than a second). This has led to the first detailed studies of the relation between accretion and ejection. It is even possible that, excluding their "soft" periods, the majority of the power in galactic sources lies in the jets and not in the accretion flows. This means that until a few years ago we were struggling with a physical problem, accretion onto compact objects, without considering one of the most important components of the system. Models that associate part of the high-energy emission and even the fast aperiodic variability to the jet itself are now being proposed and jets can no longer be ignored. At the same time, the study of the accretion/ejection connection in Active Galactic Nuclei is progressing and is being linked to the properties of small-scale systems in an attempt to find simple scaling laws that span several orders of magnitude in black hole mass. These laws are expected on basic theoretical grounds and are being found. Even neutron-star binaries are included in these unified views. In this sense, the expression "microquasar," which contains the connection to supermassive systems, appears more and more to be a synonym of X-ray binary.

Fifteen years after the discovery of the first galactic superluminal jets, the field has grown in complexity and is now spanning about eight orders of magnitude in compact-object mass. In particular, the two communities of Active Galactic Nuclei and galactic binaries are still largely orthogonal. While stellar-mass systems allow to follow the accretion/ejection connection at work in time, galactic nuclei provide a much larger sample of objects, allowing significant statistical analysis. Moreover, it is true that the variations seen in binaries are largely unaccessible to AGN observers, but on the contrary the variability observed in AGN is too fast to be observed in binaries with good accuracy. Moving up in mass means shifting the observational window to faster phenomena. Clearly, a view that encompasses all types of objects offers much more potential to understand the phenomenon which, close to the central object, is expected to be basically the same, after a suitable scaling is taken into account.

This book is aimed at covering an ample spectrum of topics and sources: from stellar-mass systems to galactic nuclei, from observational properties both in the radio and in the X-rays to theoretical models and simulations. The reader should not feel biased by the existence of separate communities and is presented a broad view of the current status of research. While there is a plethora of details which would have been impossible to cover here, the picture that is presented should provide the tools to navigate through them without losing track of one's whereabouts.

New observational facilities will become available in the near future, both ground-based radio and optical/infrared telescopes and space missions from the optical to gamma rays. The multi-wavelength coverage offered by their combination will allow us to explore in detail phenomena from both microquasars and quasars, but at the same time it will be important to have barriers as weak as possible between different classes of sources.

The book is structured as follows. Chapter 1 is an introduction to microquasars, Chaps. 2 through 5 present the current picture on the X-ray (Chaps. 2–3) and radio (4–5) emission from galactic binaries. Chapter 6 introduces jet models for the emission. Chapters 7 and 8 are on Active Galactic Nuclei. Finally, theoretical models and MHD simulations can be found in Chaps. 9 and 10.

A few words of thanks. To Springer Editor Ramon Khanna for his patience in supporting my struggle with authors repeatedly crossing deadlines. To the International Space Science Institute (ISSI) of Bern, which provided the ideal setting for discussions that have contributed to many chapters of this book. Finally, to Alice and Ilaria, who have supported me throughout this long editing process and have sacrificed some family time to allow me to finish it.

Merate, Italy,   Tomaso M. Belloni
April 2009

Contents

Chapter 1
Microquasars: Summary and Outlook

I.F. Mirabel

Abstract Microquasars are compact objects (stellar-mass black holes and neutron stars) that mimic, on a smaller scale, many of the phenomena seen in quasars. Their discovery provided new insights into the physics of relativistic jets observed elsewhere in the Universe, and in particular, the accretion–jet coupling in black holes. Microquasars are opening new horizons for the understanding of ultraluminous X-ray sources observed in external galaxies, gamma-ray bursts of long duration, and the origin of stellar black holes and neutron stars. Microquasars are one of the best laboratories to probe General Relativity in the limit of the strongest gravitational fields, and as such, have become an area of topical interest for both high energy physics and astrophysics. At present, back hole astrophysics exhibits historical and epistemological similarities with the origins of stellar astrophysics in the last century.

1.1 Introduction

Microquasars are binary stellar systems where the remnant of a star that has collapsed to form a dark and compact object (such as a neutron star or a black hole) is gravitationally linked to a star that still produces light, and around which it makes a closed orbital movement. In this cosmic dance of a dead star with a living one, the first sucks matter from the second, producing radiation and very high energetic particles (Fig. 1.1). These binary star systems in our galaxy are known under the name of "microquasars" because they are miniature versions of the quasars ("quasi-stellar radio source"), which are the nuclei of distant galaxies harboring a supermassive black hole, and are able to produce in a region as compact as the solar system, the luminosity of 100 galaxies like the Milky Way. Nowadays, the study of microquasars is one of the main scientific motivations of the space observatories that probe the X-ray and γ-ray Universe.

I.F. Mirabel (✉)

Laboratoire AIM, Irfu/Service d'Astrophysique, Bat. 709, CEA-Saclay, 91191 Gif-sur-Yvette Cedex, France and Instituto de Astronomía y Física del Espacio (IAFE), CC 67, Suc. 28, 1428 Buenos Aires, Argentina, felix.mirabel@cea.fr

Mirabel, I.F.: *Microquasars: Summary and Outlook*. Lect. Notes Phys. **794**, 1–15 (2010)

DOI 10.1007/978-3-540-76937-8_1   © Springer-Verlag Berlin Heidelberg 2010

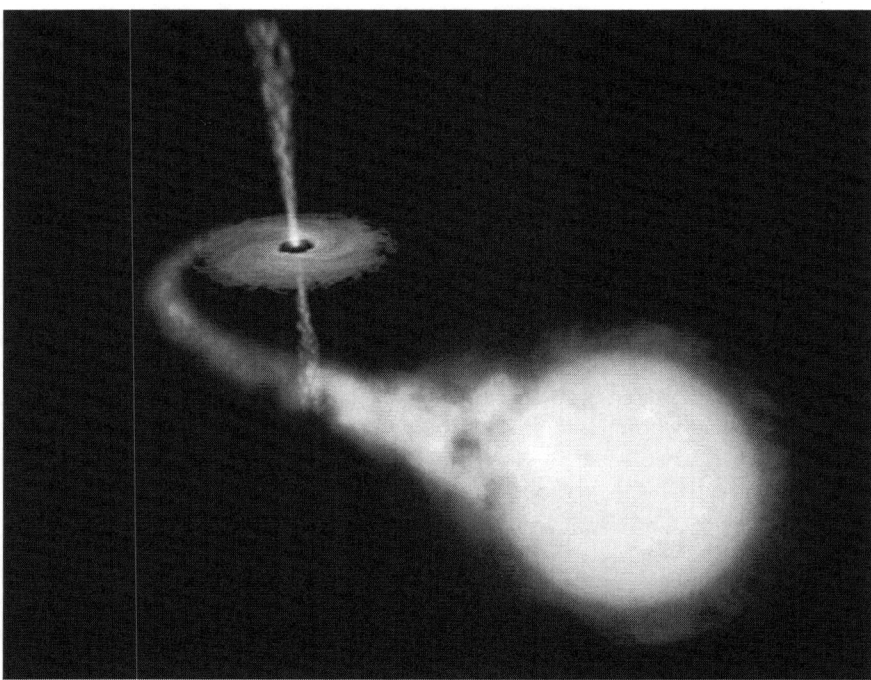

Fig. 1.1 In our galaxy there exist binary stellar systems where an ordinary star gravitates around a black hole that sucks the outer layers of the star's atmosphere. When falling out to the dense star, the matter warms and emits huge amounts of energy as X-rays and γ-rays. The accretion disk that emits this radiation also produces relativistic plasma jets all along the axis of rotation of the black hole. The physical mechanisms of accretion and ejection of matter are similar to those found in quasars, but in million times smaller scales. Those miniature versions of quasars are known under the name of "microquasars"

Despite the differences in the involved masses and in the time and length scales, the physical processes in microquasars are similar to those found in quasars. That is why the study of microquasars in our galaxy has enabled a better understanding of what happens in the distant quasars and AGN. Moreover, the study of microquasars may provide clues for the understanding of the class of gamma-ray bursts that are associated with the collapse of massive stars leading to the formation of stellar black holes, which are the most energetic phenomena in the Universe after the Big Bang.

1.2 Discovery of Microquasars

During the second half of the 18th century, John Michell and Pierre-Simon Laplace first imagined compact and dark objects in the context of the classical concept of gravitation. In the 20th century in the context of Einstein's General Relativity theory of gravitation, those compact and dark objects were named black holes. They were then identified in the sky in the 1960s as X-ray binaries. Indeed, those compact objects, when associated with other stars, are activated by the accretion of very hot gas that emits X-rays and γ-rays. In 2002, Riccardo Giacconi was awarded the Nobel Prize for the development of the X-ray Space Astronomy that led to the discovery of the first X-ray binaries [12]. Later, Margon et al. [18] found that a compact binary known as SS 433 was able to produce jets of matter. However, for a long time, people believed that SS 433 was a very rare object of the Milky Way and its relation with quasars was not clear since the jets of this object move only at 26% of the speed of light, whereas the jets of quasars can move at speeds close to the speed of light.

In the 1990s, after the launch of the Franco–Soviet satellite GRANAT, growing evidences of the relation between relativistic jets and X-ray binaries began to appear. The onboard telescope SIGMA was able to take X-ray and γ-ray images. It detected numerous black holes in the Milky Way. Moreover, thanks to the coded-mask-optics, it became possible for the first time to determine the position of γ-ray sources with arcmin precision. This is not a very high precision for astronomers who are used to dealing with other observing techniques. However, in high energy astrophysics it represented a gain of at least one order of magnitude. It consequently made possible the systematic identification of compact γ-ray sources at radio, infrared, and visible wavelengths.

With SIGMA/GRANAT it was possible to localize with an unprecedented precision the hard X-ray and γ-ray sources. In order to determine the nature of those X-ray binaries, a precision of a few tens of arc-seconds was needed. Sources that produce high energy photons should also produce high energy particles, which should then produce synchrotron radiation when accelerated in magnetic fields. Then, with Luis Felipe Rodríguez, we performed a systematic search of synchrotron emissions from X-ray binaries with the Very Large Array (VLA) of the National Radio Astronomy Observatory of the USA.

In 1992, using quasi-simultaneous observations from space with GRANAT and from the ground with the VLA, we determined the position of the radio counterpart of an X-ray source named 1E 1740.7-2942 with a precision of sub-arc-seconds. With GRANAT this object was identified as the most luminous, persistent source of soft γ-rays in the Galactic center region. Moreover, its luminosity, variability, and the X-ray spectrum were consistent with those of an accretion disk gravitating around a stellar-mass black hole, like in Cygnus X-1. The most surprising finding with the VLA was the existence of well-collimated two-sided jets that seem to arise from the compact radio counterpart of the X-ray source [22]. These jets of magnetized plasma had the same morphology as the jets observed in quasars and radio galaxies. When we published those results, we employed the term microquasar to define this new X-ray source with relativistic jets in our galaxy. This term appeared on the front page of the British journal *Nature* (see Fig. 1.2), which provoked multiple debates. Today the concept of microquasar is universally accepted and used widely in scientific publications.

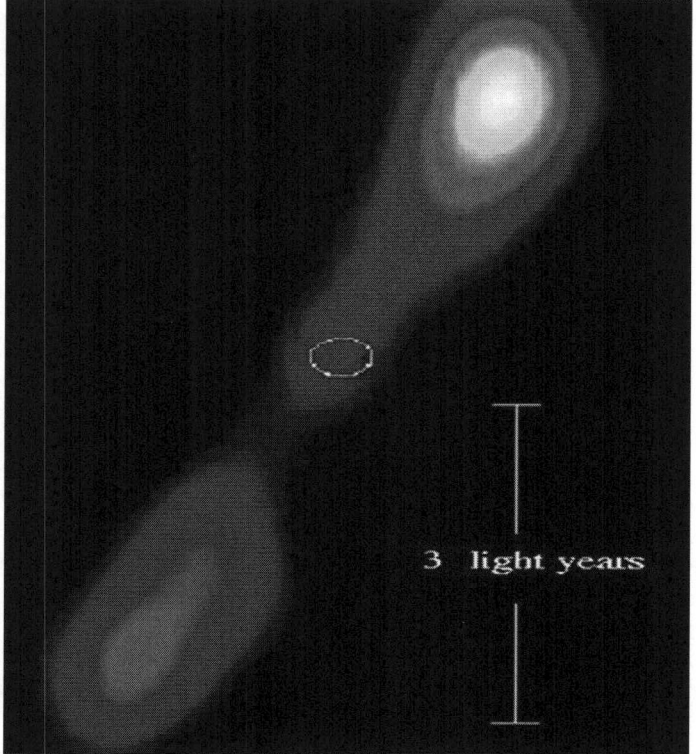

Fig. 1.2 The British journal *Nature* announced on July 16, 1992 the discovery of a microquasar in the Galactic center region [22]. The image shows the synchrotron emission at a radio wavelength of 6 cm produced by relativistic particles jets ejected from some tens of kilometers to light years of distance from the black hole binary which is located inside the small white ellipse

Before the discovery of its radio counterpart, 1E 1740.7-2942 was suspected to be a prominent source of 511 keV electron–positron annihilation radiation observed from the center of our galaxy [17], and for that reason it was nicknamed as the "Great Annihilator". It is interesting that recently it was reported [40] that the distribution in the Galactic disk of the 511 keV emission, due to positron–electron annihilation, exhibits similar asymmetric distribution as that of the hard low mass X-ray binaries, where the compact objects are believed to be stellar black holes. This finding suggests that black hole binaries may be important sources of positrons that would annihilate with electrons in the interstellar medium. Therefore, positron–electron pairs may be produced by γ–γ photon interactions in the inner accretion disks, and microquasar jets would contain positrons as well as electrons. If this recent report is confirmed, 1E 1740.7-2942 would be the most prominent compact source of anti-matter in the Galactic center region.

1.3 Discovery of Superluminal Motions

If the proposed analogy [25] between microquasars and quasars was correct, it should be possible to observe superluminal apparent motions in Galactic sources. However, superluminal apparent motions had been observed only in the neighborhood of super massive black holes in quasars. In 1E 1740.7–2942 we would not be able to discern motions, as in that persistent source of γ-rays the flow of particles is semi-continuous. The only possibility of knowing if superluminal apparent movements exist in microquasars was through the observation of a discrete and very intense ejection in an X-ray binary. This would allow us to follow the displacement in the firmament of discrete plasma clouds. Indeed, with the GRANAT satellite was discovered [3] a new source of X-rays with such characteristics denominated GRS 1915+105. Then with Rodríguez, we began with the VLA a systematic campaign of observations of that new object in the radio domain, and in collaboration with Pierre-Alain Duc (CNRS-France) and Sylvain Chaty (Paris University) we performed the follow-up of this source in the infrared with telescopes of the Southern European Observatory and telescopes at Mauna Kea, Hawaii.

Since the beginning, GRS 1915+105 exhibited unusual properties. The observations in the optical and the infrared showed that this X-ray binary was well absorbed by the interstellar dust along the line of sight in the Milky Way and that the infrared counterpart was varying rapidly as a function of time. Moreover, the radio counterpart seemed to change its position in the sky, so that at the beginning we did not know if those changes were due to radiation, reflection, or refraction in an inhomogeneous circumstellar medium ("Christmas tree effect"), or rather due to the movement at very high speeds of jets of matter. For two years we kept on watching this X-ray binary without exactly understanding its behavior. However, in March 1994, GRS 1915+105 produced a violent eruption of X-rays and γ-rays, followed by a bipolar ejection of unusually bright plasma clouds, whose displacement in the sky could be followed during two months. From the amount of atomic hydrogen absorbed in the strong continuum radiation we could infer that the X-ray binary stands at about 30,000 light-years from the Earth. This enabled us to know that the movement of the ejected clouds in the sky implies apparent speeds higher than the speed of light.

The discovery of these superluminal apparent movements in the Milky Way was announced in *Nature* [23] (Fig. 1.3). This constituted a full confirmation of the hypothesis, which we had proposed two years before, on the analogy between microquasars and quasars. With Rodríguez, we formulated and solved the system of equations that described the observed phenomenon. The apparent asymmetries in the brightness and the displacement of the two plasma clouds could naturally be explained in terms of the relativistic aberration in the radiation of twin plasma clouds ejected in an antisymmetric way at 98% of the speed of light [26]. The superluminal motions observed in 1994 with the VLA [23] were a few years later reobserved with higher angular resolution using the MERLIN array [10].

Using the Very Large Telescope of the European Southern Observatory, it was possible to determine the orbital parameters of GRS 1915+105, concluding that it

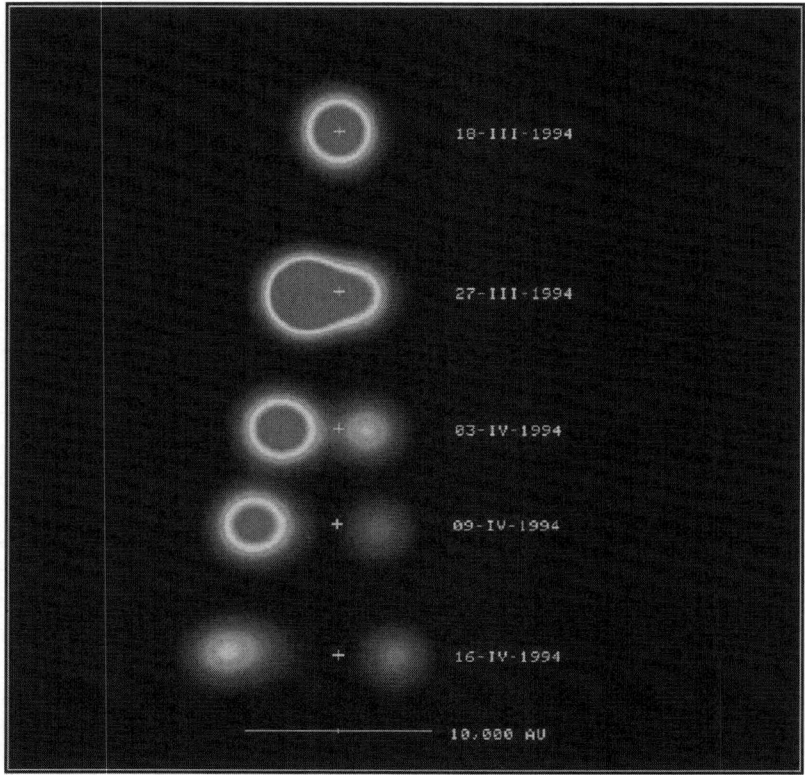

Fig. 1.3 The journal *Nature* announced on September 1, 1994 the discovery of the first Galactic source of superluminal apparent motions [23]. The sequence of images shows the temporal evolution in radio waves at a wavelength of 3.6 cm of a pair of plasma clouds ejected from black hole surroundings at a velocity of 98% the speed of the light

is a binary system constituted by a black hole of ∼14 solar masses accompanied by a star of 1 solar mass [13]. The latter has become a red giant from which the black hole sucks matter under the form of an accretion disk (see Fig. 1.1).

1.4 Disk-Jet Coupling in Microquasars

The association of bipolar jets and accretion disks seems to be a universal phenomenon in quasars and microquasars. The predominant idea is that matter jets are driven by the enormous rotation energy of the compact object and accretion disk that surrounds it. Through magneto-hydrodynamic mechanisms, the rotation energy is evacuated through the poles by means of jets, as the rest can fall toward the gravitational attraction center. In spite of the apparent universality of this relationship

between accretion disks and bipolar, highly collimated jets, the temporal sequence
of the phenomena had never been observed in real time.

Since the time scales of the phenomena around black holes are proportional
to their mass, the accretion–ejection coupling in stellar-mass black holes can be
observed in intervals of time that are millions of time smaller than in AGN and
quasars. Because of the proximity, the frequency, and the rapid variability of energetic eruptions, GRS 1915+105 became the most adequate object to study the connection between instabilities in the accretion disks and the genesis of bipolar jets.

After several attempts, finally in 1997 we could observe [24], on an interval of
time shorter than an hour, a sudden fall in the luminosity in X-rays and soft γ-rays,
followed by the ejection of jets, first observed in the infrared, then at radio frequencies (see Fig. 1.4). The abrupt fall in X-ray luminosity could be interpreted as
the silent disappearance of the warmer inner part of the accretion disk beyond the
horizon of the black hole. A few minutes later, fresh matter coming from the companion star comes to feed again the accretion disk, which must evacuate part of its
kinetic energy under the form of bipolar jets. When moving away, the plasma clouds

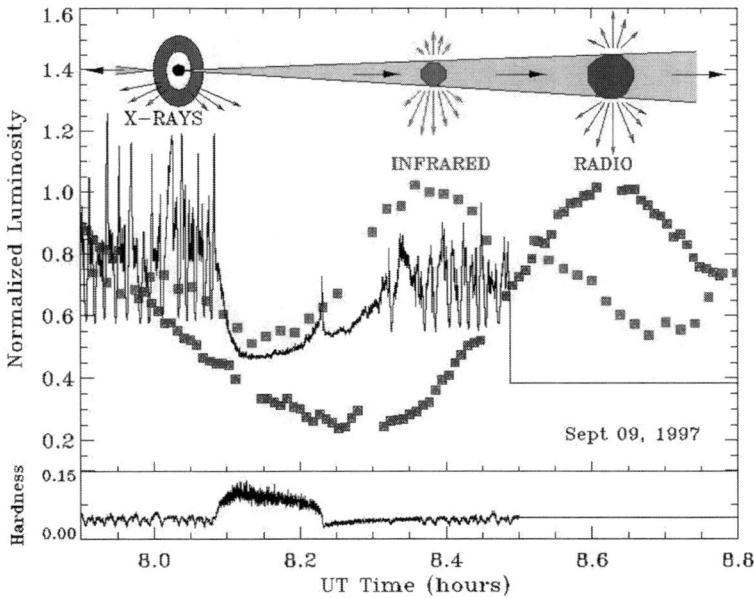

Fig. 1.4 Temporal sequence of accretion disk–jet coupling observed for the first time in real
time simultaneously in the X-rays, the infrared, and radio wavelengths in the microquasar
GRS 1915+105 [24]. The ejection of relativistic jets takes place after the evacuation and/or dissipation of matter and energy, at the time of the reconstruction of the inner side of the accretion disk,
corona, or base of the jet. A similar process has been observed years later in quasars [19], but on
timescales of years. As expected in the context of the analogy between quasars and microquasars
[25], the timescale of physical processes in the surroundings of black holes is proportional to their
masses

expand adiabatically, becoming more transparent to their own radiation, first in the infrared and then in the radio frequency. The observed interval of time between the infrared and the radio peaks is consistent with that predicted by van der Laan [39] for extragalactic radio sources.

Based on the observations of GRS 1915+105 and other X-ray binaries, it was proposed [11] a unified semiquantitative model for disk–jet coupling in black hole X-ray binary systems that relate different X-ray states with radio states, including the compact, steady jets associated with low-hard X-ray states, that had been imaged [5] using the Very Long Baseline Array of the National Radio Astronomy Observatory.

After 3 years of multi-wavelength monitoring an analogous sequence of X-ray emission dips followed by the ejection of bright superluminal knots in radio jets was reported [19] in the active galactic nucleus of the galaxy 3C 120. The mean time between X-ray dips was of the order of years, as expected from scaling with the mass of the black hole.

1.5 Can We Prove the Existence of Black Holes?

Horizon is the basic concept that defines a black hole: a massive object that consequently produces a gravitational attraction in the surrounding environment, but that has no material border. In fact, an invisible border in the space–time, which is predicted by general relativity, surrounds it. This way, matter could go through this border without being rejected, and without losing a fraction of its kinetic energy in a thermonuclear explosion, as sometimes is observed as X-ray bursts of type I when the compact object is a neutron star instead of a black hole. In fact, as shown in Fig. 1.4, the interval of time between the sudden drop of the flux and the spike in the X-ray light curve that marks the onset of the jet, signaled by the starting rise of the infrared synchrotron emission, is of a few minutes, orders of magnitude larger than the dynamical time of the plasma in the inner accretion disk. Although the drop of the X-ray luminosity could be interpreted as dissipation of matter and energy, the most popular interpretation is that the hot gas that was producing the X-ray emission falls into the black hole, leaving the observable Universe.

So, have we proved with such observations the existence of black holes? Indeed, we do not find any evidence of material borders around the compact object that creates gravitational attraction. However, the fact that we do not find any evidence for the existence of a material surface does not imply that it does not exist. In fact, such type I X-ray bursts are only observed in certain range of neutron star mass accretion rates. That means that it is not possible to prove the existence of black holes using the horizon definition. According to Saint Paul, *faith is the substance of hope for: the evidence of the not seen.* That is why for some physicists black holes are just objects of faith. Perhaps the intellectual attraction of these objects comes from the desire of discovering the limits of the Universe. In this context, studying the physical phenomena near the horizon of a black hole is a way of approaching the ultimate frontiers of the observable Universe.

1.6 The Rotation of Black Holes

For an external observer, black holes are the simplest objects in physics since they can be fully described by only three parameters: mass, rotation, and charge. Although black holes could be born with net electrical charge, it is believed that because of interaction with environmental matter, astrophysical black holes rapidly become electrically neutral. The masses of black holes gravitating in binary systems can be estimated with Newtonian physics. However, the rotation is much more difficult to estimate despite it being probably the main driver in the production of relativistic jets.

There is now the possibility of measuring the rotation of black holes by at least three different methods: (a) X-ray continuum fitting [41, 20]; (b) asymmetry of the broad component of the Fe K_α line from the inner accretion disk [38]; and (c) quasi-periodic oscillations with a maximum fixed frequency observed in the X-rays [35]. The main source of errors in the estimates of the angular momentum resides in the uncertainties of the methods employed.

The side of the accretion disk that is closer to the black hole is hotter and produces huge amounts of thermal X and γ radiations and is also affected by the strange configuration of space–time. Indeed, next to the black hole, space–time is curved by the black hole mass and dragged by its rotation. This produces vibrations that modulate the X-ray emission. Studies of those X-ray continuum and vibrations suggest that the microquasars that produce the most powerful jets are indeed those that are rotating fastest. It has been proposed that these pseudo-periodic oscillations in microquasars are, moreover, one of the best methods today to probe General Relativity theory in the limit of the strongest gravitational fields by means of observations.

Analogous oscillations in the infrared range may have been observed in the supermassive black hole at the center of the Milky Way. The quasi-periods of the oscillations (a few milliseconds for the microquasars X-ray emission and a few tens of minutes for the Galactic center black hole infrared emission) are proportionally related to the masses of the objects, as expected from the physical analogy between quasars and microquasars. Comparing the phenomenology observed in microquasars to that in black holes of all mass scales, several correlations among observables such as among the radiated fluxes in the low-hard X-ray state, quasiperiodic oscillations, and flickering frequencies, are being found and used to derive the mass and angular momentum, which are the fundamental parameters that describe astrophysical black holes.

1.7 Extragalactic Microquasars, Microblazars, and Ultraluminous X-Ray Sources

Have microquasars been observed beyond the Milky Way galaxy? X-ray satellites are detecting far away from the centers of external galaxies large numbers of compact sources called "ultraluminous X-ray sources", because their luminosities seem to be greater than the Eddington limit for a stellar-mass black hole [9]. Although a few of these sources could be black holes of intermediate masses of hundreds to thousands solar masses, it is believed that the large majority are stellar-mass black hole binaries.

Since the discovery of quasars in 1963, it was known that some quasars could be extremely bright and produce high energetic emissions in a short time. These particular quasars are called blazars and it is thought that they are simply quasars whose jets point close to the Earth's direction. The Doppler effect produces thus an amplification of the signal and a shift into higher frequencies. With Rodríguez, we imagined in 1999 the existence of microblazars, that is to say X-ray binaries where the emission is also in the Earth's direction [26]. Microblazars may have been already observed but the fast variations caused by the contraction of the timescale in the relativistic jets make their study very difficult. In fact, one question at the time of writing this chapter is whether microblazars could have been already detected as "fast black hole X-ray novae" [14]. In fact, the so-called fast black hole X-ray novae Swift J195509.6+261406 (which is the possible source of GRB 070610 [14]), and V4541 Sgr [31] are compact binaries that appeared as high energy sources with fast and intense variations of flux, as expected in microblazars [26].

Although some fast variable ultraluminous X-ray sources could be microblazars, the vast majority do not exhibit the intense, fast variations of flux expected in relativistic beaming. Therefore, it has been proposed [15] that the large majority are stellar black hole binaries where the X-ray radiation is – as the particle outflows – anisotropic, but not necessarily relativistically boosted. In fact, the jets in the Galactic microquasar SS 433, which are directed close to the plane of the sky, have kinetic luminosities of more than a few times 10^{39} erg/s, which would be super-Eddington for a black hole of 10 solar masses.

An alternative model is that ultraluminous X-ray sources may be compact binaries with black holes of more than 30 solar masses that emit largely isotropically with no beaming into the line of sight, either geometrically or relativistically [33]. This conclusion is based on the formation, evolution, and overall energetics of the ionized nebulae of several 100 pc diameter in which some ultraluminous X-ray sources are found embedded. The recent discoveries of high mass binaries with black holes of 15.7 solar masses in M 33 [32] and 23–34 solar masses in IC 10 [34] support this idea. Apparently, black holes of several tens of solar masses could be formed in starburst galaxies of relative low metal content.

1.8 Very Energetic γ-Ray Emission from Compact Binaries

Very energetic γ-rays with energies greater than 100 GeV have recently been detected with ground-based telescopes from four high mass compact binaries [21]. These have been interpreted by models proposed in the contexts represented in Fig. 1.5. In two of the four sources the γ radiation seems to be correlated with the orbital phase of the binary, and therefore may be consistent with the idea that the

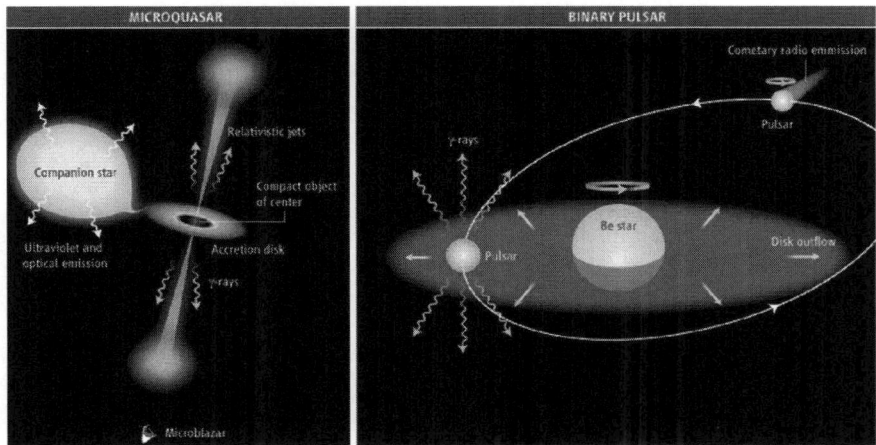

Fig. 1.5 Alternative contexts for very energetic γ-ray binaries [21]. *Left*: microquasars are powered by compact objects (neutron stars or stellar-mass black holes) via mass accretion from a companion star. The interaction of collimated jets with the massive outflow from the donor star can produce very energetic γ-rays by different alternative physical mechanisms [37], depending on whether the jets are baryonic or purely leptonic. *Right*: pulsar winds are powered by rotation of neutron stars; the wind flows away to large distances in a comet-shape tail. Interaction of this wind with the companion-star outflow may produce very energetic γ-rays [8]

very high energy radiation is produced by the interaction of pulsar winds with the mass outflow from the massive companion star [8, 6]. The detection of TeV emission from the black hole binary Cygnus X-1 [2] and the TeV intraday variability in M 87 [1] provided support to the jet models [37], which do not require relativistic Doppler boosting as in blazars and microblazars . It remains an open question whether the γ-ray binaries LS 5039 and LS I +61 303 could be microquasars where the γ radiation is produced by the interaction of the outflow from the massive donor star with jets [37] or pulsar winds [8].

1.9 Microquasars and Gamma-Ray Bursts

It is believed that gamma-ray bursts of long duration ($t > 1$ s) mark the birth of black holes by core collapse of massive stars. In this context, microquasars that contain black holes would be fossils of gamma-ray burst sources of long duration, and their study in the Milky Way and nearby galaxies can be used to gain observational inside into the physics of the much more distant sources of gamma-ray bursts. Questions of topical interest are (a) Do all black hole progenitors explode as very energetic hypernovae of type Ib/c?; and (b) What are the birth places and nature of the progenitors of stellar black holes?

The kinematics of microquasars provides clues to answer these questions. When a binary system of massive stars is still gravitationally linked after the explosion of one of its components, the mass center of the system acquires an impulse, whatever matter ejection is, symmetric or asymmetric. Then according to the microquasar movement we can investigate the origin and the formation mechanism of the compact object. Knowing the distance, proper motion, and radial velocity of the center of mass of the binary, the space velocity and past trajectory can be determined. Using multi-wavelength data obtained with a diversity of observational techniques, the kinematics of eight microquasars have so far been determined.

One interesting case is the black hole wandering in the Galactic halo, which is moving at high speed, like globular clusters [27] (Fig. 1.6). It remains an open question whether this particular halo black hole was kicked out from the Galactic plane by a natal explosion, or is the fossil of a star that was formed more than 7 billions of years ago, before the spiral disk of stars, gas, and dust of the Milky Way was formed. In this context, the study of these stellar fossils may represent the beginning of what could be called "Galactic Archaeology". Like archaeologists, studying these stellar fossils, astrophysicists can infer what was the history of the Galactic halo.

The microquasars LS 5039 [36] and GRO J1655-40 [29] which contain compact objects with less than $\sim$7 solar masses were ejected from their birth place at high speeds, and therefore the formation of these compact objects with relative small masses must have been associated with energetic supernovae. On the contrary, the binaries Cygnus X-1 [30] and GRS 1915+105 [7] which contain black holes of at least 10 solar masses do not seem to have received a sudden impulse. Preliminary results on the kinematics of the X-ray binaries suggest that low mass black holes are formed by a delayed collapse of a neutron star with energetic supernovae, whereas stellar black holes with masses equal or greater than 10 solar masses are the result of the direct collapse of massive stars, namely, they are formed in the dark. This is consistent with the recent finding of gamma-ray bursts of long duration in the near Universe without associated luminous supernovae [4].

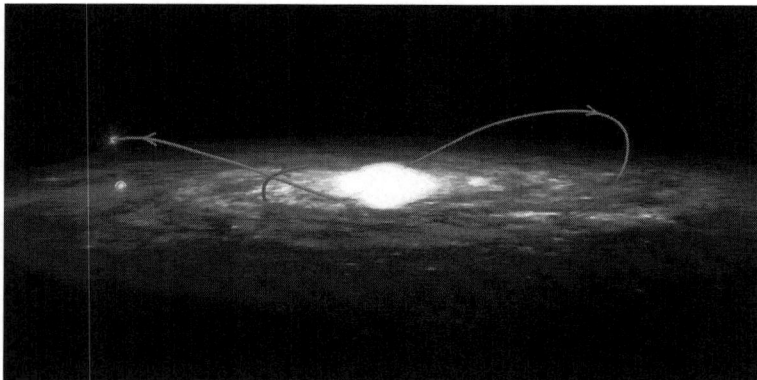

Fig. 1.6 A wandering black hole in the Galactic halo [27]. The trajectory of the black hole for the last 230 million years is represented. The *bright dot* on the *left* represents the Sun

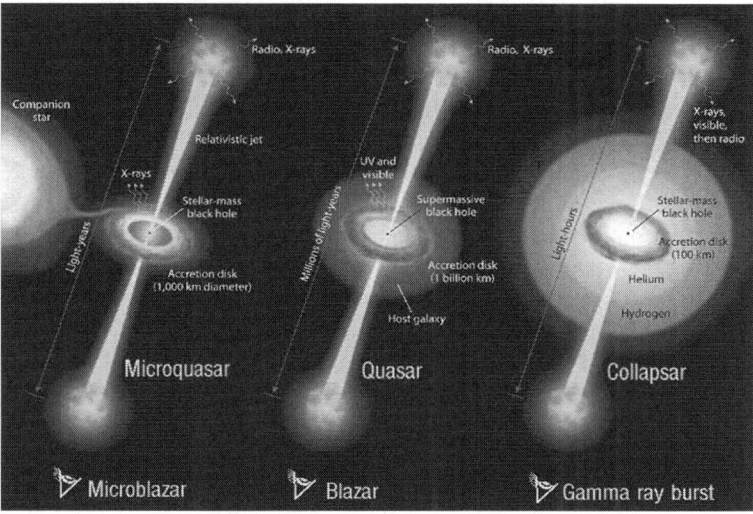

Fig. 1.7 The same physical mechanism can be responsible for three different types of objects: microquasars, quasars, and massive stars that collapse ("collapsars") to form a black hole producing gamma-ray bursts. Each one of these objects contains a black hole, an accretion disk, and relativistic particles jets. Quasars and microquasars can eject matter several times, whereas the collapsars form jets only once. When the jets are aligned with the line of sight of the observer these objects appear as microblazars, blazars, and gamma-ray bursts, respectively. Reproduced from [28]

There are indications that the mass of the resulting black hole may be a function of the metal content of the progenitor star. In fact, the black holes with 16 solar masses in M 33 [32] and more than 23 solar masses in IC 10 [34] are in small galaxies of low metal content. This is consistent with the fact that the majority of the gamma-ray bursts of long duration take place in small starburst galaxies at high redshift, namely, in Galactic hosts of low metal content [16]. Since the power and redshift of gamma-ray bursts seem to be correlated, this would imply a correlation between the mass of the collapsing stellar core and the power of the γ-ray jets.

Gamma-ray bursts of long duration are believed to be produced by ultra relativistic jets generated in a massive star nucleus when it catastrophically collapses to form a black hole. Gamma-ray bursts are highly collimated jets and it has been proposed [28] that there may be a unique universal mechanism to produce relativistic jets in the Universe, suggesting that the analogy between microquasars and quasars can be extended to the gamma-ray bursts sources, as illustrated in the diagram of Fig. 1.7.

1.10 Conclusions

Black-hole astrophysics is presently in an analogous situation as was stellar astrophysics in the first decades of the 20th century. At that time, well before the physical understanding of the interior of stars and the way by which they produce and radiate energy, empirical correlations such as the HR diagram were found and used to derive fundamental properties of the stars, such as the mass. Similarly, at present before a comprehensive understanding of black hole physics, empirical correlations between X-ray and radio luminosities and characteristic timescales are being used to derive the mass and spin of black holes of all mass scales, which are the fundamental parameters that describe astrophysical black holes. Therefore, there are historical and epistemological analogies between black hole astrophysics and stellar astrophysics. The research area on microquasars has become one of the most important areas in high energy astrophysics. In the last fourteen years there have been seven international workshops on microquasars: four in Europe, one in America, and two in Asia. They are currently attended by 100–200 young scientists who, with their work on microquasars, are contributing to open new horizons in the common ground of high energy physics and modern astronomy.

Apologies: This manuscript is based on short courses given at international schools for graduate students, intended to give an introduction to this area of research. It is biased by my own personal choice, and hence it is by no means a comprehensive review. Because references had to be minimized I apologize for incompleteness to colleagues working in the field. Part of this work was written while the author was a staff member of the European Southern Observatory in Chile.

References

1. F. Aharonian, A.G. Akhperjanian, A.R. Bazer-Bachi et al.: Science, **314**, 1424 (2006)
2. J. Albert, E. Aliu, H. Anderhub et al.: ApJ, **665**, L51 (2007)
3. A.J. Castro-Tirado, S. Brandt, N. Lund et al.: ApJ Suppl., **92**, 469 (1994)
4. M. Della Valle, G. Chincarini, N. Panagia et al.: Nature, **444**, 1050 (2006)
5. V. Dhawan, I.F. Mirabel, L.F. Rodríguez: ApJ, **543**, 373 (2000)
6. V. Dhawan, A. Mioduszewski, M. Rupen: Proceedings of the VI Microquasar Workshop: Microquasars and Beyond, PoS(MQW6)052 (2006)
7. V. Dhawan, I.F. Mirabel, M. Ribó et al.: ApJ, **668**, 430 (2007)
8. G. Dubus: A&A, **456**, 801 (2006)
9. G. Fabbiano: ARA&A, **44**, 323 (2006)
10. R.P. Fender, S.T. Garrington, D.J. McKay et al.: MNRAS, **304**, 865 (1999)
11. R.P. Fender, T.M. Belloni, E. Gallo: MNRAS, **355**, 1105 (2004)
12. R. Giacconi, H. Rursky, J.R. Waters: Nature, **204**, 981 (1964)
13. J. Greiner, J.G. Cuby, M.J. McCaughrean: Nature, **414**, 522 (2001)
14. M.M. Kasliwal, S.B. Cenko, S.R. Kulkarni et al.: ApJ, **678**, 1127 (2008)
15. A.R. King, M.B. Davies, M.J. Ward et al.: ApJ, **552**, L109 (2001)
16. E. Le Floc'h, P.-A. Duc, I.F. Mirabel et al.: A&A, **400**, 499 (2003)
17. M. Leventhal, C.J. MacCallum, S.D. Barthelmy et al.: Nature, **339**, 36 (1989)
18. B. Margon, H.C. Ford, J.I. Katz et al.: ApJ, **230**, L41 (1979)
19. A.P. Marscher, S.G. Jorstad, J.-L. Gómez: Nature, **417**, 625 (2002)
20. J.E. McClintock, R. Shafee, R. Narayan et al.: ApJ, **652**, 518 (2006)
21. I.F. Mirabel: Science, **312**, 1759 (2006)
22. I.F. Mirabel, L.F. Rodríguez, B. Cordier et al.: Nature, **358**, 215 (1992)
23. I.F. Mirabel, L.F. Rodríguez: Nature, **371**, 46 (1994)
24. I.F. Mirabel, V. Dhawan, S. Chaty et al.: A&A, **330**, L9 (1998)

25. I.F. Mirabel, L.F. Rodríguez: Nature, **392**, 673 (1998)
26. I.F. Mirabel, L.F. Rodríguez: ARA&A, **37**, 409 (1999)
27. I.F. Mirabel, V. Dhawan, R.P. Mignani et al.: Nature, **413**, 139 (2001)
28. I.F. Mirabel, L.F. Rodríguez: Sky & Telescope, 32, May (2002)
29. I.F. Mirabel, R.P. Mignani, I. Rodrigues et al.: A&A, **395**, 595 (2002)
30. I.F. Mirabel, I. Rodrigues: Science, **300**, 1119 (2003)
31. J.A. Orosz, E. Kuulkers, M. van der Klis et al.: ApJ, **555**, 489 (2001)
32. J.A. Orosz, J.E. McClintock, R. Narayan et al.: Nature, **449**, 872 (2007)
33. M.W. Pakull, L. Mirioni: Rev. Mex. Astr. & Ap., **15**, 197 (2003)
34. A.H. Prestwich, R. Kilgard, P.A. Crowther et al.: ApJ, **669**, L21 (2007)
35. R.A. Remillard, J.E. McClintock: ARA&A, **44**, 49 (2006)
36. M. Ribó, J.M. Paredes, G.E. Romero et al.: A&A, **384**, 954 (2002)
37. G.E. Romero: Gamma-ray emission from microquasars: Leptonic vs. Hadronic models. In: Relativistic Astrophysics Legacy and Cosmology – Einstein's, ESO Astrophysics Symposia, pp. 480–482. Springer-Verlag, Berlin, Heidelberg (2008)
38. Y. Tanaka: AN, **327**, 1098 (2006)
39. H. van der Laan: Nature, **211**, 1131 (1966)
40. G. Weidenspointner, G. Skinner, J. Pierre et al.: Nature, **451**, 159 (2008)
41. S.N. Zhang, W. Cui, W. Chen: ApJ, **482**, L155 (1997)

Chapter 2
X-Ray Emission from Black-Hole Binaries

M. Gilfanov

Abstract The properties of X-ray emission from accreting black holes are reviewed. The contemporary observational picture and current status of theoretical understanding of accretion and formation of X-ray radiation in the vicinity of the compact object are equally in the focus of this chapter. The emphasis is made primarily on common properties and trends rather than on peculiarities of individual objects and details of particular theoretical models. The chapter starts with discussion of the geometry of the accretion flow, spectral components in X-ray emission, and black hole spectral states. The prospects and diagnostic potential of X-ray polarimetry are emphasized. Significant attention is paid to the discussion of variability of X-ray emission in general and of different spectral components – emission of the accretion disk, Comptonized radiation, and reflected component. Correlations between spectral and timing characteristics of X-ray emission are reviewed and discussed in the context of theoretical models. Finally, a comparison with accreting neutron stars is made.

2.1 Introduction

The gravitational energy of matter dissipated in the accretion flow around a compact object of stellar mass is primarily converted into photons of X-ray wavelengths. The lower limit on the characteristic temperature of the spectral energy distribution of the emerging radiation can be estimated assuming the most radiatively efficient configuration – optically thick accretion flow. Taking into account that the size of the emitting region is $r \sim 10 r_g$ (r_g is the gravitational radius) and assuming a black body emission spectrum one obtains: $kT_{bb} \approx \left(L_X/\sigma_{SB}\pi r^2 \right)^{1/4} \approx 1.4\, L_{38}^{1/4}/M_{10}^{1/2}$ keV. It is interesting (and well known) that T_{bb} scales as directly proportional to $M^{-1/2}$. This is confirmed very well by the measurements of the disk emission temperature in stellar mass systems and around supermassive black holes in AGN. It is also

M. Gilfanov (✉)

Max-Planck-Institute for Astrophysics, Garching, Germany; Space Research Institute, Moscow, Russia, gilfanov@mpa-garching.mpg.de

Gilfanov, M.: *X-Ray Emission from Black-Hole Binaries*. Lect. Notes Phys. **794**, 17–51 (2010)
DOI 10.1007/978-3-540-76937-8_2

illustrated, albeit less dramatically, by the comparison of the soft state spectra of black holes and neutron stars, as discussed later in this chapter. The upper end of the relevant temperature range is achieved in the limit of the optically thin emission. It is not unreasonable to link it to the virial temperature of particles near the black hole, $kT_{\mathrm{vir}} = GMm/r \propto mc^2/(r/r_g)$. Unlike the black body temperature this quantity does not depend on the mass of the compact object, but does depend on the mass of the particle m. For electrons $T_{\mathrm{vir}} \sim 25(r/10r_g)^{-1}$ keV and it is correspondingly $m_p/m_e = 1836$ times higher for protons. Protons and ions are the main energy reservoir in the accretion flow, but for all plausible mechanisms of spectral formation it is the temperature of electrons that determines the spectral energy distribution of the emerging radiation. The latter depends on the poorly constrained efficiency of the energy exchange between electrons and protons in the plasma near the compact object. The values of the electron temperature typically derived from the spectral fits to the hard spectral component in accreting black holes, $kT_e \sim 50 - 150$ keV, are comfortably within the range defined by the two virial temperatures. However, the concrete value of kT_e and its universality in a diverse sample of objects and broad range of luminosity levels still remains unexplained from first principles, similar to those used in the derivation of the temperature of the optically thick soft component.

Broadly speaking, significant part of, if not the entire diversity of the spectral behavior observed in accreting black holes and neutron stars can be explained by the changes in the proportions in which the gravitational energy of the accreting matter is dissipated in the optically thick and optically thin parts of the accretion flow. The particular mechanism driving these changes is however unknown – despite significant progress in MHD simulations of the accretion disk achieved in recent years (see Chap. 10) there is no acceptable global model of accretion onto a compact object. To finish this introductory note, I will mention that non-thermal processes in optically thin media (e.g., Comptonization on the non-thermal tail of the electron distribution) may also contribute to the X-ray emission from black holes in some spectral states.

2.2 Geometry and Spectral Components

The contributions of optically thick and optically thin emission mechanisms can be easily identified in the observed spectra of accreting black holes as soft and hard spectral components (Fig. 2.1). Depending on the spectral state of the source one of these components may dominate the spectrum or they can coexist giving comparable contribution to the total emission. The soft component is believed to originate in the geometrically thin and optically thick accretion disk of the Shakura-Sunyaev type [48]. To zeroth order approximation, its spectrum can be regarded as a superposition of black body spectra with the temperatures and emitting areas distributed according to the energy release and the temperature profile of the accretion disk. The simplest example of this is the so-called multicolor disk black body model introduced by [35]. Although valid for one particular inner torque boundary condition allowing easy integration of the total flux, this model has been widely and efficiently used in

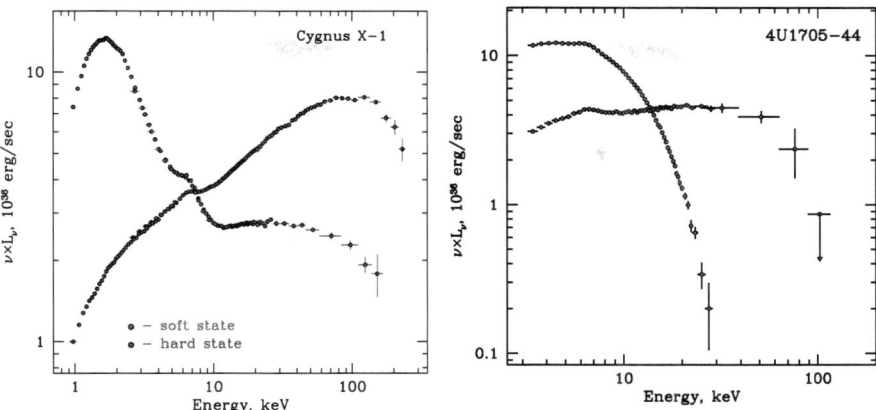

Fig. 2.1 Energy spectra of a black hole Cyg X-1 (*left*) and a neutron star 4U 1705-44 (*right*) in the soft and hard spectral states (*left panel* adopted from [18])

the era of more limited computer resources due to its simplicity, speed, and early integration in the XSPEC spectral fitting package. To achieve a higher degree of accuracy, one would need to consider more realistic inner boundary conditions, deviations of the gravitational potential from the Newtonian, and to account for such effects as distortion of the black body spectrum due to Thomson scatterings in the upper layers and in the atmosphere of the disk, Doppler effect due to rotation of the matter in the disk, etc. A number of models have been proposed to include these effects [11, 49, 46, 4], many of them currently implemented in the XSPEC package.

The inadequacy of the optically thick emission mechanism in describing the hard spectral component present in both spectra in Fig. 2.1 can be easily demonstrated. Indeed, the size of the emitting region required to achieve a luminosity of $\log(L_X) \sim 37 - 38$ with black body temperature of ~several tens of keV is $R_{em} \sim \left(L_X/\pi\sigma_{SB}T^4\right)^{1/2} \sim 20\,L_{37}^{1/2}/T_{30}^2$ m ($T_{30} = T_{bb}/30\,\mathrm{keV}$). Obviously this is much smaller than the size of the region of main energy release near an accreting black hole, $\sim 50 r_g \sim 500\text{--}1500\,\mathrm{km}$. The bremsstrahlung emission from a $\sim$ uniform cloud of hot plasma with a filling factor close to unity cannot deliver the required luminosity either, because the $\propto N_e^2$ density dependence of its emissivity makes it very inefficient in the low density regime characteristic of the optically thin accretion flow. Indeed, the optical depth of such a plasma cloud of size $\sim 10\text{--}100 r_g$ and emission measure of $N_e^2 V \sim 10^{59}\,\mathrm{cm}^{-3}$ required to explain observed hard X-ray luminosity of $\sim 10^{37}\,\mathrm{erg/s}$ would greatly exceed unity. However, bremsstrahlung emission may play a role in the advection-dominated accretion flow [37] in the low $\dot{M}$ regime of quiescent state of accreting black holes ($\log(L_X) \leq 32\text{--}33$).

It has been understood early enough that Comptonization is the most plausible process of formation of the hard spectral component [54, 55]. Thanks to its linear dependence on the gas density, Comptonization of soft photons on hot electrons in the vicinity of the compact object can efficiently radiate away the energy dissipated in the optically thin accretion flow and successfully explain the luminosity and overall spectral energy distribution observed in the hard X-ray band. Moreover, Comptonization models of varying degree of complexity do satisfactorily describe the observed broadband energy spectra of black holes to the finest detail (e.g., [64]). It is especially remarkable given the high quality of the X-ray data which became available in the last decade from observations of recent and current X-ray observatories, such as Compton GRO, RossiXTE and Chandra, XMM and Swift.

The Comptonization site – cloud of hot (thermal or non-thermal) electrons is often referred to as a "corona". Although it is generally accepted that the Comptonizing corona has to be located in the close vicinity of the compact object, there is currently no broad consensus on the detailed geometry of the region. The numerical simulations have not reached the degree of sophistication required to perform full self-consistent global simulations of the accretion flow. Two physically motivated geometries are commonly considered: (i) the "sombrero" configuration (to my knowledge first introduced in [40]) and (ii) the patchy, flaring corona above the accretion disk.

One of the variants of the "sombrero" configuration is depicted in Fig. 2.2. It is assumed that outside some truncation radius the accretion takes place predominantly

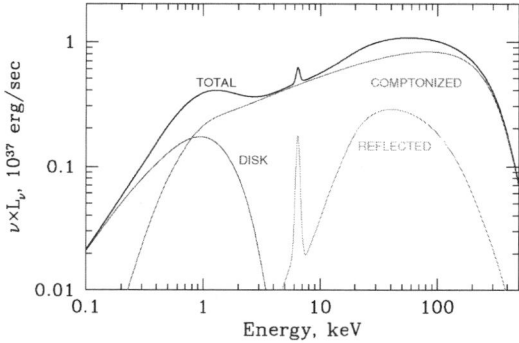

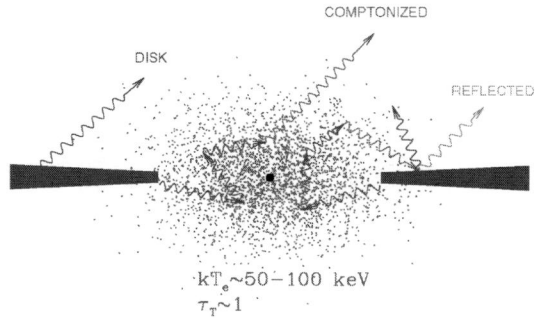

Fig. 2.2 The three main components of the X-ray emission from an accreting black hole (*top*) and a plausible geometry of the accretion flow in the hard spectral state (*bottom*)

via the optically thick and geometrically thin accretion disk, whereas closer to the compact object the accretion disk is transformed into a hot optically thin and geometrically thick flow with the aspect ratio of $H/R \sim 0.5$–1. The soft (optically thick) and hard (optically thin) spectral components are formed in the accretion disk and the hot inner flow correspondingly. The value of the truncation radius can be inferred from the observations. Although their interpretation is not unique and unambiguous, the plausible range of values is between ~ 3 and a few hundred gravitational radii. There is no commonly accepted mechanism of truncation of the disk and formation of the corona, with a number of plausible scenarios having been investigated recently. Among the more promising ones is the evaporation of the accretion disk under the effect of the heat conduction. It was initially suggested to explain quiescent X-ray emission from cataclysmic variables [31] and was later applied to the case of accretion onto black holes and neutron stars [32]. It not only provides a physically motivated picture describing the formation of the corona and the destruction of the optically thick disk but also correctly predicts the ordering of spectral states vs. the mass accretion rate. Namely, it explains the fact that hard spectra indicating prevalence of the hot optically thin flow are associated with lower $\dot{M}$ values, whereas the optically thick disk appears to dominate the photon production in the accretion flow at higher $\dot{M}$. As a historical side note I mention that in the earlier years of X-ray astronomy the presence of the hot optically thin plasma in the vicinity of the compact object was often associated with disk instabilities, therefore such a behavior appeared puzzling to many astrophysicists in the view of the $\dot{M}$ dependence of these instabilities.

Another geometrical configuration considered in the context of hard X-ray emission from black holes is a non-stationary and non-uniform (patchy) corona above the optically thick accretion disk. This scenario has been largely inspired by the suggestion by Galeev et al. [13], that a magnetic field amplified in the hot inner disk by turbulence and differential rotation may reach the equipartition value and emerge from the disk in the form of buoyant loop-like structures of solar type above its surface. These structures may lead to the formation of a hot magnetically confined structured corona similar to the solar corona, which may produce hard emission via inverse Compton and bremsstrahlung mechanisms. The model could also explain the faintness of the hard emission in the soft state as a result of efficient cooling of plasma in the magnetic loops via inverse Compton effect due to increased flux of soft photons at higher $\dot{M}$.

Although the original Galeev, Rosner, and Vaiana paper was focused on the thermal emission from hot plasma confined in buoyant magnetic loops, the latter are a plausible site of particle acceleration responsible for the non-thermal component in the electron distribution. This is expected on theoretical grounds and is illustrated very well observationally by the presence of the non-thermal emission component in the spectra of solar flares (see e.g., [9] and references therein). On the other hand, a power-law-shaped hard X-ray component is commonly observed in the spectra of black hole candidates (Fig. 2.3). It reveals itself most graphically in the soft spectral state but may be as well present in the hard state, along with the thermal Comptonized spectrum (e.g., [33]). This power-law component has a photon index

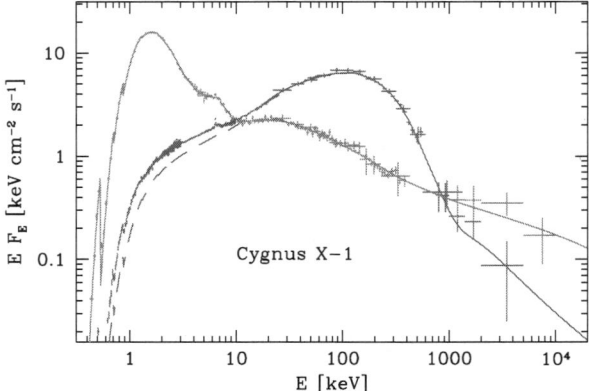

Fig. 2.3 Broadband spectra of Cyg X-1 in the soft and hard spectral state based on the data of BeppoSAX and Compton GRO missions. Adopted from [33]

of $\Gamma \sim 2$–3 and, although it may extend into the several hundred keV–MeV range (e.g., [58]), it is relatively unimportant energetically, contributing a small fraction to the total radiation output of the black hole. This is in contrast with the hard spectral component due to thermal Comptonization, observed in the hard spectral state which accounts for the most of the source luminosity.

There have been attempts to explain the power-law component as a result of Comptonization in a media with Maxwellian distribution of electrons of large temperature and small optical depth. This seems to be unlikely as in this case humps due to individual scattering will be seen in the output spectrum, in contrast with the smooth power-law spectra observed. This qualitative consideration is illustrated by results of Monte-Carlo simulations shown in Fig. 2.4.

The sombrero configuration is often associated with a predominantly thermal distribution of electrons, whereas the solar-like flares above the accretion disk may be the site of electron acceleration producing non-thermal electron distributions and power-law-like Comptonization spectra. Thermal Comptonization is believed to be the main mechanism in the hard spectral state, whereas non-thermal Comptonization is probaly relevant in the soft state (Fig. 2.3). On the other hand, observations often indicate presence of both thermal and non-thermal Comptonization components suggesting that the two types of corona may coexist (e.g. [23]).

Due to heating of the disk by Comptonized radiation from the corona and soft photon feedback, the uniform thermal corona above the optically thick accretion disk cannot explain observed hard spectra with photon index of $\Gamma \sim 1.5$–2.0. It has been first proposed by Sunyaev and Titarchuk [56] that the presence of cool optically thick media – e.g., accretion disk or surface of the neutron star, in the vicinity of the Comptonization region will affect the parameters of the latter and, consequently, the shape of the outgoing Comptonized radiation. Indeed, some fraction of the Comptonized radiation will be returned to the accretion disk increasing its temperature and, consequently, the soft photon flux to the Comptonization

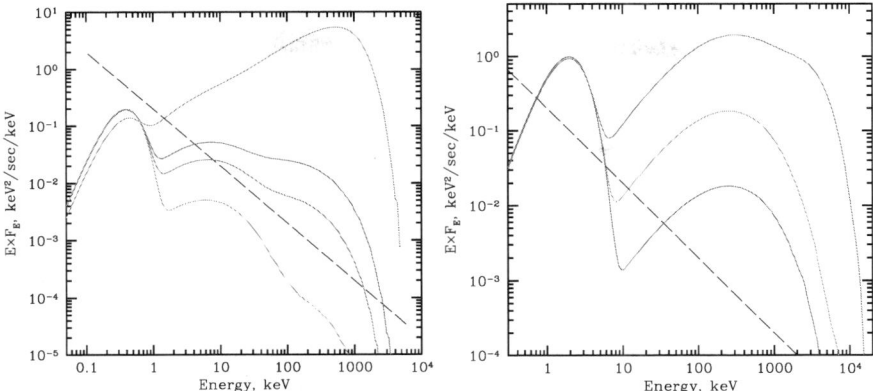

Fig. 2.4 Comptonization spectrum in the case of large temperature and small optical depth. *Left panel*: $kT_{bb} = 100\,\mathrm{eV}$, $kT_e = 300\,\mathrm{keV}$, $\tau = 0.01, 0.05, 0.1$ from the *bottom* to the *top*; *right panel*: $kT_{bb} = 500\,\mathrm{eV}$, $kT_e = 1\,\mathrm{MeV}$, $\tau = 0.001, 0.01, 0.1$. For comparison, the topmost spectrum on the *left panel* shows case of $\tau = 1$. The left-most peak in all spectra is made of seed photons which left the Comptonization region without scatterings. The *dashed line* is a power law with a photon index of $\Gamma = 3$. These spectra are the results of Monte-Carlo simulations assuming spherical geometry

region. This in turn will increase the cooling rate and will decrease the electron temperature in the Comptonization region leading to softer and steeper spectra. I will further illustrate it by the following simple quantitative consideration. In the sandwich-like geometry, assuming moderate Thompson optical depth of the corona, $\tau_T \sim 1$, a fraction $f \sim 1/2$ of the Comptonized emission will be returned to the accretion disk. Of this, a fraction of $1-\alpha$ ($\alpha \sim 0.2$ – albedo) will be absorbed and will contribute to the heating of the accretion disk, adding to its heating due to the gravitational energy release. Ignoring the latter, the luminosity enhancement factor in the Comptonization region, defined as a ratio of its total luminosity to the luminosity of the seed photons (see more detailed discussion in Sect. 2.8.2) will be $A \approx (1-\alpha)^{-1} f^{-1} < 2.5$. As well known [10], the luminosity enhancement factor is intimately related to the Comptonization parameter y and the photon index Γ of the Comptonized radiation. The above constrain on A implies $\Gamma > 2.3$ (see Fig. 2.15), which is steeper than the hard state spectra typically observed in black holes. This conclusion is confirmed by the full treatment of the Comptonization problem in "sandwich" geometry [22]. In order to produce a harder Comptonized spectrum, the value of the feedback coefficient f needs to be reduced. This is achieved, for example, in a non-uniform, patchy and/or non-stationary corona. Another example suggested by [5] involves bulk motion of the corona with mildly relativistic velocity away from the disk reducing the feedback coefficient due to the aberration effect (Sect. 2.8.2). A uniform stationary corona above the accretion disk can, in principle, be responsible for the steep power-law component often detected in the soft state, although a non-thermal electron distribution may be a more plausible explanation, as discussed above. To conclude, I also note that considerations of a similar kind involving the neutron star surface can explain the fact that the neutron star spectra are typically softer than those of black holes (e.g., [56], Sect. 2.6).

2.3 Spectral States and Geometry

The existence of different spectral states is a distinct feature of accreting X-ray sources, independently of the nature of the compact object (Fig. 2.1, 2.3). Although their phenomenology is far richer, for the purpose of this chapter I will restrict myself to the simple dichotomy between soft (high) and hard (low) spectral states and refer to the next chapter of this book for a more detailed discussion. As no global self-consistent theory/model of accretion exists, all theories explaining spectral states have to retreat to qualitative considerations. These considerations, although phenomenological in nature, usually are based on numerous observations of black hole systems, simple theoretical arguments, and some simplified solutions and simulations of the accretion problem. Described below is a plausible, although neither unique nor unanimously accepted, scenario of this kind based on the "sombrero" geometry of the accretion flow. There are a number of cartoons and geometry sketches, illustrating this and other scenarios which I will not repeat here and will refer the interested reader to the original work, e.g., [68]

As obvious from Fig. 2.1, the spectral states phenomenon is related to the redistribution of the energy released in the optically thick and optically thin components of the accretion flow. In the sombrero configuration one may associate spectral state transitions with change of the disk truncation radius – the boundary between the outer optically thick accretion disk and the inner optically thin hot flow.

In the soft (aka disk-dominated) spectral state, the optically thick accretion disk extends close to the compact object, possibly to the last marginally stable Keplerian orbit ($r = 3r_g$ for a Schwarzschild black hole), leaving no "room" for the hot optically thin flow. Therefore, the major fraction of the accretion energy is emitted in the optically thick accretion disk giving rise to a soft spectrum of the multicolor black body type. The magnetic activity at the disk surface may (or may not) produce a hard power law-like tail due to non-thermal Comptonization in the corona. As discussed in the previous section, this power-law component has a steep slope $\Gamma \sim 2$–3 and is relatively insignificant energetically.

In the hard (aka corona-dominated) spectral state, the accretion disk truncates at a distance of ~ 50–$100 r_g$ or further from the compact object. The major fraction of the gravitational energy is released in the hot inner flow. Comptonization of the soft photons emitted by the accretion disk on the hot thermal electrons of the inner flow leads to the formation of the hard spectrum of the shape characteristic for unsaturated thermal Comptonization. The typical parameters in the Comptonization region – hot inner flow, inferred from observations are: electron temperature of $T_e \sim 100\,\mathrm{keV}$ and Thompson optical depth of $\tau_e \sim 1$. The significance of the soft backbody-like emission from the optically thick disk as well as of the non-thermal emission due to magnetic flares at its surface varies depending on the disk truncation radius, increasing as the disk moves inward. There is evidence that both thermal and non-thermal hard components may coexist in the hard state in the certain range of the disk truncation radii, as suggested in [23].

2.4 Reflected Emission

After escaping the corona, a fraction of the Comptonized photons may be intercepted by the optically thick accretion disk. Part of the intercepted radiation will be dissipated in the disk, via inverse Compton effect and photoabsorption on heavy elements, contributing to its energy balance, the remaining part will be reflected due to Compton scatterings [3]. The disk albedo depends on the photon wavelength and on the chemical composition and ionization state of the disk material. Expressed in terms of energy flux it is ∼0.1–0.2 in the case of neutral matter of solar abundance [30]. Some fraction of the photobsorbed emission will be reemitted in the fluorescent lines of heavy elements and may also escape the disk. Thus, combined effects of photoabsorption, fluorescence, and Compton scattering form the complex spectrum of the reflected emission, consisting of a number of fluorescent lines and K-edges of cosmically abundant elements superimposed on the broad Compton reflection hump [3, 14] (Fig. 2.5). The peak energy of the latter depends on the shape of the incident continuum and a typical spectrum of a black hole in the hard state is located at ∼30 keV (Fig. 2.2).

Due to the dependence of the photoabsorption cross section, element abundances and fluorescent yield on the atomic number, the strongest among the spectral features associated with heavy elements are the K-edge and fluorescent K-α line of

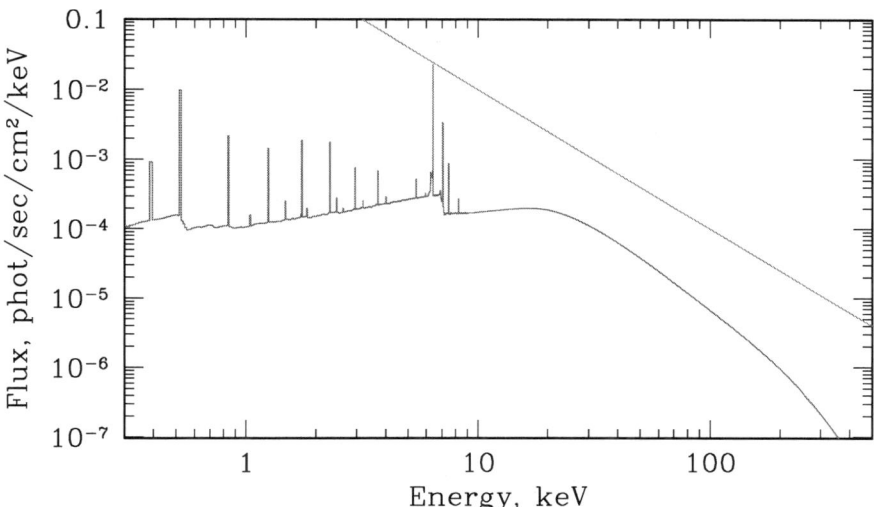

Fig. 2.5 Spectrum of emission reflected from an optically thick slab of neutral matter of solar abundance. The *solid line* shows the spectrum of the incident power-law emission

iron. For reflection from the neutral media, the centroid of the iron line is at 6.4 keV and its equivalent width computed with respect to the pure reflected continuum is ~1 keV. For a distant observer, the reflected emission is diluted with the primary Comptonized radiation and the thermal emission of the accretion disk, adding complexity to the observed spectra. It leads to the appearance of characteristic reflection features in the spectra of X-ray binaries (Fig. 2.5) – the fluorescent K_α line from iron, iron K-edge, and a broad Compton reflection hump at higher energies [3, 14]. Their shape is further modified, depending on the ionization state of the matter in the disk [47], by strong gravity effects and Doppler and aberration effects due to the Keplerian rotation of the disk, e.g., [12]. The observed line usually has non-zero intrinsic width and the K-edge is never sharp (smeared edge). Their amplitude depends on the ionization state and on the solid angle of the reflector as seen from the source of the primary radiation. This can change with the spectral state of the source. The typical value of the iron line equivalent width is ~50–300 eV.

The presence of these features makes the spectra deviate from a power-law shape, expected for Comptonized radiation with parameters typical for black holes. As their amplitude depends on the relative configurations of the corona and the accretion disk and physical conditions in the disk, they have great diagnostics potential for studying the accretion geometry in different spectral states. This is further discussed in Sect. 2.8.

2.5 Polarization of X-Ray Emission

As the distribution of matter in the accretion flow is not spherically symmetric, one may expect some degree of polarization of emerging X-ray radiation. In the conventional scenario of formation of X-ray radiation in the vicinity of a compact object the polarization is caused by Thompson scatterings of photons on free electrons in the accretion disk and hot corona and is predicted to be present at the moderate level of ~ several percent in all three main components of the X-ray emission: thermal emission of the optically thick disk, Comptonized emission, and reflected component. A considerably larger degree of polarization may be expected in some alternative scenarios, for example, in some versions of the jet scenario, discussed later in this book.

The degree of polarization of the Comptonized emission depends strongly on the geometry of the corona, the location of the sources of the soft seed radiation, and the viewing angle [57, 63]. As calculated by Sunyaev and Titarchuk [57], for the case of Comptonization in the disk it lies in the range between 0 and ~12%, depending on the optical depth of the corona and viewing angle, i.e., it can slightly exceed the maximum value of 11.7% for a pure-scattering semi-infinite atmosphere. Although the disk geometry of the corona is unlikely, this result suggests that a moderate degree of polarization of the Comptonized continuum, of the order of ~ few percent, may be expected.

Obviously, polarization should also be expected for the reflected emission from the accretion disk. The pure reflected component can be polarized to ~30% [29].

The degree of polarization drops to ∼ few percent when it is diluted with the Comptonized radiation and the thermal emission from the disk. The degree of polarization is a strong function of photon energy – it is low at low energy <10 keV and reaches its maximum of ∼5% (for a 60° viewing angle) at ∼30–50 keV, where the contribution of the Compton-reflected continuum to the overall spectrum is maximal [29].

The thermal emission generated inside the optically thick accretion disk is initially unpolarized but attains polarization as a result of Thompson scatterings in the atmosphere of the accretion disk [8, 27]. Calculations [27] show that the degree of polarization varies from 0% (face-on disk) to ∼5% (edge-on). The polarization degree depends on the photon energy, being largest in the Wien part of the thermal spectrum, where scatterings play the most important role. A remarkable property of polarized radiation from the vicinity of a compact object is that due to relativistic effects the polarization angle also becomes dependent on the photon energy [8, 27]. This is in contrast with the classical approximation, where symmetry considerations require that the polarization direction is coaligned with the minor or the major axis of the disk projection on the plane of the sky. The amplitude and shape of these dependences are sensitive to the disk inclination and the spin of the black hole. In combination with spectral information, this can be used to resolve the degeneracy between the black hole mass, spin and, disk inclination [27].

Polarization measurements are yet an unexplored area of high energy astrophysics and have great diagnostics potential. With the new generation of polarimetric detectors, X-ray polarimetry will become a powerful tool to study the geometry of the accretion flow and the properties of the compact object in accreting systems.

2.6 Variability

Variability of X-ray emission is a common and well-known property of X-ray binaries. The amplitude of flux variations depends on the timescale and the photon energy and can be as large as ∼20–30% fractional rms. The dependence of the variability amplitude on timescale is conventionally characterized by the power density spectrum, which is a square of the Fourier amplitudes of the time series, renormalized in order to give the answer in desired units, for example (fractional rms)^2/Hz. Power density spectra of X-ray sources often reveal a number of rather narrow features of various widths superimposed on a broadband continuum of aperiodic variations, suggesting that both resonances of various degrees of coherence as well as stochastic processes of aperiodic nature contribute to the observed flux variations. Some of the narrow features can be clearly associated with the spin frequency of the neutron star or orbital or precession frequency of the binary. The nature of others, broadly referred to as quasi-periodic oscillations (QPO), is poorly understood. The models range from the beat frequency model proposed by Alpar and Shaham [2] soon after the discovery of the first QPOs in neutron stars, and the relativistic precession model [53] relating QPO frequencies with fundamental frequencies of Keplerian orbits in strong gravity to models employing global oscillation modes in the accretion disk [1, 60]. QPOs were first discovered in neutron star systems, but are also commonly observed in accreting black holes. Their rich phenomenology is well documented elsewhere (e.g., [61]) and is not discussed in this chapter, which will focus on the aperiodic variability continuum.

I will conclude these introductory remarks with the following side note. The interpretation of the energy spectra of celestial X-ray sources has been greatly facilitated by the fact that a number of simple physical concepts could be employed in a straightforward manner, such as black body emission, Comptonization, photoabsorption, and fluorescence. No equivalent concepts are easily available to help in interpreting the power density spectra. This may explain why our understanding of the power density spectra lags significantly behind our understanding of the energy spectra. Indeed, spectral analysis can rely on a number of physically motivated and elaborate models which successfully describe high quality data from modern observatories. The interpretation of the power density spectra, on the contrary, has just started to advance beyond "numerology" and simple association of QPO frequencies with characteristic frequencies of test particles in the (strong) gravitational field. However, it is obvious that timing information (along with the polarization measurements) presents a completely different dimension which has to be taken into account in validating any model of formation of radiation in the vicinity of a compact object.

2.6.1 Propagation of $\dot{M}$ Fluctuations in the Accretion Disk

The remarkable characteristics of aperiodic variability is its breadth in the frequency domain. As illustrated by the power density spectra of Cyg X-1 and Cyg X-2 shown in Fig. 2.6, flux variations in the frequency range from $\log f < -8$ to $\log f \sim 2 - 3$ are present. This suggests that variations of the mass accretion rate $\dot{M}$ on an extremely broad range of timescales are present in the innermost region of the accretion flow, $r \leq 50-100\,r_g$, where X-ray emission is produced. It is to be compared with characteristic timescales in this region which are limited by two extremes – the dynamical timescale that is of the order of the Keplerian orbital time $t_K \approx 0.3\,(M/10M_\odot)\,(r/50r_g)^{3/2}$ s and the viscous timescale $t_{\mathrm{visc}} \sim \alpha^{-1}\,(h/r)^{-2}\,\Omega_K^{-1} = (2\pi\alpha)^{-1}\,(h/r)^{-2}\,t_K$ (α is viscosity parameter, h – the disk thickness). The latter is in the range between $t_{\mathrm{visc}} \sim 10^4\,t_K \sim 10^3$ s for a gas pressure-dominated Shakura-Sunyaev disk ($h/r \sim 10^{-2}$) and $t_{\mathrm{visc}} \sim 10\,t_K \sim 1-10$ s for a thicker hot flow with the aspect ratio of $h/r \sim 1/3$.

From the point of view of characteristic timescales, the high-frequency variations could potentially be produced in the vicinity of the compact object. Longer timescales, on the contrary, exceed by many orders of magnitude the longest timescales in the region of the main energy release and cannot be generated there. The low frequency $\dot{M}$ variations have to be generated in the outer parts of the accretion flow and be propagated to the region of the main energy release where they are converted into variations of the X-ray flux. The power spectra shown in

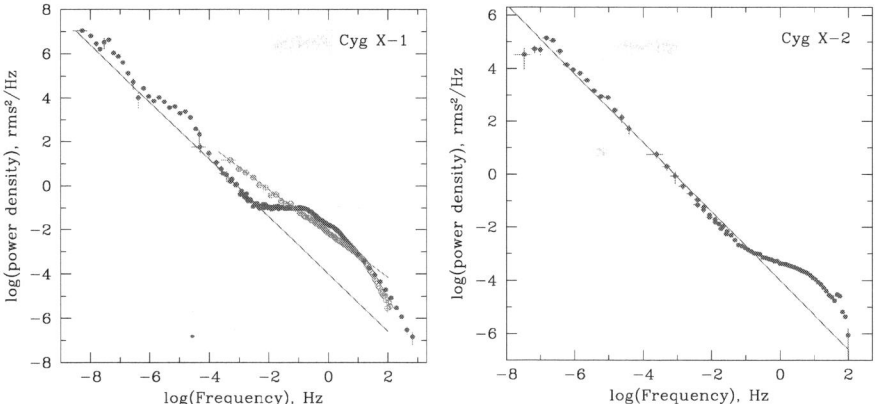

Fig. 2.6 Broad band power density spectra of Cyg X-1 and Cyg X-2. The spectra are obtained from non-simultaneous RXTE/ASM, EXOSAT/ME, and RXTE/PCA data. In the *left panel*, the *open circles* at $\log(f) > -4$ (*red* in the color version of the plot) show the 1996 high state PDS, the *filled circles* (*blue*) are hard spectral state PDS. The *solid lines* extending through entire plots in the *left* and *right panels* show a $P_\nu \propto \nu^{-1.3}$ power law; normalization is the same on both panels. The second power law in the *left panel* giving a better approximation to the soft state data of Cyg X-1 is $P_\nu \propto \nu^{-1}$ power law. The very high frequency data $\log(f) > 1$ are from [43]

Fig. 2.6 maintain the same power-law shape over a broad frequency range, suggesting that the same physical mechanism is responsible for flux variations at all frequencies. The same slope and normalization of the power-law component in the power density spectra of different sources – black holes and neutron stars – suggests that this mechanism is a property of the accretion disk and does not depend on the nature of the compact object. A plausible candidate for such mechanism may be viscosity fluctuations caused for example, by MHD turbulence in the accretion disk, as suggested in [26]. Viscosity fluctuations lead to variations of the mass accretion rate which after propagating into the innermost region of the accretion flow are transformed into variations of the X-ray flux.

However, because of the diffusive nature of disc accretion [48], fluctuations of mass accretion rate on timescales much shorter than the diffusion timescale will not be propagated inward, but will be damped close to the radius at which they originated. As demonstrated in [7], the amplitude of the fluctuations on the timescale τ will be significantly suppressed at the characteristic length scale $\Delta r / r \sim \sqrt{\tau / t_{\mathrm{visc}}}$. As the viscous timescale depends quadratically on the disk thickness, $t_{\mathrm{visc}} \sim (h/r)^{-2} \, \Omega_K^{-1}$, fluctuations on the dynamical $\tau \sim t_d \sim \Omega_K^{-1}$ or thermal $\tau \sim t_{\mathrm{th}} \sim (\alpha \Omega_K)^{-1}$ timescales will be damped in the thin disk with $h/r \ll 1$ after traveling a small distance in the radial direction, $\Delta r / r \sim h/r$, and will never reach the region of the main energy release. The inner region itself cannot generate $\dot{M}$ fluctuations on the dynamical or thermal timescales either. Indeed, the coherence length for fluctuations on a timescale $\tau \sim \Omega_K^{-1}$ is $\Delta r / r \sim h/r \sim 10^{-2}$ for the gas pressure supported Shakura-Sunyaev disk. Therefore $N \sim r/\Delta r \sim 100$ independent annuli regions will contribute to the observed flux, their variations being uncorrelated. The contribution of each region to the total flux is small, $\sim 10^{-2}$, and alone cannot cause significant variability of the total flux. Furthermore, in their combined emission uncorrelated fluctuations will be suppressed by a factor of $\sim 1/\sqrt{N} \sim 10^{-1}$ due to the averaging effect. Thus, for the geometrically thin disc, fluctuations of viscosity (or mass accretion rate) on the dynamical or thermal timescales will not contribute significantly to observed variability of the X-ray flux. In order to lead to significant modulation of the X-ray flux, viscosity or accretion rate fluctuations have to propagat inward significant radial distance and cause fluctuations of mass accretion rate in a significant range of smaller radii, including the region of the main energy release. This is only possible for fluctuations on timescales equal or longer than the viscous time of the disk at the radius, where fluctuations are "inserted" into the accretion flow [26, 7].

Thus, the standard Shakura-Sunyaev disk plays the role of a low-pass filter, at any given radius r suppressing $\dot{M}$ variations on timescales shorter than the local viscous time $t_{\mathrm{visc}}(r)$. The viscous time is, on the other hand, the longest timescale of the accretion flow; therefore, no significant fluctuations at longer timescales can be produced at any given radius. Hence, a radius r contributes to $\dot{M}$ and X-ray flux variations predominantly at a frequency $f \sim t_{\mathrm{visc}}(r)^{-1}$ [26, 7]. The broad range of variability timescales observed in the X-ray power density spectra (Fig. 2.6) is explained by the broad range of radii at which viscosity fluctuations are generated. If viscosity fluctuations at all radii have the same relative amplitude, a power-law spectrum $P_\nu \propto \nu^{-1}$ will naturally appear [26], in qualitative agreement with observations (Fig. 2.6). The picture of inward-propagating fluctuations outlined above also successfully explains the observed linear relation between rms amplitude of aperiodic variability and total X-ray flux in black holes [62] and the nearly logarithmic dependence of the time lag between time series in different energy bands on the photon energy [25].

2.6.2 Very Low Frequency Break and Accretion Disk Corona

Owing to the finite size of the accretion disk, the longest timescale in the disk is restricted by the viscous time on its outer boundary $t_{\mathrm{visc}}(R_d)$. Below this frequency, X-ray flux variations are uncorrelated; therefore, the power density spectrum should become flat at $f \leq t_{\mathrm{visc}}(R_d)^{-1}$. This explains the low frequency break clearly seen in the power spectrum of Cyg X-2 (Fig. 2.6). If there are several components in the accretion flow, for example, a geometrically thin disk and a diffuse corona above it, several breaks can appear in the power spectrum at frequencies corresponding to the inverse viscous timescale of each component. It can be easily demonstrated [21] that the break frequency is related to the orbital frequency of the binary via

$$f_{\mathrm{break}} = 3\pi\alpha(1+q)^{-1/2}\,(h/r)^2\,(R_d/a)^{-3/2}\,f_{\mathrm{orb}}$$

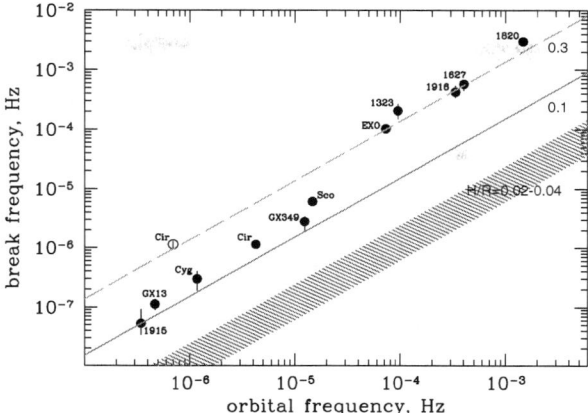

Fig. 2.7 Relation between the break frequency ($\sim t_{\mathrm{visc}}^{-1}$ at the outer edge of the accretion disk) and the orbital frequency of the binary system. The *shaded area* shows the dependence f_{visc} vs. f_{orb} expected for the standard gas pressure supported Shakura-Sunyaev disk ($h/r \sim$ few $\times 10^{-2}$). *Straight solid* and *dashed lines* are predictions for larger values of the disk thickness h/r, as indicated by the numbers on the plot. The two points for Cir X-1 correspond to the original data (*open circle*) and after correction for the eccentricity of the binary orbit (*filled circle*). Adopted from [21]

For low-mass X-ray binaries (Roche-lobe filling systems) this becomes $f_{\mathrm{break}} \propto (h/r)^2 f_{\mathrm{orb}}$. Such very low frequency breaks indeed are observed in a statistically representative sample of low mass X-ray binaries, and the break frequency is inversely proportional to the orbital period of the binary (Fig. 2.7, [21]).

However, measured values of the break frequency imply that the viscous time of the accretion flow is a factor of ≥ 10 shorter than predicted by the standard theory of accretion disks (cf. shaded area in Fig. 2.7). This suggests that significant fraction of the accretion $\dot{M}$ occurs through the geometrically thicker coronal flow above the standard thin disk. Note that the existence of the Shakura-Sunyaev disk underneath the coronal flow is required by optical and UV observations of LMXBs indicating the presence of the optically thick media of extent comparable with the Roche-lobe size [21]. The aspect ratio of the coronal flow implied by the $f_{\mathrm{br}}/f_{\mathrm{orb}}$ measurements, $h/r \sim 0.1$, corresponds to a gas temperature of $T \sim 0.01\, T_{\mathrm{vir}}$. The corona has moderate optical depth in the radial direction, $\tau_T \sim 1$, and contains $\leq 10\%$ of the total mass of the accreting matter (but the fraction of $\dot{M}$ is much larger, probably ~ 0.5). These estimates of temperature and density of the corona are in quantitative agreement with the parameters inferred by the X-ray spectroscopic observations by Chandra and XMM-Newton of complex absorption/emission features in LMXBs with large inclination angle.

In this picture, the red noise (power-law) component of the observed variability of the X-ray flux is defined by the viscosity and/or $\dot{M}$ fluctuations generated in the coronal flow rather than in the geometrically thin disk. Fluctuations produced in the standard Shakura-Sunyaev disk have much smaller amplitude due to a ~ 10 times longer viscous timescale (see discussion in Sect. 2.6.1) and do not contribute significantly to the observed variability of X-ray flux. This is further supported by the lack of variability in the soft black body component in Cyg X-1, as discussed below.

2.6.3 High and Very High Frequencies

In addition to the power-law component of red noise dominating in the larger part of the low frequency range, power density spectra usually have significant excess of power in the high frequency end, $f \geq 10^{-2} - 10^{-1}$ Hz (Fig. 2.6). This excess power is often referred to as a "band-limited noise," due its limited scope in frequency. It is this part of the power spectrum that is the subject of investigation in the majority of "standard" timing analysis projects, for example, in the ones based on a typical RXTE/PCA observation.

In black holes, this component is most prominent in the hard spectral state and usually disappears in the soft state (Fig. 2.6, 2.8). Plotted in the νP_ν units, it shows several broad humps, separated by a $\sim$0.5–1 decade in frequency (Fig. 2.8). The centroid frequencies of these bumps vary with time but they usually do correlate with each other [16, 65]. It has been proposed that they may be identified with the (much more narrow) QPO peaks observed in the power spectra of accreting neutron stars. When the proper identifications are made, black hole and neutron star systems appear to follow the same global relation between QPO frequencies [41, 65, 38, 6]. This suggests that the features on the power density spectra and their frequencies are the property of the accretion disk, their existence being unaffected by the nature of the compact object.

Based on the amplitude and timescale considerations, it is plausible to link the origin of the band-limited noise component to the inner hot optically thin flow in

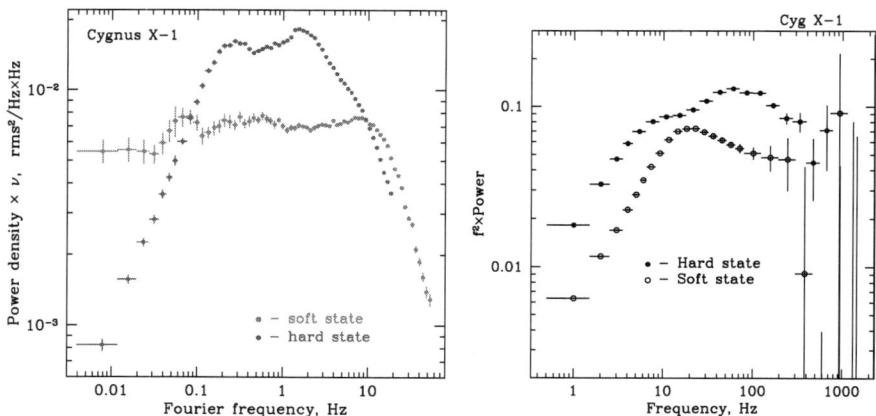

Fig. 2.8 Power density spectra of Cyg X-1 in the soft and hard spectral states. The *right hand panel* shows the very high frequency part of the power density spectrum. The power is shown in units of νP_ν (*left*) and $\nu^2 P_\nu$ (*right*). Adopted from [18] and [43] respectively

the region of the main energy release. For a $10M_\odot$ black hole, the characteristic time scales in this region ($r \sim 100r_g$) are: $t_K \sim 1$ s and $t_{\text{visc}} \sim 10$ s, the latter was computed assuming $\alpha \sim 0.2$ and $h/r \sim 0.2$. This suggest that the two main peaks observed in the νP_ν spectra may (or may not) be related to the dynamical and viscous timescales at the inner edge of the truncated accretion disk and their origin may be linked to the interaction between the geometrically thin outer accretion disk and the optically thin hot inner flow. Note that what appears to be the lower frequency hump in the νP_ν plot is in fact a break, below which the power density distribution is flat, down to the frequency where the red noise component becomes dominant (Fig. 2.6). As a flat power density distribution means lack of correlation between events on the corresponding timescales, we can draw an analogy with the very low frequency breaks associated with the viscous timescale at the outer edge of the disk (Sect. 2.6.2). This further supports the association of the low frequency hump in the νP_ν plot, Fig. 2.8 (= break in the power density spectrum) with the viscous timescale at the outer boundary of the hot inner flow. One can estimate the truncation radius of the geometrically thin disk (= outer radius of the hot inner flow in the sombrero configuration, Sect. 2.2) equating the viscous time with the inverse break frequency, $f_{\text{br}}^{-1} \sim 5$ s: $r_{\text{tr}} \sim 10^2 r_g$ assuming $\alpha = 0.2$ and the aspect ratio of the hot inner flow $h/r = 1/3$. This number is in a good agreement with the disk truncation radius inferred by other measurements, e.g., derived from the variability of the reflected component ([18], Sect. 2.7).

The right panel in Fig. 2.8 also demonstrates that there is considerable power at very high frequencies, at the level of $\nu P_\nu \sim 1 - 3\%$ at $f \sim 10^2$ Hz. The millisecond range includes a number of important characteristic timescales in the vicinity of the compact object. The Keplerian frequency at the last marginally stable orbit for a non-rotating black hole is $f_K \sim 220\,(M/10M_\odot)^{-1}$ Hz. The relativistic precession frequencies associated with Keplerian motion in the vicinity of the compact object are also in this range [53]. The typical value of the sound crossing time in the gas pressure-supported thin disk corresponds to the frequency of a few hundred Hz. The light-crossing time for the region of $\sim 10r_g$ corresponds to the frequency of ~ 1 kHz. However, the observed power spectra break somewhere between $f \sim 20$ Hz (soft state) and $f \sim 50$–100 Hz (hard state) and do not exhibit any detectable features beyond these frequencies. The upper limit on the fractional rms of aperiodic variability in the 500–1000 Hz range is $\approx 1\%$. Similar upper limit can be placed on the rms amplitude of narrow features anywhere in this range, $\approx 0.9\%$ assuming width $\nu/\Delta\nu = 20$ (all upper limits at 95% confidence) [43]. This suggests that none of the above processes leads to a significant variability and, in particular, no resonances are present at their characteristic timescales. The nature of the high frequency breaks is unclear. It may be plausible to associate them with the properties of the emission mechanism of the hard component, for example, with the details and timescales of heating and cooling of electrons in the Comptonization region. Obviously, these properties are different for the hard component in the soft and hard states as indicated by the difference in the high frequency power density spectra shown in the right panel in Fig. 2.8.

2.6.4 (Lack of) Variability in the Disk Emission

The energy spectra of black holes in the soft state often have two spectral components, a soft black body-like component emitted by the optically thick accretion disk and a hard power-law component of thermal or non-thermal origin formed in the optically thin media (Sect. 2.3, Figs. 2.1,2.3). The study of the variability properties of these two components in black hole X-ray novae in the early 1990s with the Ginga and GRANAT observatories led to the conclusion that most of the variability of the X-ray emission in the soft state is usually associated with the hard spectral component (e.g., [36, 58, 15]), the thermal disk emission generally being much more stable. This conclusion has been later confirmed by a more rigorous and quantitative analysis [7, 18, 20] whose main results are summarized below.

The energy dependence of the variability can be characterized by the frequency-resolved energy spectrum [44]. It is defined as a set of Fourier amplitudes computed from light curves in different energy channels. The Fourier amplitudes are integrated over the frequency range of interest and expressed in the units of flux. Its advantages over simple energy-dependent fractional rms are the possibility to use conventional (i.e., response folded) spectral approximations and to compare its shape with shapes of various spectral components present in the average energy spectrum of the source. However, it cannot be always regarded as the energy spectrum of the variable component. For this to be possible, certain conditions must be fulfilled [20], for example, (i) independence of the shape of the frequency-resolved spectrum on the Fourier frequency and (ii) the absence of time lags between flux variations at different energies. If these two conditions are satisfied, the spectral variability can be represented as $F(E, t) = S_0(E) + I(t)S(E)$.

These conditions are fulfilled in the soft state of Cyg X-1 (Fig. 2.9): therefore, the Fourier frequency-resolved spectrum represents the energy spectrum of the variable part of the X-ray emission. Remarkably, it coincides with the average source spectrum at $E \geq 7 - 8$ keV where the contribution of the disk emission becomes small (Fig. 2.10). Moreover, source light curves in different energy channels allow a linear decomposition in the form $F(E, t) = A(E) + I(t)B(E)$ [7]. The constant part of the source emission $A(E)$ coincides with the spectrum of the accretion disk, while the spectrum of the variable part $B(E)$ coincides with the frequency-resolved spectrum and can be described by the Comptonization model (Fig. 2.10). Thus, on timescales from ~ 100 ms to ~ 500 s (at least) the source variability is due to variations in the flux of the hard component, whose shape is kept constant in the course of these variations, the amplitude of variability of the disk emission being significantly smaller.

This can be explained in the model of inward-propagating fluctuations outlined in Sect. 2.6.1. The hard component is associated with the optically thin hot flow whose aspect ratio must be comparable to unity, $h/r \sim 0.1 - 1$. Therefore, the viscous timescale in the hot flow is $\sim 10^2$–10^4 times shorter than in the geometrically thin disk from which the soft component originates (cf. Sect. 2.6.2). This makes the geometrically thick flow more "transparent" for the high frequency perturbations. Depending on the particular value of h/r and of the α-parameter, the

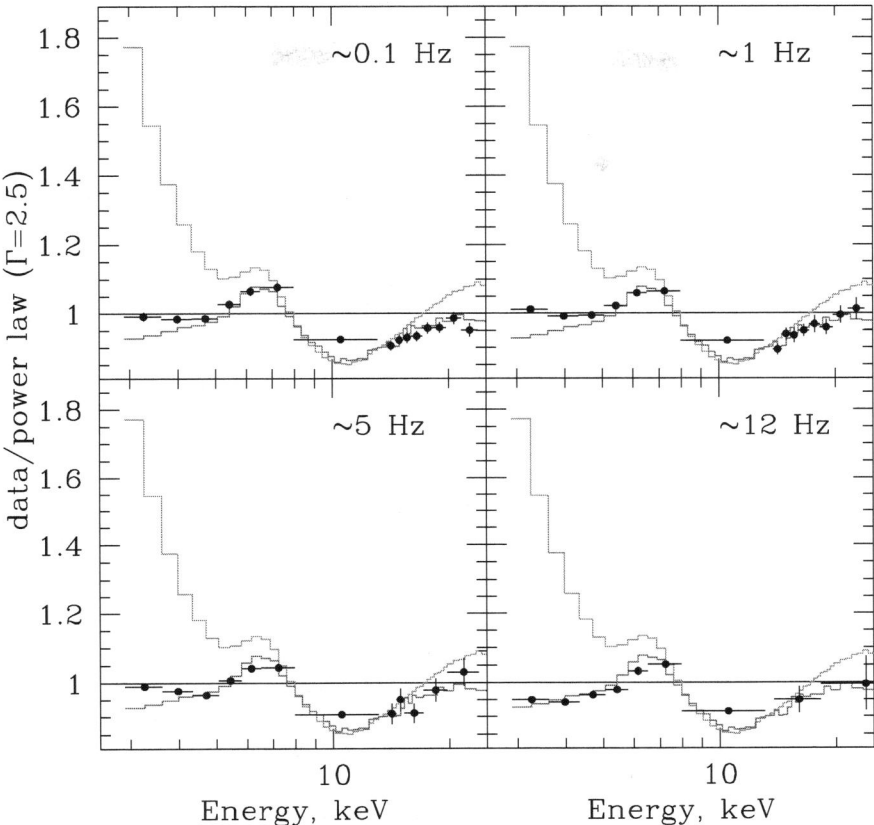

Fig. 2.9 Frequency-resolved spectra of Cyg X-1 in the soft state (June 16–18, 1996). The numbers in the *upper-right corner* of each panel indicate the median frequency. In each panel, the upper histogram shows the average spectrum, the lower histogram shows the frequency-resolved spectrum at low frequencies, 0.002–0.033 Hz (the same data as in the *left panel* of Fig. 2.10). The spectra are plotted as ratio to a power-law spectrum with photon index $\Gamma = 2.5$ and low energy absorption $N_H = 6 \cdot 10^{21}$ cm^{-2}

viscosity or $\dot{M}$ fluctuations on the thermal and even dynamical timescales will be propagated inward without significant damping and will modulate the accretion rate at all smaller radii, including the region where X-ray emission is formed, leading to a significant modulation of X-ray flux. This is not the case for the geometrically thin disk where the viscous timescale is $\sim 10^3$–10^4 times longer than the dynamical and thermal timescales; therefore, high frequency perturbations will be damped (Sect. 2.6.1). It is plausible to expect that perturbations on the thermal and dynamical timescales have larger amplitude than perturbations on the viscous timescales, thus explaining the significantly smaller variability amplitude of the disk emission.

The lack of variability of the disk emission and its interpretation are consistent with the conclusion of the Sect. 2.6.2 made from completely independent arguments.

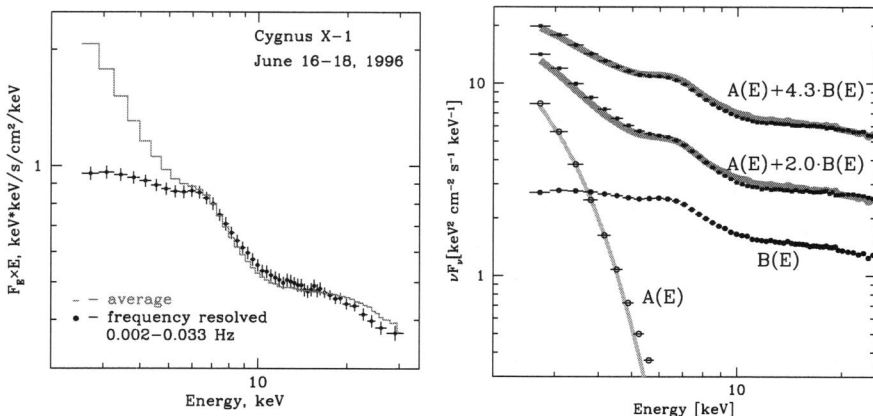

Fig. 2.10 *Left:* Average and frequency-resolved spectrum of Cyg X-1 in the 1996 soft state. *Right:* Spectra of "constant" (*open circles* – $A(E)$) and "variable" (*solid circles* – $B(E)$) components derived from the linear fits of the correlation between count rate in different channels. The normalization of the "variable" component $B(E)$ is arbitrary. For comparison, the *light grey* (*yellow* in the color version) *curve* shows the spectrum of a multicolor disk emission with a characteristic temperature of 0.5 keV. The two upper spectra (*solid squares*) were averaged over the periods of time when the count rate above 9 keV was high and low, respectively. The dark *grey* (*green*) *lines* show that these spectra can be reasonably well (within 10–15%) approximated by a model $M(E) = A(E) + I * B(E)$ consisting of the stable and variable spectral components where I (the normalization of the variable component) is the only free parameter. The *right panel* is adopted from [7]

Namely, based on the location of the low-frequency break in the power density spectra of neutron stars we concluded that the bulk of variability seen at low frequencies originates in the optically thin coronal flow with an aspect ratio of $h/r \sim 0.1$, rather than in the underlying geometrically thin disk.

Finally, it should be noted that variations of the soft component may also arise from the variations of the disk truncation radius. Such variations may be absent if the geometrically thin disk extends to the last marginally stable orbit ($3r_g$ for a Schwarzschild black hole) but appear when the disk truncation radius is larger. Indeed, detailed study of the 1996 soft state of Cyg X-1 used as an example here demonstrated that on a number of occasions, mostly at the beginning and in the end of soft state episodes, the soft component was strongly variable. Notably, the power density spectrum during these periods had a complex shape, significantly different form the simple power law shown in Figs. 2.6, 2.8.

2.7 Variability of the Reflected Emission

The reflected component (Sect. 2.4) arises from the reprocessing of the Comptonized emission in the accretion disk (Fig. 2.2), therefore, it should be expected to show some degree of variability as the Comptonized radiation is strongly

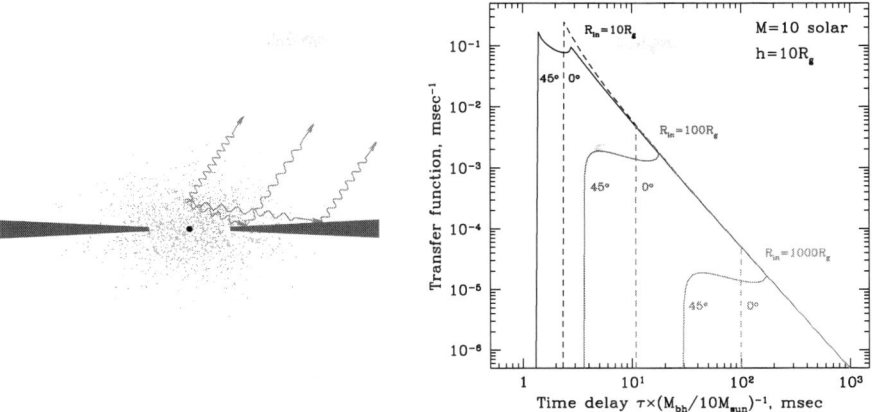

Fig. 2.11 The two effects defining the response of the reflected emission to variations of the Comptonized emission are the finite light travel time τ_d from the hot inner flow to the accretion disk and the finite size of the accretion disk itself, $\Delta r_d/c \sim \Delta \tau_d$ (*left panel*). While the first factor introduces a time delay between variations of the reflected and Comptonized components, the finite size of the disk itself leads to the suppression of high frequency variations in the reflected emission. The *right panel* shows the Green function of the time response of the geometrically thin disk around a $10 M_\odot$ black hole (from [18]). The numbers at the curves mark the disk truncation radius and inclination

variable. The characteristic times of absorption/emission processes in the accretion disk are negligibly small; therefore, the main factors defining the response of the reflected flux to variations of the Comptonized radiation are related to the light travel times (Fig. 2.11). Namely, they are (i) the finite light travel time from the source of primary radiation to the reflector, i.e., from the Comptonization region to the accretion disk and (ii) the finite size of the reflector. The first will introduce a time delay between variations of the reflected and Comptonized components. The amplitude of this delay is $\tau_d \sim r_d/c \sim 10 \left(r_d/100 r_g \right) (M/10 M_\odot)$ ms. The finite size of the disk itself will lead to suppression of the high frequency variations in the reflected component – the accretion disk acts as a low-pass filter. It seems to be possible to estimate the cut-off frequency from the size of the accretion disk making the main contribution to the reflected flux, $\Delta r_d \sim r_d$, thus leading to $f_{\text{cut}} \sim \Delta \tau_d^{-1} \sim (\Delta r_d/c)^{-1} \sim 100$ Hz, i.e., beyond the frequency range of the bulk of observed variability (Fig. 2.8). However, calculations of the transfer function [18] show that at the frequency $f_{\text{cut}} \sim \Delta \tau_d^{-1}$ the variability signal is suppressed by a significant factor of ~ 10–20, whereas a noticeable suppression of variability by a factor of ~ 2 or more occurs at frequencies ~ 10 times lower. As these effects directly depend on the mutual location and the geometry of the Comptonization region and the accretion disk, their observation is a powerful tool in studying the geometry of the accretion flow.

As suggested in [44, 18], the variability of the reflected emission can be studied using the methods of frequency-resolved spectroscopy (Sect. 2.6.4). This method is based on the analysis of Fourier amplitudes and therefore washes out phase

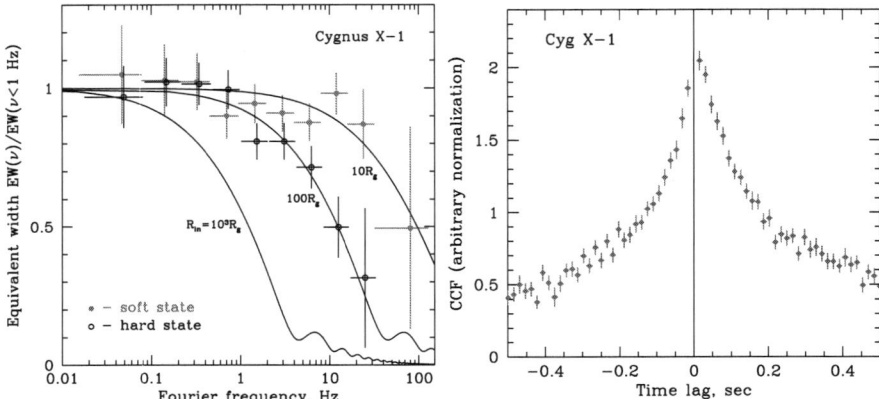

Fig. 2.12 Suppression of high frequency variations (*left*) and time delay (*right*) in the reflected emission. The *left panel* shows the equivalent width of the iron fluorescent line (= ratio of the fractional rms of the iron line flux to that of the underlying continuum) vs. Fourier frequency. The model curves are for an isotropic point source at the height $h = 10r_g$ on the axis of a flat disk with inner truncation radius of 10, 100 and $1000r_g$ (assuming a $10M_\odot$ black hole) and with an inclination angle of $50°$. A narrow line was assumed in calculations. The *right panel* shows the cross-correlation function of the iron line emission and the power-law continuum. Positive lags mean delay of the line emission with respect to the continuum

information, precluding the study of time delay effects. However, it provides a convenient way to explore the frequency dependence of the variability of the iron line flux and to compare it with that of the continuum emission which is dominated by the Comptonized radiation. Using this method, [44, 18] showed that in the soft spectral state, variations of the reflected component have the same frequency dependence of the rms amplitude as the Comptonized emission up to frequencies $\sim 30\,\mathrm{Hz}$ (Fig. 2.12). This would be expected if, for instance, the reflected flux was reproducing, with a flat response, the variations of the primary radiation down to timescales of $\sim 30\text{--}50\,\mathrm{ms}$. The sensitivity of their analysis was insufficient to study shorter timescales. In the hard spectral state, on the contrary, the variability of the reflected flux is significantly suppressed in comparison with the direct emission on timescales shorter than $\sim 0.5\text{--}1\,\mathrm{s}$. These findings are to be compared with the predictions of the simple model for the time response of the disk to variations of the primary emission. Assuming that suppression of the short-term variability of the reflected emission is caused by the finite light-crossing time of the accretion disk, one can estimate the truncation radius of the accretion disc, $r_d \sim 100r_g$ in the hard spectral state and $r_d \leq 10r_g$ in the soft spectral state (Fig. 2.12). This agrees well with the interpretation of the spectral states in black holes in the "sombrero" geometry of the accretion flow (Sect. 2.2).

In order to study the time delay in the reflected emission, one would need to separate the reflected component from the main emission in time-resolved energy spectra. This can be done most easily for the fluorescent iron line, while the reflection continuum is more difficult to separate due to its breadth in the energy domain.

An attempt to perform such an analysis based on RXTE/PCA data of Cyg X-1 in the hard state is presented in the right panel in Fig. 2.12. In each bin of the light curve with 16 ms resolution, the spectrum in the 3–20 keV band was linearly decomposed into power-law component, reflected continuum, and 6.4 keV iron line emission. The cross-correlation of the light curves of the iron line flux and the power-law component is shown in Fig. 2.12. The cross-correlation function shows a clear asymmetry, suggesting a time delay of the order of ∼10–15 ms. The amplitude of the possible time delay corresponds to a disk truncation radius of ∼$100r_g$, in good agreement with the number obtained from the analysis of high-frequency variations in the iron line flux (left panel in Fig. 2.12) and also with the numbers tentatively suggested by spectral analysis.

The results shown in Fig. 2.12 seem to suggest a rather consistent picture and favor to the "sombrero"-type configuration of the accretion flow, with the spectral state transition being related to a change of the disk truncation radius. However, a caveat is in order. While the rms amplitude behavior of the iron line emission is a rather robust observational result, the search of the time delay of the reflected component requires separation of line and continuum emission components. On ∼10 ms timescales, this cannot be done through direct spectral fitting because of insufficient statistics, even with the large collecting area of the PCA instrument aboard RXTE and requires more sophisticated data analysis techniques, for example, the one used to produce the cross-correlation function shown in Fig. 2.12. Second, although the interpretation of these results in terms of finite light travel times is the most simple and straightforward, alternative scenarios are also possible as discussed in length in [18]. Nevertheless, the former interpretation is, in my view, the simplest and most attractive one and is further supported by the results of the spectral analysis, as described in the next section.

2.8 $R - \Gamma$ and Other Correlations

Observations show that spectral and timing parameters of accreting black holes often change in a correlated way, e.g., [16, 42, 39]. One of the most significant correlations is the one between the photon index of the Comptonized spectrum, the amplitude of the reflected component, and the characteristic frequencies of aperiodic variability (Figs. 2.13, 2.16). The correlation between spectral slope and reflection amplitude is also known as $R - \Gamma$ correlation [16, 66]. Its importance is further amplified by the fact that it is also valid for supermassive black holes (Fig. 2.14) [67].

2.8.1 $R - \Gamma$ Correlation

The spectrum formed by the unsaturated Comptonization of low frequency seed photons with characteristic temperature T_{bb} on hot electrons with temperature T_e has a nearly power-law shape in the energy range from ∼$3kT_{bb}$ to ∼kT_e [55]. For

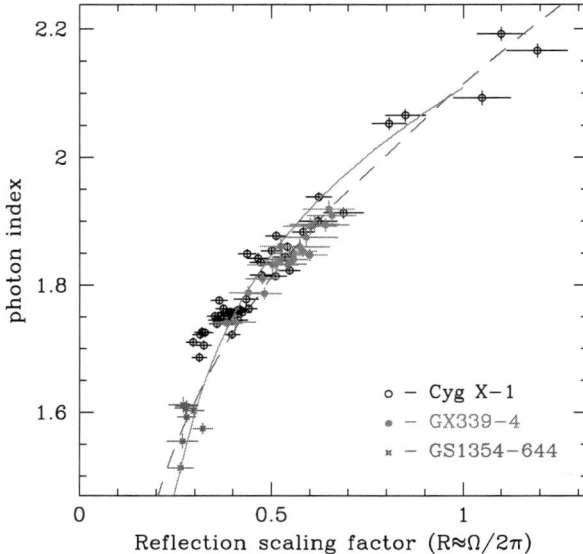

Fig. 2.13 The $R - \Gamma$ correlations between the photon index Γ of the Comptonized radiation and the relative amplitude of the reflected component R in three "well-behaved" black hole systems Cyg X-1, GX 339-4, and GS 1354-644. The *solid* and *dashed lines* show the the dependence $\Gamma(R)$ expected in the disk-spheroid model and in the plasma ejection model discussed in the text. Adopted from [17]

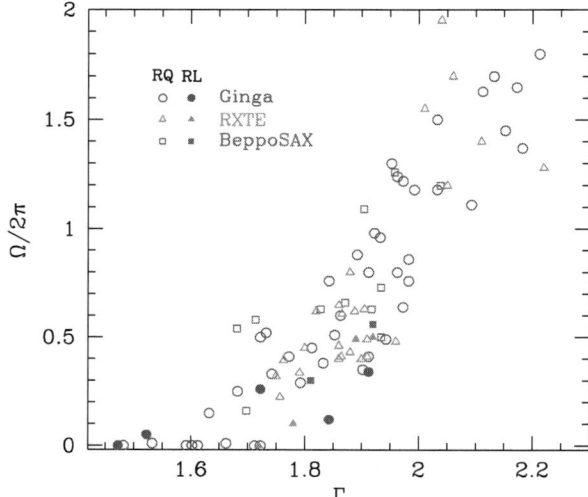

Fig. 2.14 $R - \Gamma$ relation for supermassive black holes, from [67]

the parameters typical for black hole X-ray binaries in the hard spectral state, this corresponds to the energy range from ~0.5–1 keV to ~50–100 keV (e.g., Fig. 2.2). The photon index Γ of the Comptonized spectrum depends in a rather complicated way on the parameters of the Comptonizing media, primarily on the electron temperature and the Thompson optical depth [55]. It is more meaningful to relate Γ to the Comptonization parameter y or, nearly equivalently, to the Compton amplification factor A. The latter describes the energy balance in the corona and is defined as the ratio of the energy deposition rate into hot electrons and the energy flux brought into the Comptonization region by soft seed photons. The concrete shape of the $\Gamma(A)$ relation depends on the ratio T_{bb}/T_e of the temperatures of the seed photons and the electrons, the Thomson optical depth, and the geometry, but broadly speaking, the higher the Compton amplification factor, the harder is the Comptonized spectrum [56, 10, 22, 15]. This is illustrated by the results of Monte-Carlo simulations shown in Fig. 2.15.

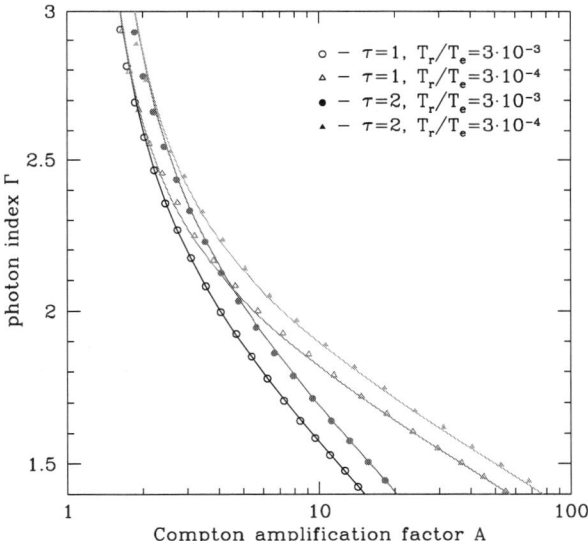

Fig. 2.15 Relation between the photon index of Comptonized radiation Γ and the Compton amplification factor A. The symbols show results of Monte-Carlo simulations assuming spherical geometry for different values of parameters of the Comptonization region and soft seed photons. The solid lines are calculated using Eq. (2.1). Adopted from [17]

The strength of the reflected component in the spectrum depends on the fraction of the Comptonized radiation intercepted by the accretion disk (Sect. 2.4). The latter is defined by the geometry of the accretion flow, namely, by the solid angle Ω_{disk} subtended by the accretion disk as seen from the corona. In addition, the spectrum of the reflected emission depends on the ionization state of the disk, in particular, its low energy part which is formed by the interplay between Thomson scattering and photoabsorption and fluorescence by metals. The problem is further complicated by the fact that the ionization state of the disk can be modified by the Comptonized radiation.

Observations show that there is a clear correlation between the photon index of the Comptonized radiation Γ (i.e., the Comptonization parameter) and the relative amplitude of the reflected component R (Fig. 2.13). Softer spectra (lower value of the Comptonization parameter y and of the Compton amplification factor A) have stronger reflected component, revealing itself, for example, via a larger equivalent width of the iron fluorescent line. The existence of this correlation suggests that there is a positive correlation between the fraction of the Comptonized radiation intercepted by the accretion disk and the energy flux of the soft seed photons to the Comptonization region [66]. This is a strong argument in favor of the accretion disk being the primary source of soft seed photons to the Comptonization region. Indeed, in the absence of strong beaming effects a correlation between Ω_{disk} and the seed photons flux should be expected since an increase in the solid angle of the disk seen by the hot electrons ($= \Omega_{\mathrm{disk}}$) should generally lead to the increase of the fraction of the disk emission reaching the Comptonization region.

2.8.2 Toy Models

We illustrate the above considerations with two simple and idealized models having a different cause of change of Ω_{disk}. In the first, the disk-spheroid model (cf. "sombrero" configuration, Sect. 2.2), an optically thin uniform hot sphere with radius r_{sph}, the source of the hard Comptonized radiation is surrounded by an optically thick cold disk with an inner radius r_{disk}, Ω_{disk} depending on the ratio $r_{\mathrm{disk}}/r_{\mathrm{sph}}$. Propagation of the the disk toward/inward the hot sphere (decrease of $r_{\mathrm{disk}}/r_{\mathrm{sph}}$) leads to an increase of the reflection scaling factor R, a decrease of the the Compton amplification factor A, and a steepening of the Comptonized spectrum. In such a context the model was first studied by Zdziarski et al. [66]. In the second, the plasma ejection model, the value of the Ω_{disk} is defined by the intrinsic properties of the emitting hot plasma, particularly by its bulk motion with mildly relativistic velocity toward or away from the disk, which itself remains unchanged [5]. In the case of an infinite disk, values of the reflection scaling factor R below and above unity correspond to the hot plasma moving, respectively, away from and toward the disk.

Both models predict a relation between reflection R and Compton amplification factor A which can be translated to $\Gamma(R)$ given a dependence $\Gamma(A)$ of the photon index of the Comptonized spectrum on the amplification factor. The relation between Γ and A can be approximated by

$$A = (1 - e^{-\tau_T}) \cdot \frac{1 - \Gamma}{2 - \Gamma} \cdot \frac{(T_e/T_{bb})^{2-\Gamma} - 1}{(T_e T_{bb})^{1-\Gamma} - 1} + e^{-\tau_T}. \qquad (2.1)$$

This formula is based on the representation of the Comptonized spectrum by a power law in the energy range $kT_{bb} - kT_e$ and takes into account that a fraction $e^{-\tau_T}$ of the soft radiation will leave the Comptonization region unmodified. Despite its simplicity, it agrees with the results of the Monte-Carlo calculations with reasonable accuracy for optical depth $\tau_T \sim 1$ and $T_{bb}/T_e \sim 10^{-5} - 10^{-3}$ (Fig. 2.15).

The expected $\Gamma(R)$ relations are shown in Fig. 2.13. With a proper tuning of the parameters, both models can reproduce the observed shape of the $\Gamma(R)$ dependence and in this respect are virtually indistinguishable. The models plotted in Fig. 2.13 were calculated with the following parameters: the disk-spheroid model assumes the disk albedo $a = 0.1$, Thomson optical depth of the cloud $\tau_T = 1$, and the ratio of the temperature of the seed photons to the electron temperature $T_{bb}/T_e = 10^{-4}$. I note that the latter value is too small for stellar mass black holes, a more realistic one being in the range $T_{bb}/T_e = 10^{-3}$. However, this does not invalidate the model as a number of significant effects are ignored in this calculation which can modify the $\Gamma(R)$ dependence, for example, disk inclination, gravitational energy release in the disk, etc. The exact importance of these factors is yet to be determined. The plasma ejection model parameters are $a = 0.15$, $\tau_T = 1$, $T_{bb}/T_e = 3 \cdot 10^{-3}$, and $\mu_s = 0.3$. The observed range of the reflection $R \sim 0.3 - 1$ and the slope $\Gamma \sim 1.5 - 2.2$ can be explained assuming variation of the disk radius from $r_{\text{disk}} \sim r_{\text{sph}}$ to $r_{\text{disk}} \sim 0$ in the disk-spheroid model or variation of the bulk motion velocity from $v \sim 0.4c$ away from the disk to $v \sim 0$ in the plasma ejection model.

Of course these models are very simple and schematic and the real configuration of the accretion flow is likely to be far more complex. They are presented here with the only purpose to demonstrate that simple geometrical considerations can successfully explain the observed correlation between parameters of the Comptonized and reflected emission in black holes. More sophisticated scenarios are considered, for example, in [28].

2.8.3 Characteristic Frequencies of Variability

As discussed in Sect. 2.6.3, the power density spectra of accreting black holes above $f \geq 10^{-2}$ Hz have a number of bumps and peaks which define several frequencies characterizing the variability timescales in the accretion flow. These frequencies usually correlate with each other, therefore, almost any of them may be used to represent characteristic variability timescales. As illustrated in Fig. 2.16, a tight correlation exists between the reflection amplitude R and the characteristic variability frequencies – an increase in the amplitude of the reflected component in the energy spectrum is accompanied by an increase in the variability frequencies. This correlation covers a remarkably broad dynamical range, nearly two orders of magnitude in frequency.

Although the precise nature of the characteristic noise frequencies is still unknown, it is plausible that they are associated with the Keplerian and viscous timescales of the disk and corona at various characteristic radii, for example, at the truncation

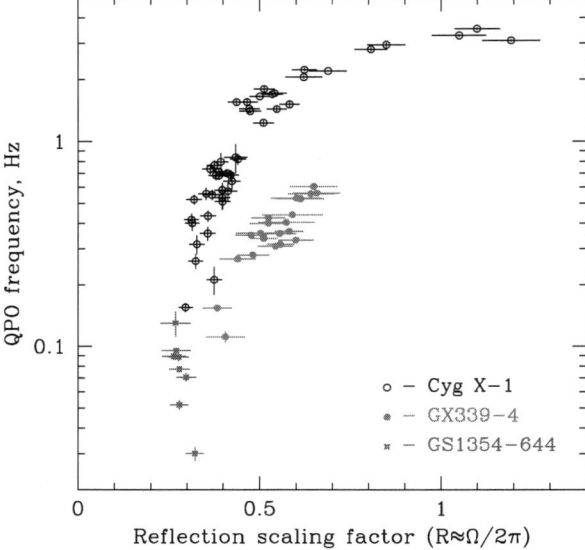

Fig. 2.16 The correlations between the amplitude of the reflected component R and the characteristic frequencies of aperiodic variability. The frequency of the second peak in the νP_ν plot (Fig. 2.8) was used to represent the latter

radius of the disk (Sect. 2.6.3). If this is true, the correlation between R and ν can be easily understood, at least qualitatively. Indeed, in the truncated disk picture the increase in reflection is caused by the inward propagation of the inner disk boundary, hence, it is accompanied by an increase in the Keplerian frequency at the disk truncation radius and the corresponding increase in the characteristic noise frequencies.

2.8.4 Doppler Broadening of the Iron Line

The spectrum of the emission, reflected from a Keplerian accretion disk, is modified by special and general relativity effects [12], in particular, the width of the fluorescent line of iron is affected by the Doppler effect due to Keplerian motion in the disk (Sect. 2.4). If the increase in the reflection amplitude is caused by the decrease in the inner radius of the accretion disk, a correlation should be expected between the amplitude of reflected emission and its Doppler broadening, in particular, the Doppler width of the iron line. Such a correlation is a generic prediction of the truncated disk models and might be used to discriminate between different geometries of the accretion flow. The energy resolution of RXTE/PCA, whose data have been used for this study [19], is not entirely adequate for the task to accurately measure the relativistic smearing of the reflection features. However, the data shown in Fig. 2.17 suggest a correlated behavior of the reflection and the Doppler broadening of the fluorescent line of iron.

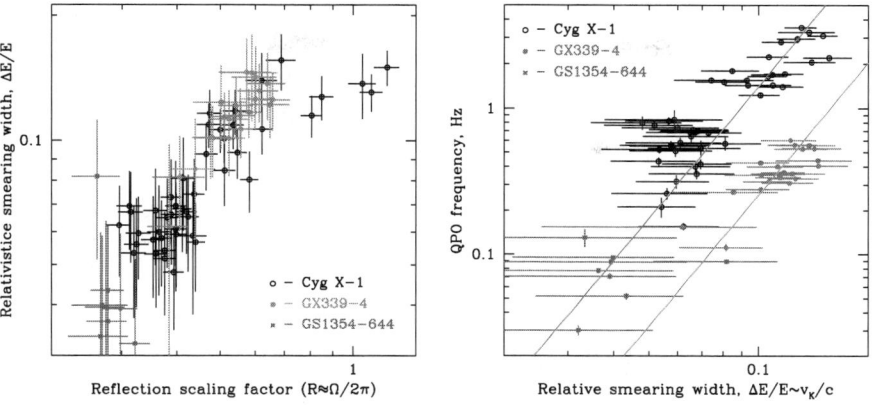

Fig. 2.17 Relation between the Doppler broadening of the iron line and the reflection scaling factor (*left*) and the characteristic frequency of variability (*right*). The *straight lines* in the right panel show the dependence $\Delta E/E \propto \nu_{QPO}^{1/3}$ expected in the truncated disk picture if the characteristic frequencies of variability were proportional to the Keplerian frequency at the inner boundary of the accretion disk

Speculating further, if the characteristic frequencies of variability are proportional to the Keplerian frequency at the inner boundary of the disk, they should scale as

$$\nu_{QPO} \propto \omega_K \propto r_{\mathrm{disk}}^{-3/2}.$$

As the reflected emission is likely to originate primarily from the innermost parts of the accretion disk, closest to the source of Comptonized radiation, the effect of the Doppler broadening should be proportional to the Keplerian linear velocity at the inner edge of the disk:

$$\frac{\Delta E}{E} \propto \frac{\mathrm{v}_K}{c} \sin i \propto r_{\mathrm{disk}}^{-1/2} \sin i.$$

Therefore, one might expect that the characteristic frequencies of variability and the Doppler broadening of the fluorescent line should be related via:

$$\nu_{QPO} \propto \frac{(\Delta E/E)^3}{M_{\mathrm{BH}} \sin^3 i},$$

where M_{BH} is the black hole mass and i is inclination of the binary system. The PCA data indicate that such dependence might indeed be the case (Fig. 2.17, right panel). However, an independent confirmation by observations with higher energy resolution instruments is still needed.

Thus, observations speak in favor of the truncated disk scenario. However, this point of view is not universally accepted and alternative interpretations and scenarios are being investigated. The counter arguments are based, for example, on the possible detection of the relativistic broad Fe K line and cool disk emission component in the hard state spectra of several black holes, suggesting that the optically thick disk might be present in the vicinity of the compact object in the hard state as well. For the detailed discussion the interested reader is referred to the original work, e.g., [34] and references therein.

2.9 Comparison with Neutron Star Binaries

Neutron star radii are most likely in the range $\sim$10–15 km, i.e., of the order of $\sim$3–4r_g for a $\sim$1.4$M_\odot$ object. This is comparable to the radius of the last marginally stable Keplerian orbit around a non-rotating black hole; hence, the efficiency $\eta = L_X/\dot{M}c^2$ of accretion onto a neutron star is not much different from that on a black hole. Therefore at comparable $\dot{M}$ accreting black holes and neutron stars would have comparable X-ray luminosities (but see below regarding the contribution of the boundary layer). The Eddington luminosity limit, however, is proportional to the mass of the central objects, and the typical maximum luminosity is by $M_{BH}/M_{NS} \sim 5$–10 times smaller for neutron stars. This is in general agreement with observations of peak luminosities of black hole transients and their comparison to luminosities of persistent and transient neutron star systems in the Milky Way.

The fact that the size of a neutron star is of the order of $\sim$3r_g, i.e., is comparable to the radius of the last marginally stable Keplerian orbit around a black hole also suggests that the structure of the accretion disk in both cases may be similar. This is indirectly confirmed by the existence of two spectral states in accreting neutron stars, whose properties are qualitatively similar to black holes (Fig. 2.1). A more direct argument is presented by the similarity of the accretion disk spectra in the soft state, which can be described by the same spectral models in black holes and neutron stars (e.g. [20]). Moreover, the power density spectra of the accretion disk in both types of systems follow a $P_\nu \propto \nu^{-\alpha}$ power law with α close to $\sim$1 (Sect. 2.6 and [20]). The strong magnetic field which may exist around young neutron stars can change the picture, causing disruption of the accretion disk at a large distance from the neutron star and modifying dramatically the structure of the accretion flow inside the magnetospheric radius $r_m \gg r_{NS}$. This may lead to the phenomenon of X-ray pulsations common in high-mass X-ray binaries and is not considered in this chapter.

The main qualitative difference between the two types of compact objects is obviously the existence of a solid surface of the neutron star which is absent in the case of a black hole. Neutron star rotation frequencies are typically in the few hundred Hz range, i.e., $\sim$ several times smaller than the Keplerian frequency near its surface, $\nu_K \sim$ kHz. Therefore, a boundary or spreading layer will appear near the surface of the neutron star where accreting matter decelerates from Keplerian rotation in the accretion disk down to the neutron star spin frequency and settles onto its surface (Fig. 2.18, left panels). In Newtonian approximation, half of the energy of the test particle on a Keplerian orbit is in the form of kinetic energy of the Keplerian

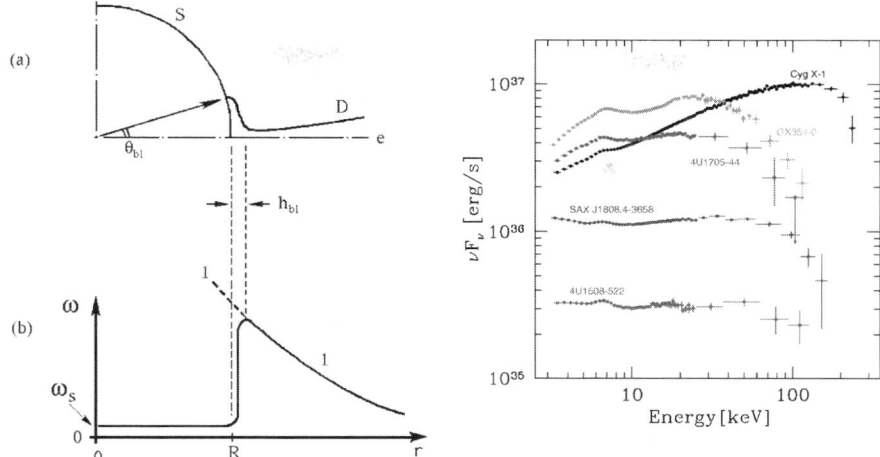

Fig. 2.18 The *left panels* show the geometry of the spreading layer on the surface of the neutron star (*upper*) and radial dependence of the angular velocity of accreting matter (*lower*). Adopted from [24]. On the *right*: Hard state spectra of several weakly magnetized accreting neutron stars and of a black hole Cyg X-1. Based on the data from RXTE observations

rotation; hence, for a non-rotating neutron star half of the energy release due to accretion would take place in the boundary/spreading layer. The effects of general relativity can increase this fraction, e.g., up to $\sim 2/3$ in the case of a neutron star with radius $r_{NS} = 3r_g$ [52, 50]. Rotation of the neutron star and deviations of the space–time geometry from Schwarzschild metric further modify the fraction of the energy released on the star's surface. A luminous spectral component emitted by the boundary layer will exist in the X-ray spectrum of an accreting neutron star, in addition to the emission from the accretion disk. The shape of the spectra of luminous neutron stars (soft state spectrum in Fig. 2.1) suggests that both the accretion disk and the boundary layer are in the optically thick regime. As the luminosities of the two components are comparable, but the emitting area of the boundary layer is smaller than that of the accretion disk, its spectrum is expected to be harder.

Looking from a different angle we may say that because of the presence of the stellar surface, at the same $\dot{M}$ neutron stars are approximately twice more luminous than black holes. Indeed, in the case of the black hole accretion the kinetic energy of Keplerian motion at the inner edge of the accretion disk is nearly all advected into the black hole. If the central object is a neutron star, this energy is released in the boundary/spreading layer on its surface, approximately doubling the luminosity of the source.[1] An interesting consequence of this is that the ratio of L_{Edd}/M is larger for neutron stars than for black holes.

---

[1] A more accurate consideration should take into account geometry of the problem, namely the emission diagrams and the orientation with respect to the observer of the emitting surfaces of the boundary layer and accretion disk. This makes the boundary layer contribution to the observed emission dependent on the inclination of the binary system, see, for example, [20].

Due to the similarity of the spectra of the accretion disk and the boundary layer, the total spectrum has a smooth curved shape, which is difficult to decompose into separate spectral components. This complicates analysis and interpretation of the neutron star spectra and in spite of a very significant increase in the sensitivity of X-ray instruments made in the last decade, still often leads to ambiguous and contradicting results, even based on physically motivated spectral models. This degeneracy may be removed with the help of timing information. It has been noticed already in the early 1980s that the variability patterns may be different for the boundary layer and accretion disk [35]. Further progress has been made almost 20 years later, thanks to the large collecting area of the PCA instrument aboard RXTE and the use of novel data analysis techniques. Using the method of frequency-resolved spectral analysis it has been shown that aperiodic and quasi-periodic variability of bright LMXBs – atoll and Z-sources, on $\sim$ s – ms timescales is caused primarily by variations of the luminosity of the boundary layer [20]. It was also shown that the boundary layer spectrum remains nearly constant in the course of the luminosity variations and is represented to certain accuracy by the Fourier frequency-resolved spectrum. This permits to resolve the degeneracy in the spectral fitting and to separate contributions of the boundary layer and disk emission (Fig. 2.19, left panel). Interestingly, the spectrum of the boundary layer emission has the same shape in different objects and

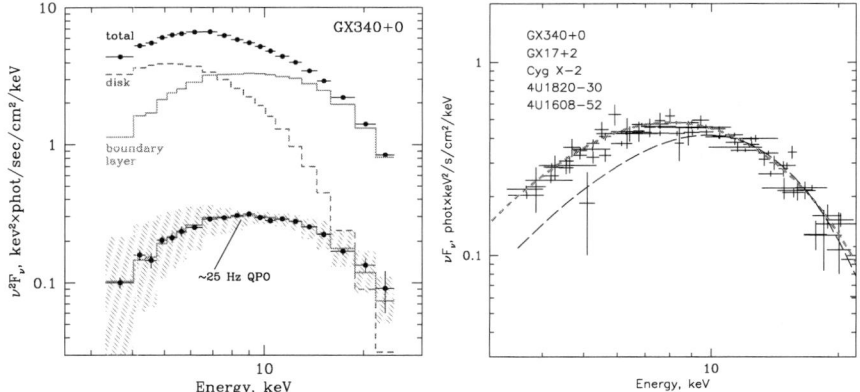

Fig. 2.19 *Left*: Points with error bars showing total and frequency-resolved spectra of GX 340+0. The *dashed* (*blue* in the color version of the plot) and *upper solid* (*red*) histograms show the disk and boundary layer spectra, the latter computed as a difference between the (observed) total and (predicted) accretion disk spectrum. The lower solid histogram is obtained from the upper scaling it to the total energy flux of the frequency-resolved spectrum. Adopted from [20]. *Right*: Fourier frequency-resolved spectra, corrected for the interstellar absorption ($\approx$boundary layer spectra) of five luminous accreting neutron stars. For 4U 1608-52, the frequency-resolved spectrum of the lower kHz QPO is shown. All spectra were corrected for the interstellar absorption. The *thick short-dashed line* shows the best fit Comptonization model with $kT_s = 1.5$, $kT_e = 3.3$ keV, $\tau = 5$. The *thin long-dashed line* shows a black body spectrum with temperature $kT_{bb} = 2.4$ keV. Adopted from [45]

is nearly independent of the global mass accretion rate in the investigated range of $\dot{M} \sim (0.1 - 1)\dot{M}_{Edd}$ and in the limit of $\dot{M} \sim \dot{M}_{Edd}$ is close to a Wien spectrum with $kT \sim 2.4\,\text{keV}$ (Fig. 2.19, right panel). Its independence on the global value of $\dot{M}$ lends support to the theoretical suggestion by [24] that the boundary layer is radiation pressure supported. With this assumption, one can attempt to measure gravity on the neutron star surface and hence M/R^2, from the shape of the boundary layer spectrum, similarly to the photospheric expansion (i.e., Eddington limited) X-ray bursts. This gives results within the range of values obtained by other methods [45, 51].

As neutron stars have smaller mass than black holes, the linear scale corresponding to the gravitational radius is smaller, $r_g \propto M$. This has two important consequences. First, the surface area of the emitting region is directly proportional to M^2, therefore at the same luminosity the temperature of black body emission will be larger, $T_{bb} \propto M^{-1/2}$. The spectrum is further modified by the contribution of the emission from the boundary layer, which has yet smaller emitting region and harder spectrum. This is illustrated by Fig. 2.1 – the soft state spectrum of the neutron star 4U 1705-44 is noticeably harder than that of the black hole Cyg X-1.

Second, the smaller linear scale in neutron star accretion shifts the characteristic frequency scales by a factor $\propto M^{-1}$, suggesting ~ 5 times higher frequencies of variability in neutron star systems. This is confirmed by observations as demonstrated in the extensive comparison of black hole and neutron star power density spectra in [59]. In addition, characteristic timescales in the boundary or spreading layer are significantly shorter than those in the accretion disk and are in the tens of kHz frequency domain. It has been suggested that a very high frequency component may exist in the power density spectra of neutron stars associated with the turbulence in the spreading layer [59, 24]. If detected, this may become a unique diagnostics tool of physical conditions in the spreading layer on the surface of the neutron star. So far, only upper limits have been obtained; they are at the level of $\sim 10^{-2}$ fractional rms [59].

The relatively cold surface of the neutron star is a source of copious soft photons. This is mostly relevant in the hard spectral state – the low energy photons emitted by the neutron star surface result in a more efficient cooling of electrons in the Comptonization region, thus changing its energy balance and Comptonization parameter (luminosity enhancement factor, cf. Sect. 2.8.1). Consequently, the energy spectra of neutron stars in the hard state are significantly softer than those of black holes – they have larger spectral index (smaller Comptonization parameter) and a smaller value of the high energy cut-off (lower electron temperature). Both these effects are clearly seen in the spectra shown in Figs. 2.1, 2.18.

The existence of soft and hard spectral states in neutron stars suggests that, similarly to black holes, the accretion flow can change its configuration from optically thin to optically thick. A remarkable fact is that this change seems to occur to both accretion disk and boundary layer (quasi-) simultaneously. Indeed, to my knowledge, no two-component (soft + hard) spectrum has been observed in the case of accreting neutron stars so far.

References

1. M. Abramowicz: Astron. Nachr., **326**, 782 (2005)
2. M.A. Alpar, J. Shaham: Nature, **316**, 239 (1985)
3. M. Basko, R. Sunyaev, L. Titarchuk: A&A, **31**, 249 (1974)
4. T. Belloni, D. Psaltis, M. van der Klis: ApJ, **572**, 392 (2002)
5. A.M. Beloborodov: ApJ, **510**, L123 (1999)
6. S. Bhattacharyya, D. Bhattacharya, A. Thampan: MNRAS, **325**, 989 (2001)
7. E. Churazov, M. Gilfanov, M. Revnivtsev: MNRAS, **321**, 759 (2001)
8. P.A. Connors, R.F. Stark, T. Piran: ApJ, **235**, 224 (1980)
9. P. Coppi: In: J. Poutanen, R. Svensson (eds.) High Energy Processes in Accreting Black Holes, ASP Conference Series 161, p. 375 ISBN 1-886733-81-3, (1999)
10. C.D. Dermer, E.P. Liang, E. Canfield: ApJ, **369**, 410 (1991)
11. K. Ebisawa, K. Mitsuda, T. Hanawa: ApJ, **367**, 213 (1991)
12. A.C. Fabian: MNRAS, **238**, 729 (1989)
13. A.A. Galeev, R. Rosner, G.S. Vaiana: ApJ, **229**, 318 (1979)
14. I.M. George, A.C. Fabian: MNRAS, **249**, 352 (1991)
15. M. Gilfanov, E. Churazov, R. Sunyaev et al.: In: M.A. Alpar, U. Kiziloglu, J. van Paradijs (eds.) The Lives of the Neutron Stars. Proceedings of the NATO Advanced Study Institute on the Lives of the Neutron Stars, Kemer, Turkey, August 29–September 12, Kluwer Academic, p. 331 (1993)
16. M. Gilfanov, E. Churazov, M. Revnivtsev: A&A, **352**, 182 (1999)
17. M. Gilfanov, E. Churazov, M. Revnivtsev: In: G. Zhao, J.-J. Wang, H. M. Qiu, G. Boerner (eds.) Proceedings of 5-th Sino-German workshop on Astrohpysics, SGSC Conference Series, vol. 1, p. 114 (1999)
18. M. Gilfanov, E. Churazov, M. Revnivtsev: MNRAS, **316**, 923 (2000)
19. M. Gilfanov, E. Churazov, M. Revnivtsev: In: P. Kaaret, F. K. Lamb, Jean H. Swank (eds.) X-ray Timing 2003: Rossi and Beyond. AIP Conference Proceedings, vol. 714, p. 97 (2004)
20. M. Gilfanov, M. Revnivtsev, S. Molkov: A&A, **410**, 217 (2003)
21. M. Gilfanov, V. Arefiev: in preparation (astro-ph/0501215)
22. F. Haardt, L. Maraschi: ApJ, **413**, 507 (1993)
23. A. Ibragimov, J. Poutanen, M. Gilfanov et al.: MNRAS, **362**, 1435 (2005)
24. N. Inogamov, R. Sunyaev: Astr. Lett., **25**, 269 (1999)
25. O. Kotov, E. Churazov, M. Gilfanov: MNRAS, **327**, 799 (2001)
26. Yu. E. Lyubarskii: MNRAS, **292**, 679 (1997)
27. L.-X. Li, R. Narayan, J.E. McClintock: ApJ, in press, eprint arXiv:0809.0866 (2009)
28. J. Malzac, A. Beloborodov, J. Poutanen: MNRAS, **326**, 417 (2001)
29. G. Matt, A.C. Fabian, R.R. Ross: MNRAS, **264**, 839 (1993)
30. P. Magdziarz, A.A. Zdziarski: MNRAS, **273**, 837 (1995)
31. F. Mayer, E. Meyer-Hofmeister: A&A, **288**, 175 (1994)
32. F. Meyer, B.F. Liu, E. Meyer-Hofmeister: A&A, **354**, 67 (2000)
33. M.L. McConnell, A.A. Zdziarski, K. Bennett et al.: ApJ, **572**, 984 (2002)
34. J.M. Miller, J. Homan, D. Steeghs: ApJ, **653**, 525 (2006)
35. K. Mitsuda, H. Inoue, K. Koyama et al.: PASJ, **36**, 741 (1984)
36. S. Miyamoto, S. Kitamoto, S. Iga et al.: ApJ, **435**, 398 (1994)
37. R. Narayan, I. Yi: ApJ, **428**, L13 (1994)
38. M. Nowak: MNRAS, **318**, 361 (2000)
39. K. Pottschmidt, J. Wilms, M.A. Nowak et al.: A&A, **407**, 1039 (2003)
40. J. Poutanen, J.H. Krolik, F. Ryde: MNRAS, **292**, L21 (1997)
41. D. Psaltis, T. Belloni, M. van der Klis: ApJ, **520**, 262 (1999)
42. M. Revnivtsev, M. Gilfanov, E. Churazov: A&A, **380**, 520 (2001)
43. M. Revnivtsev, M. Gilfanov, E. Churazov: A&A, **363**, 1013 (2000)
44. M. Revnivtsev, M. Gilfanov, E. Churazov: A&A, **347**, L23 (1999)

45. M. Revnivtsev, M. Gilfanov: A&A, **453**, 253 (2006)
46. R.R. Ross, A.C. Fabian: MNRAS, **281**, 637 (1996)
47. R.R. Ross, A.C. Fabian, A.J. Young: MNRAS, **306**, 461 (1999)
48. N. Shakura, R. Sunyaev: A&A, **24**, 337 (1973)
49. T. Shimura, F. Takahara: ApJ, **331**, 780 (1995)
50. N. Sibgatullin, R. Sunyaev: Astr. Lett., **26**, 699 (2000)
51. V. Suleimanov, J. Poutanen: MNRAS, **369**, 2036 (2006)
52. R. Sunyaev, N. Shakura: SvAL, **12**, 117 (1986)
53. L. Stella, M. Vietri: ApJ, **492**, L59 (1998)
54. R. Sunyaev, J. Trümper: Nature, **279**, 506 (1979)
55. R. Sunyaev, L. Titarchuk: A&A, **86**, 121 (1980)
56. R. Sunyaev, L. Titarchuk: In: The 23rd ESLAB Symposium on Two Topics in X Ray Astron-
    omy, Volume 1: X Ray Binaries, pp. 627–631 (SEE N90-25711 19-89) (1989)
57. R. Sunyaev, L. Titarchuk: A&A, **143**, 374 (1985)
58. R. Sunyaev, E. Churazov, M. Gilfanov et al.: ApJ, **389**, 75 (1992)
59. R. Sunyaev, M. Revnivtsev: A&A, **358**, 617 (2000)
60. L. Titarchuk, N. Shaposhnikov: ApJ, **626**, 298 (2005)
61. M. van der Klis: In: W. Lewin, M. van der Klis (eds.) Compact Stellar X-ray Sources, Cam-
    bridge University Press, Cambridge (2006)
62. P.M. Uttley, I.M. McHardy: MNRAS, **323**, L26 (2001)
63. K. Viironen, J. Poutanen: A&A, **426**, 985 (2004)
64. G. Wardziński, A.A. Zdziarski, M. Gierliński et al.: MNRAS, **337**, 829 (2002)
65. R. Wijnands, M. van der Klis: ApJ, **514**, 939 (1999)
66. A.A. Zdziarski, P. Lubiński, D. Smith: MNRAS, **303**, L11 (1999)
67. A.A. Zdziarski, P. Lubiński, M. Gilfanov et al.: MNRAS **342**, 355 (2003)
68. A.A. Zdziarski, M. Gierliński: Prog. Theor. Phys. Suppl., **155**, 99 (2004)

Chapter 3
States and Transitions in Black-Hole Binaries

T. M. Belloni

Abstract With the availability of the large database of black-hole transients from the Rossi X-Ray Timing Explorer, the observed phenomenology has become very complex. The original classification of the properties of these systems in a series of static states sorted by mass accretion rate proved not to be able to encompass the new picture. I outline here a summary of the current situation and show that a coherent picture emerges when simple properties such as X-ray spectral hardness and fractional variability are considered. In particular, fast transition in the properties of the fast time variability appear to be crucial to describe the evolution of black-hole transients. Based on this picture, I present a state classification which takes into account the observed transitions. I show that, in addition to transients systems, other black-hole binaries and Active Galactic Nuclei can be interpreted within this framework. The association between these states and the physics of the accretion flow around black holes will be possible only through modeling of the full time evolution of galactic transient systems.

3.1 Introduction

The presence of two different "states" in the X-ray emission of the first black-hole candidate, Cygnus X-1, was realized in the early 1970s with the Uhuru satellite [90]. A transition lasting less than 1 month was observed from a soft spectrum to a hard spectrum: the source intensity decreased by a factor of 4 in the 2–6 keV band and increased by a factor of 2 in the 10–20 keV band. The associated radio source was not detected before the transition and appeared at a flux at least three times higher after the transition. Because of the large flux swing below 10 keV, where most of the instrument response was, this led to the identification of a "high" state (with soft spectrum and no radio detection) and a "low" state (with hard spectrum and associated radio source). In its "low" state, Cyg X-1 was observed to show strong

T. M. Belloni (✉)
INAF – Osservatorio Astronomico di Brera, Via E. Bianchi 46, I-23807 Merate, Italy,
tomaso.belloni@brera.inaf.it

Belloni, T.M.: *States and Transitions in Black-Hole Binaries*. Lect. Notes Phys. **794**, 53–84 (2010)
DOI 10.1007/978-3-540-76937-8_3

aperiodic noise [91, 71], while the soft state is characterized by a much reduced noise level.

As we know now, most black-hole binaries (BHB) are transient sources (black-hole transients, BHT), whose detection depends strongly on the availability of all-sky monitors and wide-field instruments. The first BHT was A 0620-00 discovered with the Ariel satellite in 1975 [23]. As its flux reached 50 Crab at peak, the spectral distribution could be followed throughout the outburst [79]. The spectrum was hard at the start of the outburst, while the flux peak was observed to be caused by a strong soft (<10 keV) enhancement, while the hard flux dropped. The two states recognized in Cyg X-1 appeared to be present also here, as recognized by [17].

Other sources appeared to follow this bi-modal state classification. The two persistent systems LMC X-1 and LMC X-3 were always observed in the high state (although it is now recognized that LMC X-3 shows occasional transitions to a harder state (see [100, 36]). The bright source GX 339-4 was not detected with Uhuru and was discovered with OSO-7 [54]. The early observations showed that the source was extremely variable over timescales of 100 days, alternating between three states: a "high" state with a soft spectrum, a harder "low" state, and an "off" state, which was later recognized as a low-flux extension of the low state (see [44]). The similarity of GX 339-4 and Cyg X-1 was also strengthened by the observation of similar strong aperiodic variability [82]. Since the names "low" and "high" derive from the 1–10 keV flux and the fluxes at higher energies reverse, I will refer to them as "low-hard" (LHS) and "high-soft" (HSS), respectively (see Chap. 2).

The all-sky monitor on board the Ginga satellite allowed the discovery of new X-ray transients, which could then be followed-up with extensive observations with its large proportional counters. Three major bright transients were observed: GS 1124-684, GS 2000+251, and GS 2023+338 (see [89]). The latter showed very unusual properties and dramatic variability caused by variable intrinsic absorption. GS 2000+251 showed low and high states similar to those already known. More interesting was the case of GS 1124-684. Here, in addition to the two known states, an additional state with different properties was found [47]. This new state had already been recognized in earlier observations of GX 339-4 [64]: since it appeared at the brightest flux levels, it was dubbed "very-high" state (VHS). Its spectral properties were a mixture of those observed in the high and low states, while the timing properties appeared very complex. Low-frequency quasi-periodic oscillations (QPO) were observed, but not at all times, and fast transitions could be seen (see [67, 88]). Properties similar to those of the "very-high" state were seen in the same two sources, but during observations at a much lower luminosity [4, 60] a new "intermediate" state (IMS) was therefore proposed.

Overall, the amount of information available until 1995 was scarce and it was difficult to derive a coherent picture (see [89] for a review). At the very end of 1995, the Rossi X-Ray Timing Explorer was launched [13]. The presence of an all-sky monitor (ASM), a large proportional counter (PCA) and a high-energy instrument (HEXTE), together with an extreme flexibility of operation, make it an ideal mission for the study of bright black-hole binaries,[1] most of which are of transient nature. A vast database is available, which of course is enhanced by the presence of (sparser) observations by other X-ray missions. The large amount of new information has naturally resulted in a burst of publications. Since the observed phenomenology is complex, it resulted in a number of different classifications in terms of source states, often evolving with time and difficult to compare with each other. In the following, I present the current situation, with the aim of guiding the reader through the jungle of source properties and states. My approach touches the basic properties, intentionally ignoring all inevitable complications, in order to give an overview of the general picture that has emerged, pointing to selected publications where the subject can be examined in more detail. I concentrate on the X-ray properties, since the connection with other energy bands and with the jet ejection is discussed in other chapters. The use of the term BHB instead of microquasar is intentional and is aimed at denoting the complete class of sources, whether relativistic jet ejections have been observed or not. The ubiquitous presence of radio emission, at least in some states, seems to indicate that the two definitions probably point to the same set of objects.

3.2 The Fundamental Diagrams

While the HSS and the LHS are relatively well defined and identified (see Chap. 2), everything else is still a matter of discussion. The light curves of black-hole transients are quite varied and even the same source can exhibit very different time evolutions of their flux during different outbursts (see [33, 56]). However, the spectral hardness (defined as the ratio of observed counts in two energy bands) and the integrated fractional rms variability are quantities which exhibit considerable regularity [8, 39] and have the additional advantage of being model independent. Since RXTE provides a large database obtained with the same high-area instrument (the PCA), after correction for time variations of the instrument gain, it is possible to compare a large number of sources, a comparison which would be impossible combining different instruments. Clearly, the interpretation of hardness is not obvious and need to be supported by spectral fitting, also considering that the observed spectrum consists of the superposition of different components. However, a careful selection of the energy bands allows to separate the major components and avoid complex degeneracies.

Two diagrams are particularly useful for characterizing the behavior of BHT: the hardness–intensity diagram (HID), where the total count rate is plotted as a function of hardness, and the hardness–rms diagram (HRD), where the fractional rms, integrated over a broad range of frequencies, is plotted versus hardness. The HID

---

[1] Notice that the original denomination of "black-hole candidate" (BHC) has at some point been replaced with "black-hole binary" (BHB) without any qualitative breakthrough to justify the change in nomenclature.

is equivalent to the color–magnitude diagram of optical astronomy, with the major
difference that for BHT we can follow the movement of a single source on short
timescales. A number of diagrams can be produced from parameters obtained from
X-ray spectral fitting: these can be considered the equivalent of the Herzsprung–
Russell diagram (see e.g., [31, 77]). However, for stellar astronomy the emission
model is better understood, justifying the loss of model independence, while theoretical assumptions in the case of BHT are usually far from being agreed upon.

Work on the complete RXTE/PCA sample of BHT shows that, although the time
evolution of the parameters can be quite different, these diagrams allow the identification of a surprising number of common properties [39, 40]. A full comparative
analysis of the BHT in the RXTE archive is presented by [40]. Here I will base my
description on the 2002–2003 outburst of GX 339-4, as published in two earlier
papers [70, 8]. This source is clearly the clearest example, as confirmed by the
fact that its HID and HRD are extremely similar also for all outbursts observed
with RXTE, but the general statements that can be drawn from it apply to all other
transients observed by RXTE [40].

The HID and HRD of GX 339-4 are shown in Fig. 3.1. From the q-shaped
HID (also referred to as "turtle head"), four distinct branches can be identified,

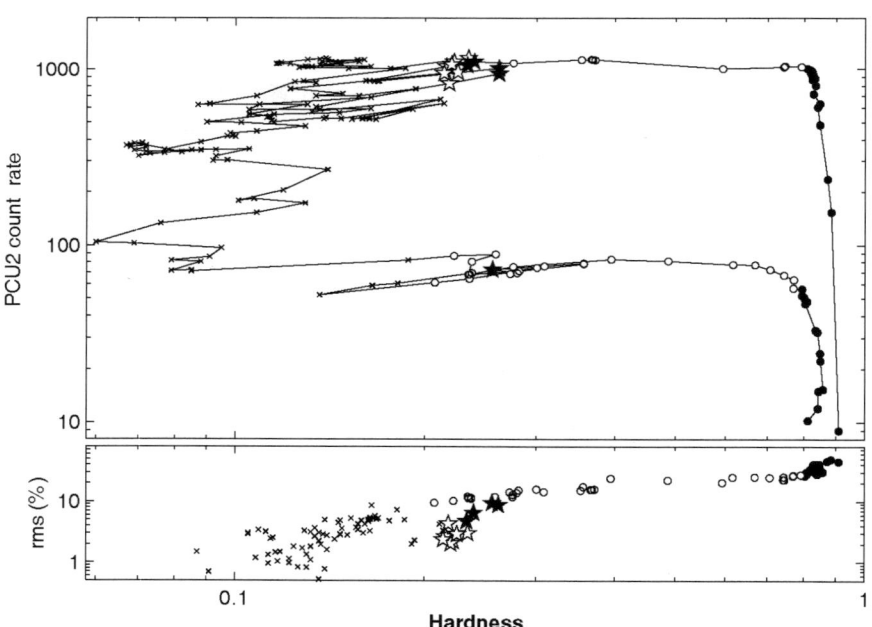

Fig. 3.1 Hardness–intensity diagram (HID: top panel) and hardness–rms diagram (HRD: bottom
panel) for the 2002–2003 outburst of GX 339-4 (adapted from [8]). PCA count rate is in the energy
range 3.8–21.2 keV. Hardness is defined as the ratio of counts in the energy bands 6.3–10.5 and
3.8–6.3 keV. Fractional percentage rms corresponds to the frequency range 0.1–64 Hz and to the
full energy range. Symbols correspond to different types of power density spectra (see Sect. 3.3):
type 1 (*filled circles*), type 2 (*empty circles*), type 3 (*filled stars*), type 4 (*empty stars*), type 5 (*dots*)

corresponding to the four sides of the "q." These could be associated with the original four pre-RXTE states: a LHS on the hard branch to the right, a HSS on the soft branch to the left, a VHS on the branch on the top and a IMS on the bottom, the last two with large hardness variations. From a complete analysis of the 1998 outburst of XTE J1550-564, it was shown that the properties of VHS and IMS are very similar in all respects and can take place at different luminosity levels, leading to a unification of these into one likely physical state [41]. This leaves us with three separate states. However, when timing properties are taken into account, the HRD (bottom panel in Fig. 3.1) gives a different picture: here most of the points follow a single correlation: the LHS points have a high level of variability (more than 20%), the HSS points vary much less (<10%), and the intermediate points are, of course, intermediate. Interestingly, there is a narrow band in hardness where many points have a lower variability than those on the main correlation (shown as stars in Fig. 3.1). This diagram suggests the presence of two separate states only, corresponding to the points on the main correlation (spanning a large range in hardness) and to those deviating from it.

The diagrams of all other BHT observed by RXTE, although the shape of the HID deviate from the "q" shown here, are qualitatively in agreement with this picture [39, 40, 27]. Three examples can be seen in Fig. 3.2.

The general time sequence of the points in the HID is clear: the "q" diagram is followed in a counterclockwise direction, although there are clearly more complex movements on the left branch. This will be discussed in Sect. 3.4.

3.3 Aperiodic Variability

The next step is to examine the properties of fast aperiodic variability (see [95]). I will concentrate on the power density spectra (PDS): although important information can be extracted from higher-order timing tools, they are not essential for the basic determination of states and state transitions.

3.3.1 Classification of Power Density Spectra

The original paper on GX 339-4 reports a number of different PDS shapes, but it is now clear that we can classify them as belonging to few basic types (see [40, 16, 15]). Furthermore, as we will see below, these types are closely related to the position on the HID and HRD. All the basic types but two were observed in the PDS of the 2002–2003 outburst of GX 339-4: they are shown in Fig. 3.3.

- PDS labeled 1 in Fig. 3.3 corresponds to the hard points in the HID. Its shape can be fitted with a small number, 3–4, of very broad Lorentzian components, plus in some case a low-frequency QPO peak [72, 7, 8]. The characteristic frequencies of all these components increase with source flux, while since the HID branch is almost vertical, the hardness remains almost unchanged, with only a minor

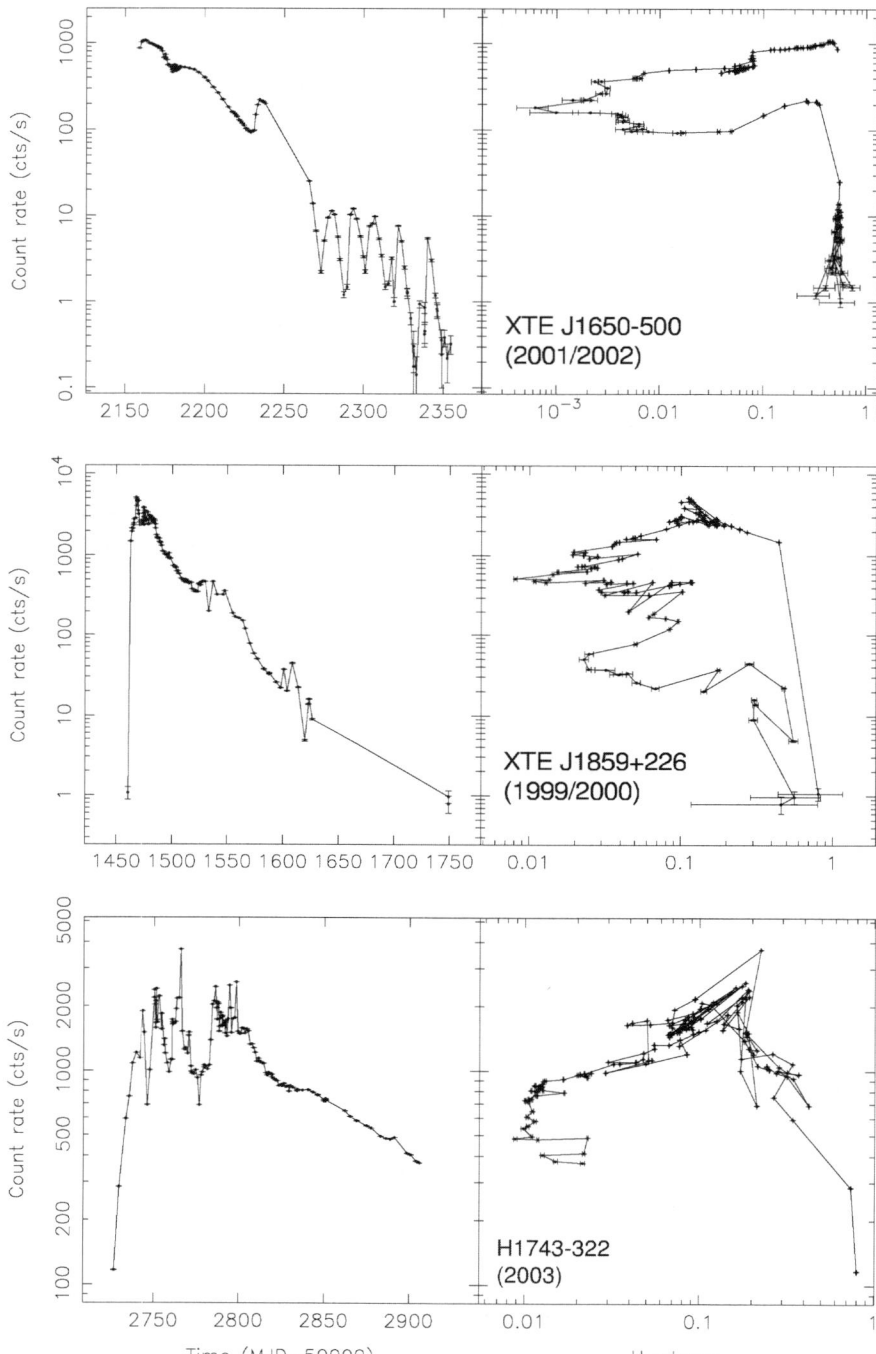

Fig. 3.2 Light curves (*left*) and HIDs (*right*) of three transients observed with the RXTE/PCA (from [39])

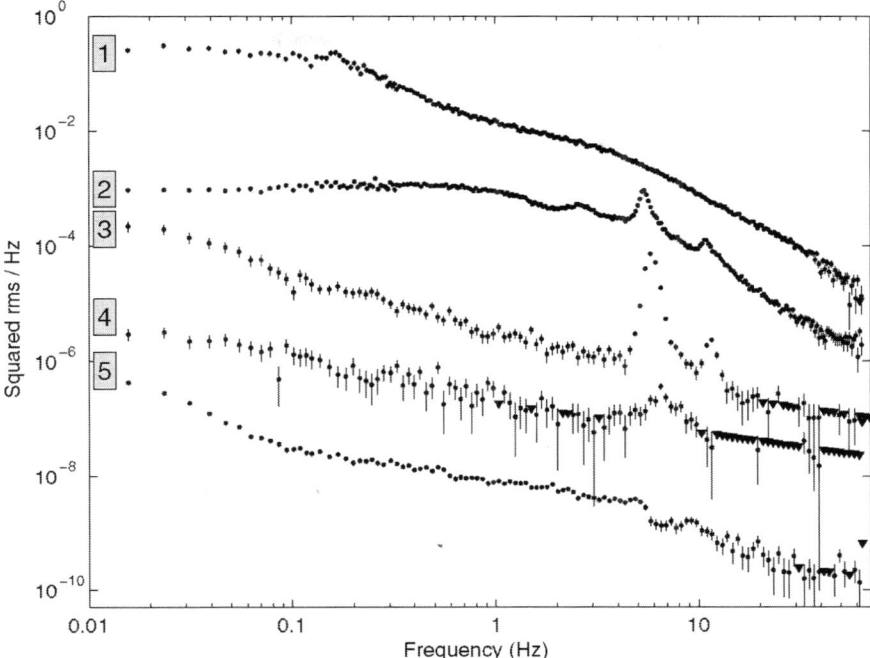

Fig. 3.3 Examples of the five types of PDS described in the text (labeled on the *left*). The data are from the 2002/2003 outburst GX 339-4 [8] and correspond to the energy range 3.8–21.2 keV. The PDS are shifted in power for clarity

softening at high flux. As it can be seen in the HRD, the total fractional rms on the right branch is rather high, larger than 20%. As a function of energy, the fractional rms shows a slight decrease (see e.g., [32]). This is the band-limited noise typical of the LHS [72, 74].

• PDS 2 in Fig. 3.3 can be considered a high-frequency extension of PDS 1 and can be found at intermediate hardness values, notably along the horizontal branches. It can be decomposed into a number of Lorentzian components which correspond to those found in the LHS (see e.g., [8]). The most prominent feature is a QPO with centroid frequency varying between ∼0.01 and ∼20 Hz. All Lorentzian components vary together, including the QPO. Unlike the previous case, they are strongly correlated with hardness: softer spectra correspond to higher frequencies and also to lower integrated rms variability (see Fig. 3.1). The low-frequency QPO (LFQPO) observed here is termed "type-C" QPO (see [16] for precise definitions of QPO types): its most important property for our description is that it always appears together with moderately strong (∼5–20% rms) band-limited noise. The total fractional rms *increases* with energy, in marked difference with PDS 1 [32]. It is usually accompanied by at least two peaks harmonically related: one at half the frequency and one at twice the frequency. Although the QPO frequency is correlated with spectral hardness, on short (<10 s) timescales, it is rather stable.

The frequency of the type-C QPO is very strongly correlated with the characteristic (break) frequency of the underlying broad-band noise components [99, 7], a correlation which also extends to neutron-star binaries (see [95]). The two main band-limited noise components have break frequencies which are one at the same frequency of the QPO and the other a factor of 5 lower [99, 7]. Often, only one broad-band component is detected, depending on the energy band considered, as their energy dependence is rather different (see Fig. 3.4 in [18]).

- PDS 3 in Fig. 3.3 (found over a rather narrow range of intermediate hardnesses) also shows a QPO (called "type-B" QPO), but with very different characteristics than the one discussed above. A detailed discussion of the difference between QPO types can be found in [16]. The fact that type-B QPOs are not simply an evolution of type-C QPOs is demonstrated by the fast transitions observed between them (see [15]) and by the few cases of simultaneous detection (i.e., GRO J1655-40 observed by RXTE on May 18, 2005). The total fractional variability is lower, due mainly to the fact that the band-limited components are replaced by steeper and weaker components. As for PDS 2, the fractional rms

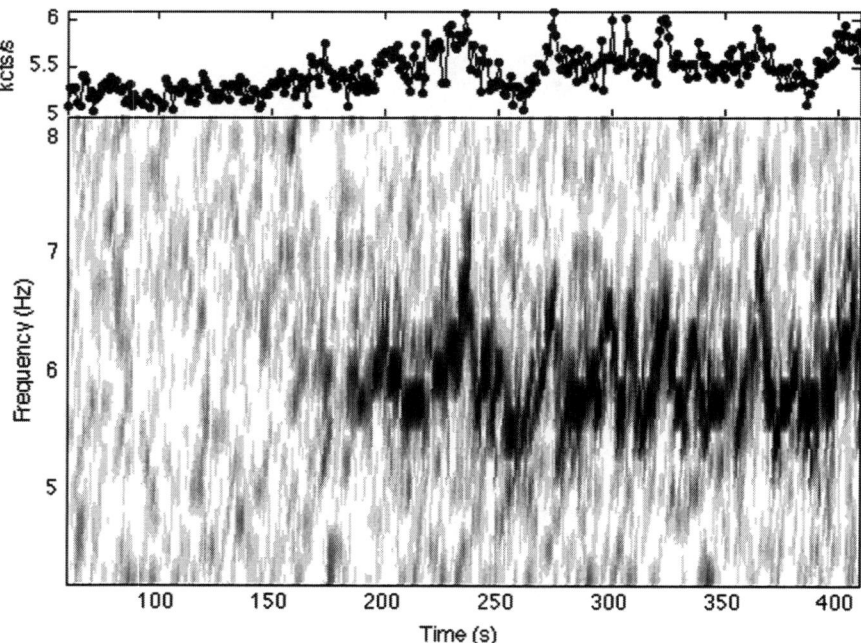

Fig. 3.4 The first 6 min of the PCA observation of GX 339-4 from May 17, 2002. *Top panel*: light curve with 1-s binning. *Bottom panel*: corresponding spectrogram in the 4–8 Hz range; darker regions correspond to higher power. The onset of a sharp and strong type-B QPO at around 6 Hz is evident, as well as the variability of its centroid frequency. In the first 2 min, a type-A QPO is present, but it is too weak to show in the spectrogram. Adapted from [70]

increases with energy. Type-B QPOs show a harmonic structure similar to that of type-C QPOs (two harmonic peaks are visible in Fig. 3.3, while often a peak at half the frequency of the main peak is observed). While type-C QPOs span large range in frequency, type-B QPOs are limited to the range 1–6 Hz, but detections during high-flux intervals are concentrated in the narrow 4–6 Hz range [16]. The centroid frequency appears positively correlated with source intensity rather than hardness (as they are associated with a very narrow range of hardness, see below).

A closer look at Fig. 3.3 will reveal that the shape of the type-B peak is different from that of type-C. While the latter can be fitted with a Lorentzian model, type-B QPOs are consistent with a Gaussian shape, with broader wings. This is due to the fact that type-B QPOs jitter in time on short timescales to the effect that in the average PDS its peak is smoothed [70]. This can be seen in Fig. 3.4, where a spectrogram of 6 min of PCA observation of GX 339-4 is shown: the jitter in frequency takes place on a timescale of 10 s [70].

- The last PDS in Fig. 3.3 showing a QPO is #4, found at hardness values systematically slightly lower than those of PDS 3. Here, a so-called type-A QPO is shown. Being a much weaker and broader feature, we know less details about this oscillation; sometimes it is only detected by averaging observations. Its frequency is always in the very narrow range 6–8 Hz and it is associated with an even lower level of noise than type B. In fact, the three types of QPO can be separated by plotting them against the integrated fractional rms of the PDS in which they appear [16].

- Finally, PDS 5 in Fig. 3.3 shows a weak steep component, which often needs a long integration time for a detection. Weak QPOs at frequencies >10 Hz are sometimes observed, as well as a steepening/break at high frequencies. The total fractional rms can be as low as 1%, increasing with energy [32]. This PDS corresponds to the soft points at the extreme left of the HID.

- As mentioned above, in addition to the PDS shapes in Fig. 3.3, there are two types of PDS which were not observed in GX 339-4. Examples from the 2005 outburst of GRO J1655-40 can be seen in Fig. 3.5, together with PDS 1 and 5 from Fig. 3.3. The integrated fractional rms of both is intermediate. One PDS has a featureless curved shape, the other is similar below a few Hz, but shows an additional bump and a QPO. As we will see, these "anomalous" PDS shapes are associated with anomalous "flaring states."

3.3.2 Fast Transitions

The PDS described in the previous section have been obtained as averages over hundreds or thousands of seconds of observations. While for most cases this averaging procedure is justified, there are observations where very fast transitions are observed, requiring a time-resolved analysis. An example is shown in Fig. 3.4. Not only the centroid frequency of the QPO jitters with time, but the oscillation is clearly not visible in the first 2 min of data. Averaging the first part, a type-A QPO appears [70]. These transitions are common and have been observed with RXTE in many

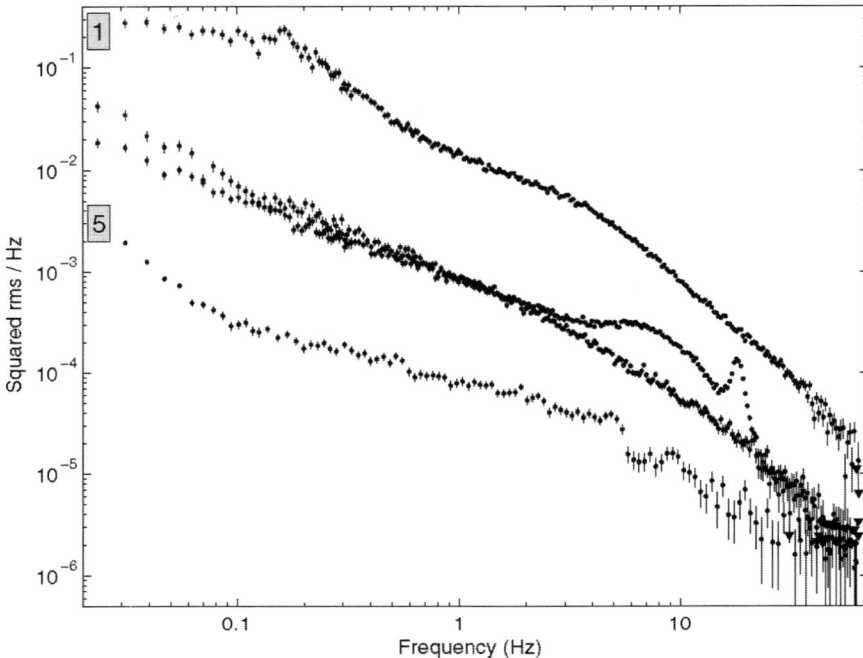

Fig. 3.5 Examples of the two types of PDS which have not been found in GX 339-4 (see text). The *top* and *bottom* PDS are # 1 and 5 from Fig. 3.3, shifted in power for clarity. The *middle* ones correspond to two observations of GRO J1655-40 from its 2005 outburst: the one with a QPO from May 24, 2005 the other from May 8, 2005. The middle ones are not shifted in power

sources [8, 15, 40]: they correspond to those already discovered with Ginga [67, 88]. All of them, with no exception, involve a transition between a type-B and a type-A QPO, or between a type-B and a type-C QPO. A number of detailed transitions from XTE J1859+226 can be found in [15]: there, an exponentially time-decaying threshold in count rate was found for the occurrence of the three types of QPO. This is clearly not a common case, since from Fig. 3.4 it is evident that the count rate corresponding to the type-A QPO for GX 339-4 is lower than that of the type-B QPO, opposite to the case of XTE J1859+226.

3.4 The Time Evolution

In what presented above, the time evolution of the X-ray properties was not shown, as the attention was focused on the different classes of hardness and variability properties. The light curves of transient BHB can be quite diverse (see e.g., [39, 56, 40]). GX 339-4 has had three major outbursts in the past 5 years. In the bottom panel of Fig. 3.6 one can see that the time evolution of the three events is very different. Despite these differences, a general behavior can be identified, leading to basic facts

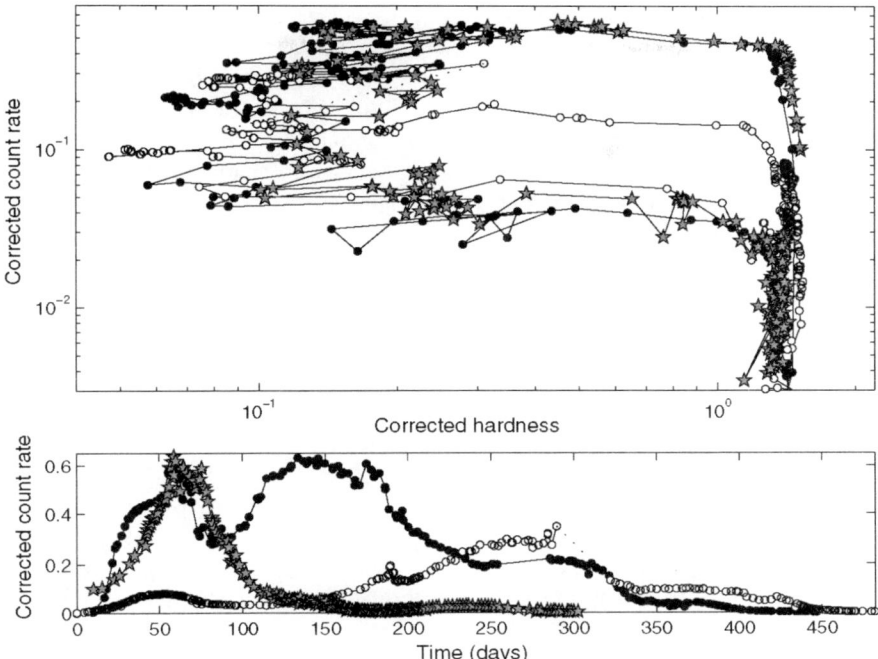

Fig. 3.6 *Top panel*: HIDs for the three recent outbursts of GX 339-4. In order to compare different epochs the spectra have been divided by a simulated Crab spectrum for each observation date. Therefore, the count rate is in Crab units in the 3.8–21.2 keV band and the hardness of 1 corresponds to a Crab spectrum. The symbols identify the three outbursts (2002/2003: *filled circles*; 2004/2005: *empty circles*; 2007: *gray stars*). The *dashed line* indicates long time gaps. *Bottom panel*: corresponding light curves plotted on the same scale. Symbols are the same

shared by all sources. The top panel of Fig. 3.6 shows that the three outbursts of GX 339-4 had a very similar evolution, despite the differences in the time domain. Other sources show a HID qualitatively similar to the q-shaped diagram of GX 339-4 (see [39] and Fig. 3.2), while others behaved more erratically (see [56, 40]).

The basic properties of the time evolution of outbursts are the following:

- All sources start from quiescence, become bright for a period typically of several months, then faint again and return to quiescence (with the exception of the three persistent bright BHB Cyg X-1, LMC X-3, and LMC X-1, and of GRS 1915+105 which is still very bright after 15 years). This is the definition of transient X-ray sources.
- For some sources, the initial rising part of the outburst was not observed (see for instance XTE J1650-500 in Fig. 3.2). The first pointed RXTE observations already show a high flux level. All sources are observed in the later parts of the outburst to become faint down to the minimum detectable levels with the RXTE PCA. This is obviously due to the fact that an unknown source or outburst must be discovered, leading to a delay in pointed observations. However, the typical

reaction time is at maximum a few days. This means that the initial rise can be as fast as a couple of days in these sources.

- In all sources for which the initial rise is observed, the start of the outburst is hard and so is the end of the outburst (see Fig. 3.6). There are a number of sources for which pointed observations started late and that were initially observed with a low hardness. Since the initial rise was fast, this is compatible with a hard rise.
- Most sources do show significant changes in hardness, extending to the left branch in the HID, corresponding to a soft spectrum. Intermediate values of hardness are observed. A few sources, such as XTE J1118+480, did not leave the hard branch and move in the HID on a roughly vertical hard track (see [28, 55]). This also applies to some outbursts of BHB which have shown other outbursts with a complex structure. An example is XTE J1550-564: its first two outbursts were complex and showed spectral variations (see [41, 80]), while the next (fainter) two did not (see [6, 87]).
- A good number of sources followed an evolution similar to that of GX 339-4 (Figs. 3.2 and 3.6): the q-shaped pattern was followed in counterclockwise direction. While there can be many state transitions, the q shape identifies two main ones: a hard-to-soft one at high flux in the first part of the outburst, and a soft-to-hard one at the end of the outburst (see below). The first one is particularly important since it has been shown to be associated with main ejections of relativistic jets (see with [25]).
- From the analysis of data from the Ginga satellite, the presence of a hysteretic behavior in the evolution of transient BHB was noticed [68]: the transition from the right to the left branch, namely from hard to soft, takes place at higher luminosities than the reverse transition later in the outburst. This was also found with RXTE data of transient BHB and the neutron-star transient Aql X-1 [52] and is evident from Fig. 3.6 for the case of GX 339–4. Other cases can be seen in [41, 39, 77, 40]. In the case of GX 339-4, which showed three outbursts qualitatively similar to each other, it appears as the luminosity difference between the bright and the faint transition (the two horizontal branches in the HID) is related to the level of the first. It was suggested that the flux at the initial hard peak (roughly corresponding to the top of the right branch in the HID) is correlated with the time from the final hard peak (the return to the LHS at the end of the outburst) of the previous outburst [104]. This correlation is also followed by the 2007 outburst.

3.5 Definitions of Source States

From what discussed in the previous sections, we can identify a number of source states that must be examined separately. As shown above, many observed properties change smoothly throughout the basic diagrams, but some do not. In particular, it is the inspection of the fast-variability properties which indicates the presence of abrupt variations. It is these sharp changes that must be taken as landmarks to separate different states. Whether these correspond to actual physical changes in the accretion flow is obviously not a priori clear and must be determined with the application of models. This approach has the advantage of being phenomenological and completely model independent and is meant to provide theoreticians with a solid observational framework on which to base their investigation. Below, I present a revised state classification, together with the observed transitions between them. This classification is similar but not identical to that presented in [39]. A graphical summary can be seen in Fig. 3.7.

- *Low-Hard State (LHS)*: It is associated *only* with the early and late phases of an outburst and it corresponds to the vertical branch at the extreme right of the HID. At the end of an outburst, once this state is reached, no more transitions have ever been observed. In the spectral domain, it is characterized by a hard spectrum, with a power-law index of 1.6–1.7 (in the 2–20 keV band) and little moderate variations (right-most zone in the HID). A high level of aperiodic variability is seen in the form of strong band-limited noise components (PDS 1 in Fig. 3.3), with typical rms values of ~30%, anti-correlated with flux and positively correlated with hardness. The PDS can be decomposed into a number of Lorentzian components (see [72, 7, 74, 48]), whose characteristic frequencies increase with flux. One of these components can take the form of a type-C QPO peak. The time spent in this state, both at the beginning and at the end of the outburst, can be quite variable.

- *Transition to the Hard-Intermediate State*: From the HID and HRD, it is difficult to mark a precise end of the LHS and the start of the HIMS. As shown in the next bullet, the properties of the HIMS are compatible with being an extension of those of the LHS. Changes in the speed of increase of the timing components, or in the frequency–frequency correlations can be seen [8], but the clearest marker for the transition has been observed in the IR/X-ray correlation for GX 339-4: the change in the correlation was very fast and marked and provided a precise marker for the transition [42]. Without making use of observations at other wavelengths, the time of the transition can only be identified with an uncertainty of a few days.

- *Hard-Intermediate State (HIMS)*: At the start of the outburst, if the source ever leaves the LHS, it enters the HIMS, moving along the top horizontal branch in the HID. The spectrum is softer, the softening being due to two simultaneous effects: an increase of the power-law index that can reach 2.4–2.5 and the appearance of a thermal disk component. The fast aperiodic variability corresponds to PDS B in Fig. 3.3, with a band-limited noise and a strong type-C QPO. The total fractional rms is lower than in the LHS (10–20%) and decreases with softening. The PDS can be decomposed into the same Lorentzian components as in the LHS, with characteristic frequencies that are higher than in the LHS. In the LHS, the frequencies increased with flux, with little spectral variations, in the HIMS they increase with softening of the energy spectrum. The top branch in the HID is traveled rather fast and lasts at most a dozen days. This state is also observed in the central and final parts of the outburst (see below).

- *Transition to the Soft-Intermediate State (Jet line)*: This transition is marked *only* by the timing properties. The overall level of noise drops (see HRD) and a type-B

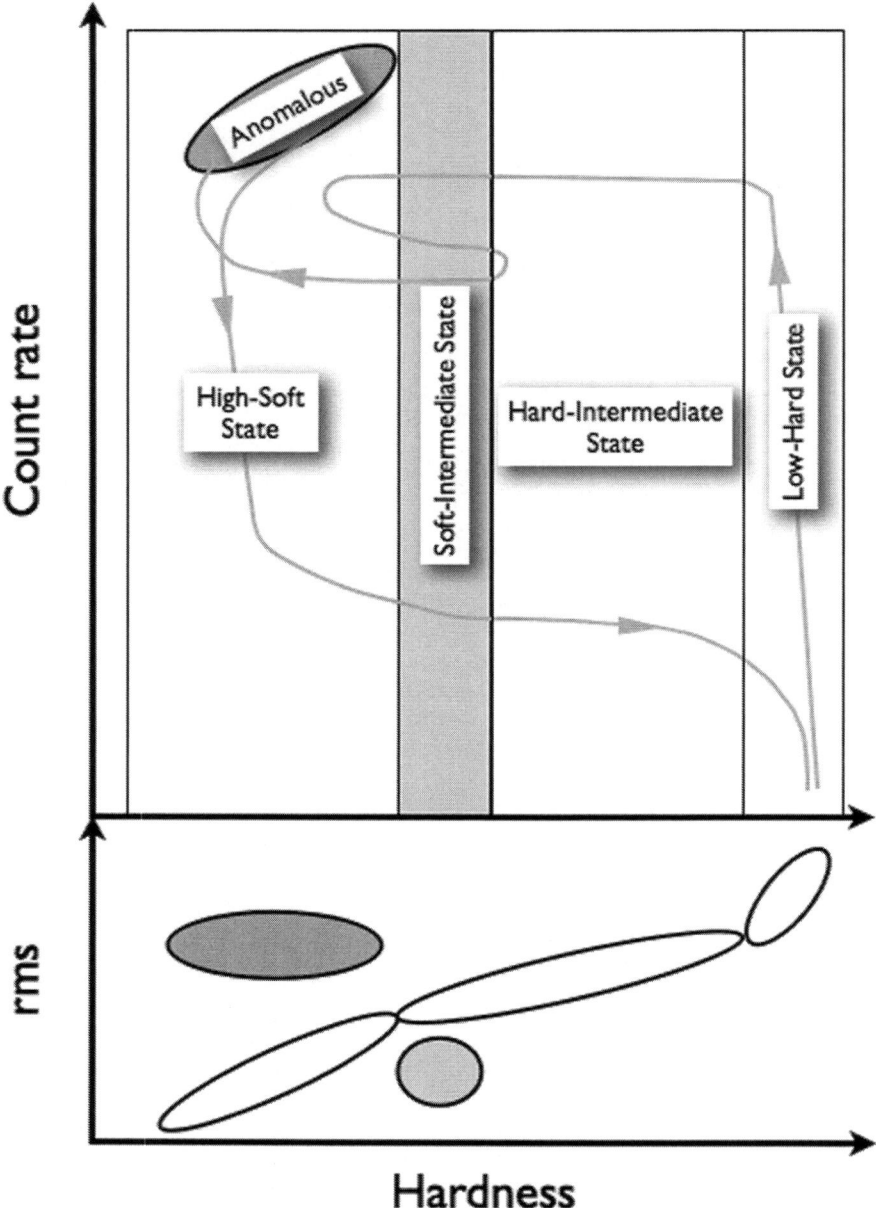

Fig. 3.7 Sketch of the general behavior of black-hole transients in the HID (*top*) and HRD (*bottom*), with the HID regions corresponding to the states described in the text

QPO appears in the PDS (PDS 3). All this takes place with a minor softening of the spectrum: this abrupt transition associated with a small change in hardness is associated with the "jet line" in the HID [8, 25]. The time coincidence of the crossing of the jet line and the HIMS–SIMS transition is however not exact [27].

- *Soft-Intermediate State (SIMS)*: Although spectrally below 20 keV the energy spectrum is only slightly softer than in the HIMS, as shown by the similar hardness, the timing properties are radically different and mark a clearly different state. The lower level of variability, which can be as low as a few percent is a clear indication of this state (see the HRD). A QPO is often present, either of type A or of type B (PDS 3 and 4 in Fig. 3.3). It is not clear whether it is always present, since observations without clear QPO peaks could feature an undetected type-A QPO.
- *Transitions from and to the Soft-Intermediate State*: Once the SIMS is reached, a number of transitions can be seen. All of them involve moving from or to the SIMS. In Fig. 3.7 this is exemplified with a short return to the HIMS, but the situation can be much more complex. Very fast transitions such as that shown in Fig. 3.4 can be observed, all of them involving a type-B QPO (see also [15]). In particular, the jet line can be crossed more than once.
- *High-Soft State (HSS)*: This soft state corresponds to the softest spectrum, dominated by a thermal disk component, with only a small contribution to the flux by a hard component. The variability is in the form of a weak (down to 1% fractional rms) steep component. Weak QPOs are sometimes detected in the 10–30 Hz range. The hardness variations shown in Fig. 3.1 are due to changes in the hard component.
- *"Anomalous" State*: In some sources, a different class of PDS has been seen at high flux (see Fig. 3.5). Although the hardness corresponds to that of the HSS (and sometimes to that of the SIMS), there is evidence that the energy spectrum is different (see below). The anomalous PDS are also accompanied by a higher integrated variability (see HRD).
- *Transitions back to the Low-Hard State*: The HIMS presents itself again at the end of the outburst, with very similar properties, preceding the final LHS. The timing properties along this transition are smoother, also because of the lower statistics associated with the lower flux of these observations. The range in hardness of the HIMS is similar at low fluxes, but not identical, as the SIMS appears to cover a smaller extent in hardness. These transitions during outburst decay are extensively presented in [46] and references therein.

As mentioned above, the source states outlined above stem from the hardness/timing properties observed in all systems observed with RossiXTE. They do not necessarily correspond to markedly different physical conditions. Indeed, the only transition which marks sharp changes in the timing properties is the HIMS–SIMS one. Moreover, looking at the HRD, it appears that LHS, HIMS, and HSS follow a single path, suggesting that they have something in common. As mentioned above, the HIMS can be seen as an extension of the LHS, but also the HSS shows similar timing properties (a low-frequency QPO, aperiodic noise decreasing as the source softens). The only states which are radically different are the SIMS and the anomalous one. In Sect. 3.9, we will see what the situation is in terms of broad-band emission and spectral models.

3.5.1 Comparison with Other State Classifications

Although different authors use different terms, even mixing nomenclatures, there are only two other state classifications in addition to the one presented here. It is important to compare them in order to clarify similarities and differences.

3.5.1.1 Ginga Canonical States

The original "canonical" states introduced on the basis of observations made by Ginga were the following. A hard (called low) state and a soft (called high) state were defined in a similar fashion as was done before. The hard state [65] was identified with strong variability in the form of a band-limited noise whose break frequency is correlated with the flat-top level (see [2, 60]), with hard phase lags increasing with energy and a 2–20 keV energy spectrum characterized by a power law with photon index ~1.5. The soft state by a strong thermal accretion disk energy spectrum associated with a very low level of aperiodic variability. The new "very-high" state (VHS) was originally identified in the brightest observations of GS 1124-684 and GX 339-4 on the basis of a different PDS, with lower variability, higher characteristic frequencies, and a QPO [65]. Subsequently, a second "flavor" of VHS was reported, with a different PDS shape and a lower integrated fractional rms [66]. This was obviously associated with a different type of QPO [88]. Finally, all the characteristics of the original VHS were discovered in both GX 339-4 and GS 1124-684 at much lower fluxes, observed after a long interval of soft state. This led to the tentative inclusion of an intermediate state taking place between the soft and the hard states [4, 60].

Given the relatively sparse time coverage obtained by Ginga, it was not possible to follow in detail transitions between states (with the exception of fast "flip–flops", which were not recognized as transitions [64]). This prevents a precise comparison. However, in addition to the obvious identification of the LHS and HSS, the two VHS flavors can be identified with the HIMS and the SIMS. The need for a separate intermediate state was removed by the discovery that mass accretion rate is not the only parameter driving the evolution of the outbursts [41].

3.5.1.2 Quantitative State Classification

More recently, a state classification based on the determination of timing and spectral parameters has been proposed (see [56, 77]). The underlying idea is to have a definition based on instrument-independent parameters, with a precise definition of states which makes possible a comparison between sources. An updated definition of states can be found in [57]. Three states are presented: hard, thermal, and steep power law (SPL), which are identified with the LHS, HSS, and VHS of the Ginga classification, respectively. Their definition is based on the precise boundaries of a number of parameters such as integrated fractional rms and the presence of a QPO in the PDS, power-law photon index, and disk fraction in the energy spectra. These three states do not fill the complete parameter space and all observations which do not qualify are generically classified as "intermediate." Among the RXTE observations of the 2003 outburst of H 1743-322, about 17% of the observations were intermediate between two states. A large fraction of observations of 4U 1630-47 also did not fit the three-state classification [94]. Therefore, this classification is not meant to be exhaustive like the one presented above, but rather to guide through general parameters of emission, such as flux ratio between main spectral components, spectral indices, and fractional rms of QPOs. Whether to adopt one or the other is dependent on what one's final aim is.

As mentioned above, the LHS and HSS are normally undisputed (but see below for the HSS): what is difficult to identify is their boundary. As I have shown in Sect. 3.5, there are sharp changes in some observable, which can be taken as markers of state transitions (IR/X correlation for LHS–HIMS, rms drop for other transitions). It is not clear at this stage whether the boundaries in this classification can reproduce these transitions. In particular, the SPL is defined in a rather complex way which involves the presence of a QPO of either type and the presence of a power-law component with photon index >2.4. The HIMS–SIMS transitions of GX 339-4 were observed at different power-law indices: in 2002 the index was 2.44 before *and* after the transition [70], in 2004 it varied from 1.9 to 2.3 [10]. It appears as this transition falls well within a single state. On the other hand, the spectral analysis of 5 years of RXTE observations of Cyg X-1 showed that a photon index of 2.1 indeed marks a transition between the LHS and a softer state [101].

As to the comparison with the system presented here, independent of the precise parameters adopted for the states, which have evolved with time, it is clear that the Hard and Soft can be identified with the LHS and HSS, respectively. The SPL and the different intermediate states would then correspond (roughly) to the SIMS and HIMS.

3.6 High-Frequency QPOs

Particular importance is attached to the high-frequency (>30 Hz) QPOs (HFQPO) detected in the PDS of some BHB. There are few instances of such detections, yet the situation here is entangled and even though few publications are available, it is difficult to derive a clear pattern. Here I summarize the basic information available, relying on significant detections of narrow ($Q>2$) features with a centroid frequency above 20 Hz (references can be found in [9]). Single QPO peaks have been detected from three sources: XTE J1650-500 (250 Hz),4U 1630–47 (variable frequency), and

XTE J1859+226 (90 Hz). From four other sources, pairs of QPO peaks have been detected: GRO J1655-40 (300/400 Hz), XTE J1550-564 (184/276 Hz), H 1743-322 (165/241 Hz), and GRS 1915+105 (41/69 Hz). The case of GRS 1915+105 is more complex and will be discussed below. All detected peaks are weak, with a fractional rms of a few percent and strongly dependent on energy. In some cases, only a few detections are available, but in others many peaks have been found: the centroid frequencies are not always constant, as one can see in the case of XTE J1550-564 and 4U 1630-47. However, it has been first noticed by [1] that when two simultaneous peaks are detected, their frequencies are in a 3:2 ratio in three cases out of four, while for GRS 1915+105 the ratio is 5:3. It is interesting to note that for XTE J1550-564, a double peak was discovered averaging a few observations from the 2000 outburst [61], while for the previous outburst in 1998 many single detections were reported [78]: those associated with type-B QPOs cluster around 180 Hz, while those associated with type-A QPOs are around 280 Hz. The fact that some sources appear to have preferred frequencies suggests that they are associated with basic parameters of the black hole. Indeed, a degree of anti-correlation with the dynamically determined black-hole mass has been found (see [77] but also [9]).

The comparison with kHz QPOs in neutron-star X-ray binaries shows that these are most likely different features: unlike the BHB HFQPO, which are mostly observed at the same frequency and have never been detected together with type-C WPOs, kHz QPOs span a large range in frequency and are correlated with the low-frequency QPOs which are the NS counterparts of type-C (see [95, 7]). Moreover, the kHz QPOs have been shown not to have preferred frequencies [11, 12].

For this paper, we are interested in where observations with an HFQPO are located in the HID and HRD and with which PDS shape they are associated. From the literature, many HFQPOs correspond to observations in the SIMS, i.e., associated with power-law noise and often type-A/B QPOs. There exist a a few exceptions, notably the case of XTE J1550-564, where one detection was reported together with a type-C QPO [41]; in this case, the frequency of the type-C QPO was high, indicating that the source was close to the SIMS.

3.7 Other Sources

The paradigm presented above has been derived from black-hole transients. These constitute the large majority of known black-hole binaries, but there are a few other systems that need to be examined. There are persistent systems, such as Cyg X-1, LMC X-1, and LMC X-3. Because of their large distance, the latter two are weak sources in our instruments, which means that the statistical level of the signal is much lower. LMC X-1 appears to be locked in the HSS, as does LMC X-3, which however has shown brief transitions to a harder state, even so extreme as not to be detected with RossiXTE. It is difficult to say more about them [100, 73, 36]. Moreover, it is impossible to ignore the prototypical microquasar GRS 1915+105 (see [24]) which, despite its peculiarity, needs to be compared with other systems. Here

I present these two sources, Cyg X-1 and GRS 1915+105 in a similar framework, analyzing the differences and similarities with those outlined above.

3.7.1 Cygnus X-1

Cygnus X-1 is not a transient system: it is found most of the time in the LHS (the original definitions of the states come from this source), with occasional transitions to a softer state, usually interpreted as a HSS, and a series of "failed state transitions" (see [74, 101]). However, already the first state-transition observed with RossiXTE in 1996 showed properties which were not compatible with those of a full-fledged HSS [3]. The total integrated fractional rms is rather high even as the source softens and does not go down to a few percent as in transients (see [34]). Spectrally, the hard component does become rather soft [101], but no sharp QPO like those shown above has been observed (see e.g., [74]). In the energy spectrum, a change of properties has been identified corresponding to a low-energy photon index of 2.1 (see above), which was then taken as a marker of state transitions [101].

We can compare the properties of Cyg X-1 with those of transient systems by analyzing in the same way, through the production of HID/HRD. The result can be seen in Fig. 3.8, where the points corresponding to 1065 RossiXTE observations ranging from February, 1996 to October, 2005 are plotted over the points from GX 339-4. The count rate here has been corrected also for the difference in distance between the two sources. Most of the observations found Cyg X-1 in the LHS, as shown by the hardness histogram. Clearly, the histogram features a second peak at softer values, which can also be identified with a bend in the HID. However, Cyg X-1 does not show type-C QPOs as would be expected on that branch. It is interesting to note that the SIMS region is reached only for a few observations, termed "failed state-transitions", but also no type-B QPOs were seen [74]. In correspondence to most of these events, a small radio flare was observed [74]. The HRD shows a strong similarity with GX 339-4: the points are well correlated and match those for GX 339-4, besides the few soft points which do not show evidence for a rms drop. The rms difference between the first and the subsequent LHS observations reported by [74] is present but invisible in the HRD due to the number of points. Interestingly, a transient relativistic jet has been observed from Cyg X-1 [26]. It corresponds in time to one of the excursions to a soft spectrum, but only observed down to hardness ∼0.3 and not to the extreme values in Fig. 3.8.

Another major difference is the absence of observable hysteresis (see [52]). Only a single branch is seen, which travels in both directions. However, GX 339-4 has shown that the weaker the outburst is, the smaller is the difference in flux between the hard-to-soft and soft-to-hard transitions (see Fig. 3.6). As Cyg X-1 never reaches very high accretion rate values, its behavior would be compatible with a very small unmeasurable level of hysteresis. Clearly, this system has somewhat different properties, probably related to the fact that it is not a transient. Nevertheless, its behavior in the HID and the HRD is not very different from that of other sources.

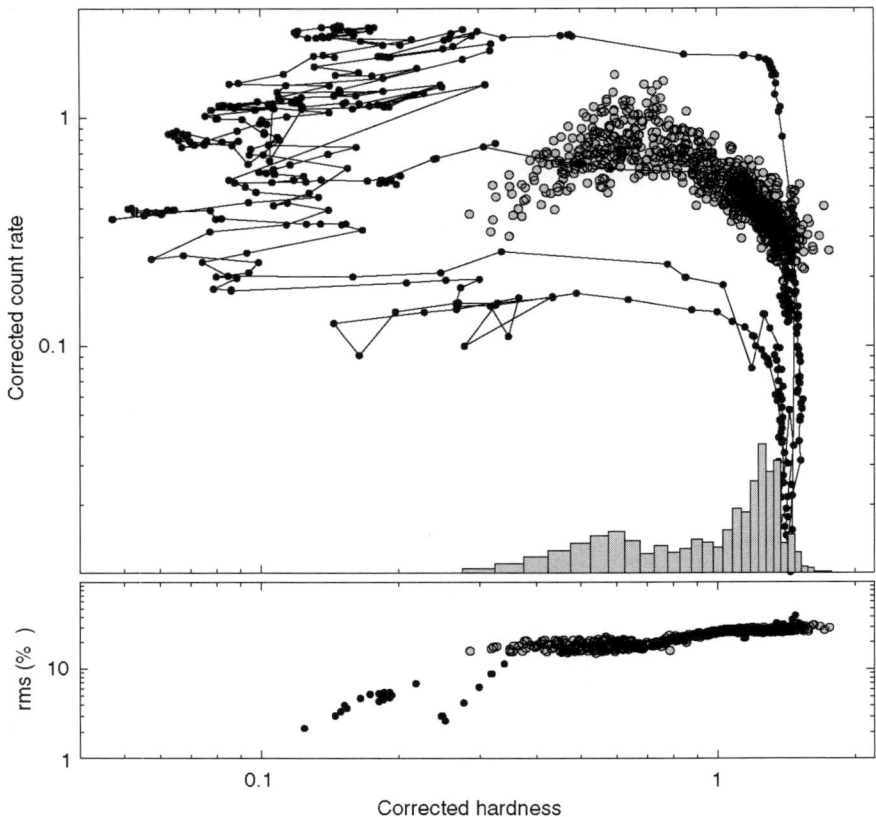

Fig. 3.8 *Top panel*: HID of all the RXTE/PCA observation of Cygnus X-1 (*gray circles*), plotted over the HID of the 2002/2003 and 2004/2005 outbursts of GX 339-4 (*black circles*). The *gray* histogram shows the distribution of hardness values on the same *X*-axis. The presence of two separate peaks is evident. Note that the GX 339-4 points have been shifted to bring the source to the same distance as Cygnus X-1. *Bottom panel*: HRD of the same observations, plotted over a few of the GX 339-4 from the 2002/2003 outburst (see *bottom panel* of Fig. 3.1)

3.7.2 GRS 1915+105

On the topic of microquasars, it is not possible not to discuss GRS 1915+105. This system shows very peculiar properties, which however can be successfully compared with those of other sources [24]. Three separate states were defined for this source, with a naming convention that in this context appears particularly unfortunate: states A, B, and C [5]. Both from the energy spectra and the PDS, it is clear that GRS 1915+105 never reaches a LHS: the high-energy end of the spectrum does not reach the typical 1.6 photon index, the thermal accretion disk is always present with a measured inner temperature >0.5 keV, and a strong type-C QPO is present during state C. A comparative analysis of the PDS in the three states of GRS 1915+105 not only shows that the LHS is never reached, but also suggests that states A and B correspond to the "anomalous state" described above [76].

Until recently, only a type-C QPO was observed from GRS 1915+105 in its hardest state, state C. However, a time-resolved analysis of fast varying light curves has led to the discovery of a type-B QPO in correspondence to state transitions, confirming the association between this oscillation and spectral changes [86]. It would be interesting to compare the HID and HRD of GRS 1915+105 with those of other more conventional transient sources in order to ascertain whether its peculiar states can be associated with the states outlined above and whether their evolution is compatible. The presence of type-C QPOs in state C suggests that it would correspond to the HIMS, while the softer states could be associated with the HSS. The fast (<1 s) transitions would include the SIMS, as shown by the detection of a

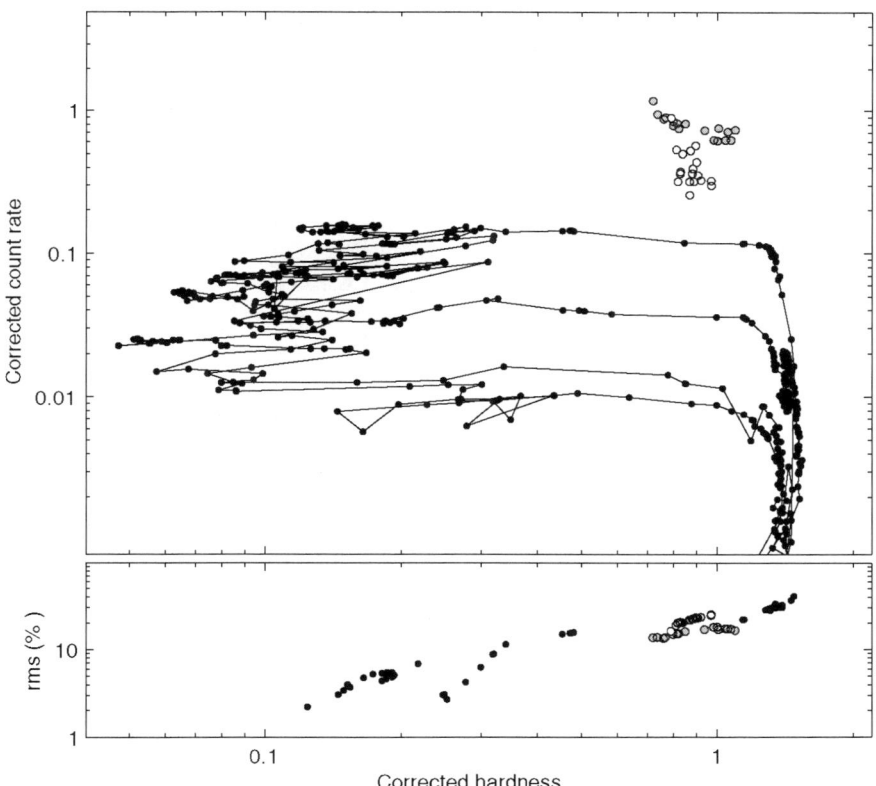

Fig. 3.9 *Top panel*: HID of two "plateau" observations of GRS 1915+105, plotted over the HID of the 2002/2003 and 2004/2005 outbursts of GX 339–4 (*black circles*). The GRS 1915+105 points correspond to the observations of classes χ_1 and χ_3 (*gray circles*) and class χ_2 (*empty circles*) from [5]. Note that the GX 339-4 points have been shifted to bring the source to the same distance as GRS 1915+105. *Bottom panel*: HRD of the same observations, plotted over a few of the GX 339-4 from the 2002/2003 outburst (see *bottom panel* of Fig. 3.1)

type-B QPO. However, the typical choice of X-ray energy bands for the production of hardness fails for this system: in its A and B states, the thermal accretion disk component is so hot that it extends well into the high-energy band. As a result, the hardness of these states is higher than that of state C. Moreover, the fast transitions would require a large amount of additional analysis effort.

However, one can accumulate the intensity/hardness/rms of observations during the so-called plateaux, which are long (one to few months) intervals of state C, usually followed by a major radio ejection [24]. During these observations, the flux only shows white noise on timescales longer than 1 s and the hardness can be compared with other sources as the thermal component is not dominant in the energy spectrum. Figure 3.9 shows the points of the observations corresponding to two of the plateaux examined in detail by [93]: the long one Oct 1996–Apr 1997 and the shorter during October 1997. The first one corresponds to variability class χ_2, the second to classes χ_1 and χ_3 (see [5]). The HID shows points which lie, as expected, in the HIMS region, with the χ_1 points at a higher count rate, probably as a result of a higher accretion rate. As in the case of Cyg X-1, the count rate has been corrected also for the difference in distance between the two sources and the high brightness of GRS 1915+105 even during plateaux is evident. Their distribution is elongated in count rate more than in hardness, at variance with the points of GX 339-4. In the HRD, the points from classes χ_1 and χ_3 overlap with those from GX 339-4, while χ_2 points lie above. This was also noted by [93], who showed that the difference is due to the presence of an additional broad component at high frequencies (50–100 Hz). This is the same component that was associated with lower-kHz QPO oscillations from neutron-star X-ray binaries (see [7]). Moreover, the χ_3 points correspond to a plateau with higher radio flux than that observed for χ_2 [24, 93]. GRS 1915+105 does therefore show some differences with other black-hole transients while in its HIMS, but overall the emerging picture is compatible.

3.8 Neutron-Star Binaries

Many neutron-star low-mass X-ray binaries display a transient nature similar to that of black-hole systems (see [95]). Their outbursts can also be characterized by the presence of two states, a hard and a soft one, with the hard state associated with low-flux intervals at the beginning and at the end of the outburst. Once a HID is produced, the similarity appears even stronger. Examples of HIDs for Aquila X-1 can be found in [95, 50] (see also Chaps. 4 and 5 of this book). A HID produced in the same way as the ones shown above can be seen in Fig. 3.10 (left panel). It represents three well-sampled outbursts of Aql X-1 as observed by RXTE. The similarity with Fig. 3.6 is evident. The similarities between X-ray outbursts of black-hole and neutron-star transients were explored by [51, 102, 103] and clearly suggest that the overall phenomenon of outburst evolution is similar. A similar evolution was recently found in the X-ray emission of a persistent neutron-star X-ray binary, 4U 1636-53 [12]. Systems of this class, called "atoll sources" alternate two states, characterized by spectral and timing features, a soft and a hard one, and accretion

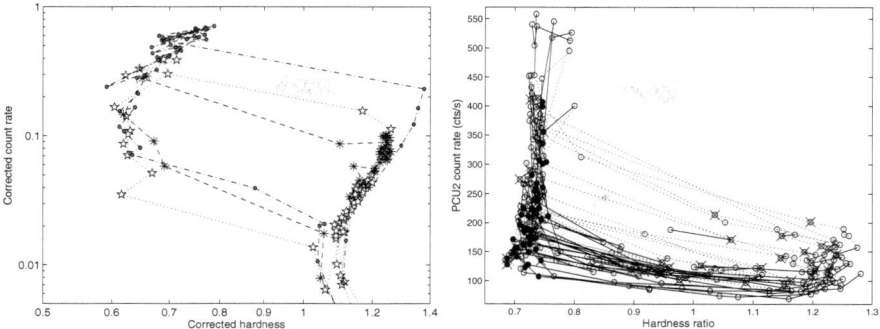

Fig. 3.10 *Left panel*: HID for three outbursts of the transient neutron-star binary Aquila X-1 as observed by RXTE/PCA. *Stars, circles,* and *asterisks* indicate outbursts from 1999, 2000, and 2004, respectively. *Right panel*: HID from 305 RXTE/PCA pointings of the persistent neutron-star binary 4U 1636-53. The lines connect the observations in time sequence: *solid line* means time evolution from soft to hard, *dotted line* the reverse. From [12]

rate is thought to be higher in the soft state than in the high state [95]. During the RXTE lifetime, 4U 1636-53 started long-term oscillations with a period of ~45 days, which decreased with time to about ~30 days [83, 12]. These oscillations correspond to regular state transitions: the production of a HID showed that the source travels a counterclockwise path very similar to that of transients (see Fig. 3.10, right panel) [12]. The similarities were also extended to cataclysmic variables (see [50] and Chap. 5).

3.9 Active Galactic Nuclei

The similarities between galactic X-ray binaries and AGN are an important tool for the study of both systems. The inner region of the accretion disk is expected to be independent of the nature of the accreting system. When appropriate scaling laws are applied, comparison must be possible (see e.g., [29]). The aperiodic timing properties of AGN have been studied in detail with long-term projects, needed because of the long timescales involved (see Chap. 8). From the analysis of 12 years of monitoring campaigns of AGN, two Seyfert 1 systems can be identified displaying variability properties similar to those of the HIMS. These are Ark 564 and Ton S180 [59, 22].

A similar approach can be attempted through spectral/hardness analysis. However, there is a fundamental problem to be addressed before producing an AGN HID: the temperature of the thermal component from the inner radius of an optically thick accretion disk scales with the black-hole mass as $M^{-1/4}$. Therefore, for AGN this component does not appear in the X-ray band and the resulting HID would not be comparable to that of a galactic system. Recently, a comparison has been proposed based on "disk-fraction luminosity diagrams" (DFLDs), where in place of an X-ray color the fraction of the overall spectral distribution attributed to the accretion disk is used [49]. Although this is a promising approach, it requires a very different type of analysis and is not directly comparable with BHB.

The HID of a galactic system is made of hard observations, soft observations, and transitions between them. The softening at the beginning of a transient outburst is due to two combined effects: the appearance of a strong thermal disk component and the steepening of the hard component. The hardening at the end of the outburst is the reverse effect. This means that even ignoring the disk component, the HID would have a similar shape, although the source would never reach very soft values of hardness. In other words, after the thermal disk is removed, the LHS points would not change (no strong disk component is detected there) and the HSS points would become considerably harder (and weaker, since here the flux is almost all thermal disk). In this respect, it appears meaningful to produce HIDs for AGN using the same procedure used for binaries. From the analysis of all Seyfert 1 AGN in the RXTE database, two systems emerge as rather different in their HID: not surprisingly, they are Ark 564 and Ton S180, the same ones singled out by the timing analysis. As an example, Fig. 3.11 shows the results for Ark 564. The top panel shows the light curve (one point per RXTE observation) over the period 1999–2003. The flux here is in the 3.8–21.2 keV, renormalized to the flux at Eddington accretion rate measured in the same band (i.e., an Eddington-normalized flux without a bolometric correction). The top axis shows the same $\sim$4-year scale linearly scaled from the mass of the black hole in Ark 564 to 10 $M_{\odot}$: 4 years for the AGN correspond to 500 s for a typical galactic black hole. The bottom panel is the HID with the same flux as in the top panel and a hardness corrected for the spectrum of the Crab nebula at the time of the observation, in order to compensate for gain changes throughout the RXTE lifetime. The dots show the points for the 2002/2003 outburst of GX 339-4, where the hardness is similarly corrected. It is evident that the points of Ark 564 are distributed along the horizontal HIMS line of the galactic binary. As explained above, their range is naturally limited and cannot reach very low values of hardness as those of the HSS of GX 339–4.

Notice however that GX 339-4 moved through the HID branch on a timescale of days and in a right-to-left direction. Ark 564 moves erratically and its points span a mere 500 s after correction for the black-hole mass. This means that, while the average value of hardness indicates Ark 564 as a good candidate for an AGN in the HIMS, the elongated shape of the distribution represents change on a different timescale than what is observed from GX 339-4. In this sense, AGN offer a unique chance to explore short timescales for which galactic sources do not have sufficient statistics.

While the two "intermediate" AGN can be identified both from timing and hardness analysis, there is a problem for the others. All Seyfert 1 AGN are located on the hard branch, corresponding to the LHS, while time variability suggests that they are all in the HSS (see e.g., [58]).

Recently, a 1-hour QPO was discovered in the Narrow-Line Seyfert 1 RE J1034+ 396 through a long XMM observation (see Chap. 8), opening an important window for comparison of timing properties of supermassive and stellar-mass systems

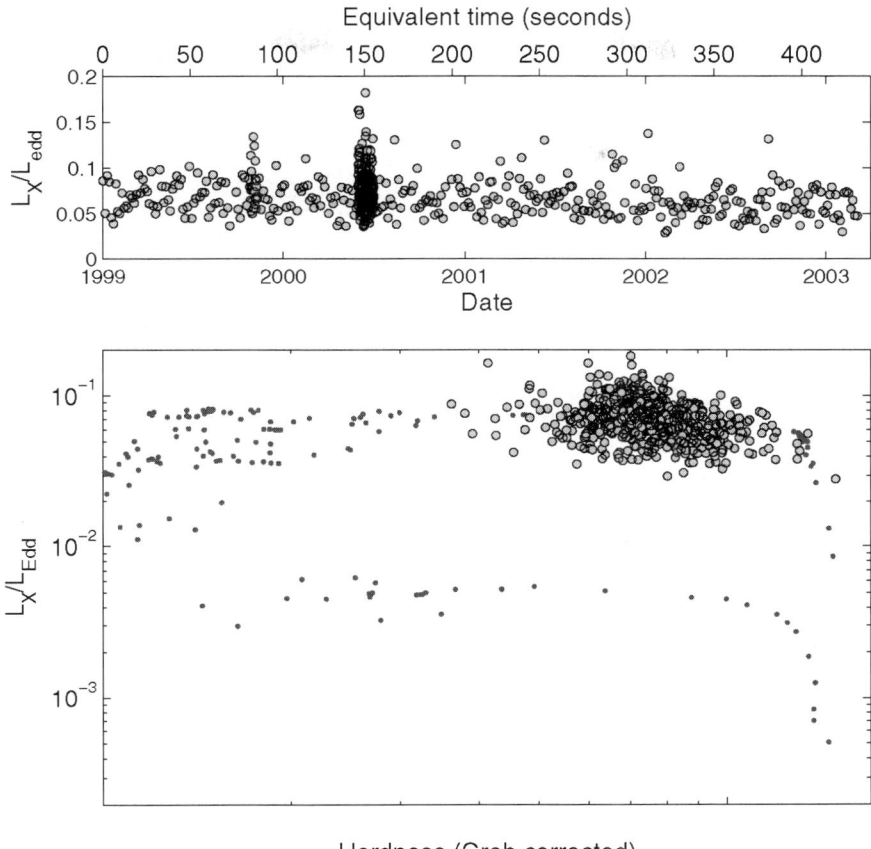

Fig. 3.11 *Top panel*: RXTE/PCA light curve (one point per observation) of the Seyfert 1 AGN Ark 465. The Y-axis is in units of Eddington luminosity for a mass of $3.2 \times 10^6\,M_\odot$, with no bolometric correction from the 3.8–21.2 keV band. The top axis indicates the corresponding dynamical timescale if the mass of the black hole was $10\,M_\odot$. *Bottom panel*: corresponding HID, plotted over the 2002/2003 points of GX 339-4

[33]. Given its relatively high frequency, this oscillation would correspond to high-frequency QPOs in BHBs.

3.10 Models and Interpretation

In the previous sections, I have presented a phenomenological picture for the evolution of hardness and timing properties in galactic black-hole transients and have compared it with those of other stellar-mass systems and AGN. In order to derive a physical picture, energy spectra must be extracted and theoretical models applied. It appears as the timing properties are crucial, but to date we do not have a complete theoretical framework for them (see [95] for a review). The emission mechanisms associated with the LHS and the HSS were discussed in Chap. 2, together with the geometry of the accretion flow in those states. Here, I will once again concentrate on the variations along the diagram, with particular emphasis on the state transitions. The picture presented in the previous sections is based on the emission in the 4–20 keV energy band, while a physical interpretation cannot be attempted without information on a much broader energy range.

3.10.1 Energy Spectra

All outbursts probably start with a LHS period, although in some case we do not observe it because it is too short. The general direction of accretion rate along the path in Fig. 3.7 is vertical: higher points correspond to a higher accretion rate. When a source moves horizontally, accretion rate should be higher on the left, but this is clear only along the main transitions at the top and bottom of the HID. Since accretion rate drives the movement along in the HID, a different physical parameter must be responsible for the position, i.e., accretion rate threshold, of the LHS–HIMS transition, which can vary between outbursts of the same source. The nature of this second parameter is very important, but still unclear (see Chap. 9).

Over the right branch in the HID, the broad-band energy spectrum is hard, with a high-energy cutoff around 100 keV, see [37]. However, an INTEGRAL observation of GRO J1655-40 has failed detecting a high-energy cutoff [14]. In the few cases when the level of interstellar absorption is low, a soft thermal disk component is also detected, with a large disk inner radius [28, 55]. Recently, a hotter thermal disk has been detected from a few sources, suggesting that the inner disk radius might always be close to the innermost stable orbit [62, 63]. The issue of the truncation of the inner disk radius is still unsolved. A Suzaku observation of Cygnus X-1 in the LHS from 2005 shows that the inner disk radius is indeed large, larger than that measured in the HSS [53]. Moreover, a reanalysis of the RXTE and XMM data for Swift 1753.5-0127 showed that a receded disk cannot be ruled out [38]. The hard component is usually interpreted as the result of thermal Comptonization or a combination of thermal/non-thermal Comptonization (e.g., [43]). Over the left branch, in the HSS, the energy spectrum is dominated by a hot thermal accretion disk, together with a (variable) hard component. This hard component has been observed as a power law without a cutoff up to 1 MeV [37].

What is the shape of the broad-band energy spectrum in the HIMS/SIMS and does anything change during the fast transitions? In other words: how does the energy spectrum change from the LSS to the HSS? The energy spectrum in these states is a combination of those of LSS and HSS. The thermal disk is clearly present and it is responsible for part of the horizontal excursion in the HID. Whether the inner disk radius moves inward as a source softens through the HIMS is still debated (see above). Significant radius changes were observed from GRS 1915+105 [5]. The hard component steepens as the disk fraction increases. Across transitions, the 2–20 keV spectrum does not change appreciably. For transients, a broader-band spectrum of good quality is only possible for the hard-to-soft transitions at high luminosities.

Recent RXTE/INTEGRAL observations of GX 339-4 have given conflicting results as to whether the high-energy cutoff increases/decreases/disappears across a HIMS–SIMS transition [10, 19]. Recent observations of GX 339-4 [69] and GRO J1655-40 [45] show that the situation is more complex than a simple increase or disappearance of the cutoff. More observations are needed. These observations are difficult as the transition is fast and can only be predicted with an accuracy of weeks. More information is available from GRS 1915+105, which as we saw above is only found in intermediate states. Simultaneous RXTE/OSSE observations showed spectra without a measurable cutoff up to ~600 keV, which could be interpreted with a hybrid thermal/non-thermal Componization model [105]. An RXTE/INTEGRAL campaign on GRS 1915+105 showed that the energy dependence of the type-C QPO (a HIMS signature) can be explained also with a hybrid emission model [81]. In addition, quasi-simultaneous ASCA/RXTE/OSSE data of XTE J1550-564 during the HIMS were found also to require a hybrid thermal/non-thermal model [30]. The RXTE database of GX 339-4, including the 2002/2003 outburst shown in the previous sections, has been analyzed with spectral models by [106].

The HIMS–SIMS transition corresponds to the crossing of the "jet line", which approximately marks the time of the ejection of a fast relativistic jet from the system (see the following chapters). This correspondence is not precise, as shown by [27], see Chap. 5. However, this event takes place in a short interval of time (minutes to hours). It is possible that the transient annihilation line detected from GS 1124-684, which was seen only over an interval of ~12 h, was indeed associated with such a transition, since the observation took place well within the Very-High State of the source, corresponding to HIMS/SIMS [35]. However, no evidence for such a spectral feature was found in the INTEGRAL data of the 2004 transition of GX 339-4.

As to the anomalous state, detailed spectral fitting needs to be performed. Here the flux is high, associated with high luminosity and, since the energy spectrum is soft, mass accretion rate is probably also high. GRO J1655-40 showed a well populated anomalous branch in both its outburst covered by RXTE. XTE J1550-564 in 1998 showed a very high flux in a single observation, exceeding a flux of 6 Crab [85], which could be interpreted as belonging to the same "state" [25]. What these observations have in common are a high inner-disk temperature, a steep hard component, a very small derived inner disk radius, and a small percentage contribution of the disk to the overall flux [84, 85]. The same effect was measured in 4U 1630-47 in a number of RXTE observations [94]. These were interpreted as a signature to the transition to a slim disk regime [94, 98]. The issue of the shape of the soft component has been discussed by various authors; in particular, the distortion effects due to Comptonization have been investigated [20]. A comprehensive physical picture of the accretion flow onto black holes and neutron stars has been proposed [21], but it is not yet clear how this corresponds to the states presented above and the phenomenology is mixed with the modeling.

3.10.2 Time Variability

The fast aperiodic variability appears to be a crucial element for our understanding of the accretion (and jet ejection) properties in black-hole binaries. In particular, the most important state-transition (HIMS–SIMS) associated with the crossing of the jet line is identified only through the changes in the timing properties.

The PDS of the LSS and the HIMS is compatible with a smooth evolution of power-spectral components. The band-limited noise components and the type-C QPO follow very tight correlations which can be extended to neutron-star binaries and possibly to cataclysmic variables [99, 7, 97, 95]. It is clear now that for these features, the distinction between noise and QPO components is somewhat arbitrary (see [7]). An interesting theoretical approach that does not discriminate between them has been presented, based on a dynamical model that investigates the filtering effect of the presence of a transition radius in the accretion disk [75]. Moreover, the frequencies of these components correlate with spectral parameters. A strong general correlation was found between the frequency of the type-C QPO and the photon index of the hard power-law component in the energy spectrum [96]. This correlation has been suggested as a possible way to estimate the compact-object mass from X-ray data (see e.g., [92]). The fact that these correlations appear to extend also to the HSS indicates that the nature of the fundamental frequencies driving the variability is the same from the LHS across the HIMS to the HSS. It is also possible that the actual emission process is the same in the hard and soft states. This would indicate that the "corona" which some models associate with the hard-state emission would be present in the soft state. More analysis is needed in order to establish this fact.

Definitely different is the case of the SIMS. Here the timing properties are different: there is no band-limited noise and the QPOs have different characteristics, notably are rather stable in frequency. At the same time, high-frequency QPOs are at times observed. What happens during this state, the onset of which is associated with the launch of fast relativistic jets, is still unexplored. The sharp and fast transitions observed from and to this state show that probably some type of instability is at work here, but no modeling has been attempted.

Unlike in the HID, the source track in the HRD does not show evidence of hysteresis (see Figs. 3.1 and 3.7). The main difference between the high-flux and low-flux branches is related to the small extent of the SIMS in the latter. However, detailed analysis does show that the two HRD paths are different [8]. Therefore, the timing properties seem to depend only weakly on the flux, provided that the energy spectrum is the same. This contributes to indicate that time variability provides a powerful and direct approach to basic parameters in the accretion flow.[2] This also applies to the elusive HFQPOs, which are observed in or near the SIMS. Unfortunately, the scarcity of detections does not allow more thorough investigations on the nature of these oscillations and their relation to the spectral characteristics of the associated emission.

3.11 Conclusions: How Many States?

The phenomenology of the evolution of outbursts of black-hole transients appears complex, but not impossible to treat. The use of diagrams such as HID and HRD allows to disentangle most of the properties and to compare the behavior of different systems. This paradigm can also be applied to other stellar-mass systems and even to supermassive objects. Although the state classification presented in the previous sections manages to capture all the essential timing/hardness properties of these systems, a major question which still remains open is that of physical states. The original picture, still widely applied today, interprets observations in terms of two separate physical states: a hard one dominated by thermal Comptonization and a soft one dominated by an optically thick accretion disk with a contribution by a non-thermal hard component. The study of state transitions and global timing properties suggests that the differences between hard and soft states are not so marked. They could be considered as evolving smoothly from the hardest energy spectra associated with strong variability to the softest energy spectra with only a few percent variability. The energy spectra also do not seem to show strong transitions: the INTEGRAL observation of the HIMS–SIMS transition of 2004 shows a hint of a change in the high-energy cutoff, but the spectral index at low energies is unchanged (see also [70]). The only interruption is the presence of the SIMS, which is marked by a fast drop in variability and the appearance of specific types of QPOs. This picture suggests that the only physically different state is the SIMS, which is also associated with the ejection of relativistic jets. As mentioned at the beginning, this picture and the corresponding state classification are meant only to present a coherent picture: the development and application of theoretical models will identify physically meaningful states and lead to a full classification. This will be possible only if the complete evolution of transient black-hole binaries is considered.

Acknowledgments A large number of colleagues have contributed to the shaping of the concepts which I have put in this chapter, and of course also to the results. Impossible to list them here, but easy to identify them as coauthors of many papers (and of other chapters of this book). The main one is certainly Jeroen Homan, whose work has started the new wave in black-hole states and is now consolidating it.

---

[2] Notice that the different energy dependence of the fractional integrated variability corresponding to the different states [32] means that the HRD looks different at different energies. In particular, the different energy dependence on the LHS and HIMS breaks the continuity of the path in the HRD at high energies.

References

1. M. Abramowicz, W. Kluzniak: A&A, **374**, L19 (2001)
2. T. Belloni, G. Hasinger: A&A, **227**, L33 (1990)
3. T. Belloni, M. Méndez, M. van der Klis et al.: ApJ, **472**, L107 (1996)
4. T. Belloni, M. van der Klis, W.H.G. Lewin et al.: A&A, **322**, 857 (1997)
5. T. Belloni, M. Klein-Wolt, M. Méndez, M. van der Klis et al.: A&A, **355**, 271 (2000)
6. T. Belloni, A.P. Colombo, J. Homan: A&A, **390**, 199 (2002)
7. T. Belloni, D. Psaltis, M. van der Klis: ApJ, **572**, 392 (2002)
8. T. Belloni, J. Homan, P. Casella et al.: A&A, **440**, 207 (2005)
9. T. Belloni, P. Soleri, P. Casella et al.: MNRAS, **369**, 305 (2006)
10. T. Belloni, I. Parolin, M. Del Santo et al.: MNRAS, **1113**, 367 (2006)
11. T. Belloni, M. Méndez, J. Homan: MNRAS, **376**, 1133 (2007)
12. T. Belloni, J. Homan, S. Motta, E. Ratti: MNRAS, **379**, 247 (2007)
13. H.V. Bradt, R.E. Rothschild, J.H. Swank: A&A Suppl., **97**, 335 (1993)
14. M.D. Caballero García, J.M. Miller, E. Kuulkers et al.: ApJ, **669**, 534 (2007)
15. P. Casella, T. Belloni, J. Homan et al.: A&A, **426**, 687 (2004)
16. P. Casella, T. Belloni, L. Stella: ApJ, **629**, 403 (2005)
17. M.J. Coe, A.R. Engel, J.J. Quenby: Nature, **259**, 544 (1976)
18. W. Cui, S.N. Zhang, W. Chen, et al.: ApJ, **512**, L43 (1999)
19. M. Del Santo, T.M. Belloni, J. Homan et al.: MNRAS, **392**, 992 (2009).
20. C. Done, A. Kubota: MNRAS, **371**, 1216 (2006)
21. C. Done, M. Gierliński, A. Kubota: Astr. Ap. Rev., **15**, 1 (2007)
22. R. Edelson, T.J. Turner, K. Pounds et al.: ApJ, **568**, 610 (2002)
23. M. Elvis, C.G. Page, K.A. Pounds et al.: Nature, **257**, 656 (1975)
24. R.P. Fender, T. Belloni, ARA&A, **42**, 317 (2004)
25. R.P. Fender, T. Belloni, E. Gallo: MNRAS, **355**, 1105 (2004)
26. R.P. Fender, A.M. Stirling, R.E. Spencer et al.: MNRAS, **369**, 603 (2006)
27. R.P. Fender, J. Homan, T.M. Belloni: MNRAS, **396**, 1370 (2009)
28. F. Frontera, L. Amati, A. Zdziarski et al.: ApJ, **592**, 1110 (2003)
29. R.P. Fender, E. Körding, T. Belloni et al.: Proc. of the VI Microquasar Workshop: Microquasars and Beyond, PoS(MQW6)011 (2006)
30. M. Gierliński, C. Done: MNRAS, **342**, 1083 (2003)
31. M. Gierliński, C. Done: MNRAS, **347**, 885 (2004)
32. M. Gierliński, A. Zdziarski: MNRAS, **363**, 1349 (2005)
33. M. Gierliński, M. Middleton, M. Ward et al.: Nature, **455**, 369 (2008)
34. T. Gleissner, J. Wilms, K. Pottschmidt et al.: A&A, **414**, 1091 (2004)
35. A. Goldwurm, J. Ballet, B. Cordier et al.: ApJ, **389**, L79 (1992)
36. D. Götz, S. Mereghetti, D. Merlini et al.: A&A, **448**, 873 (2006)
37. J.E. Grove, W.N. Johnson, R.A. Kroeger et al.: ApJ, **500**, 899 (1998)
38. B. Hiemstra, P. Soleri, M. Mendez et al.: MNRAS, **394**, 2080 (2009)
39. J. Homan, T. Belloni: Ap&SS, **300**, 107 (2005)
40. J. Homan, T. Belloni: in preparation (2009)
41. J. Homan, R. Wijnands, M. van der Klis et al.: ApJ Suppl, **132**, 377 (2001)
42. J. Homan, M. Buxton, S. Markoff et al.: ApJ, **624**, 295 (2005)
43. A. Ibragimov, J. Poutanen, M. Gilfanov et al.: MNRAS, **362**, 1435 (2005)
44. S.A. Ilovaisky, C. Chevalier, P.A. Charles et al.: A&A, **164**, 67 (1986)
45. A. Joinet, E. Kalemci, F. Senziani: ApJ, **679**, 655 (2008).
46. E. Kalemci, J.A. Tomsick, R.E. Rothschid et al.: ApJ, **603**, 231 (2004)
47. S. Kitamoto, H. Tsunemi, S. Miyamoto et al.: ApJ, **394**, 609 (1992)
48. M. Klein-Wolt, M. van der Klis: ApJ, **675**, 1407 (2008)
49. E.G. Körding, S. Jester, R. Fender: MNRAS, **372**, 1366 (2006)
50. E.G. Körding, M. Rupen, C. Knigge et al.: Science, **320**, 1318 (2008)

51. T.J. Maccarone: A&A, **409**, 697 (2003)
52. T.J. Maccarone, P.S. Coppi: MNRAS, **338**, 189 (2003)
53. K. Makishima, H. Takahashi, S. Yamada et al.: PASJ, **60**, 585 (2008)
54. T.H. Markert, C.R. Canizares, G.W. Clark et al.: ApJ, **184**, L67 (1973)
55. J.E. McClintock, C.A. Haswell, M.R. Garcia, J.J. Drake et al: ApJ, **555**, 477 (2001)
56. J.E. McClintock, R.A. Remillard: Black-hole binaries. In: W.H.G. Lewin, M. van der Klis (eds.) Compact Stellar X-Ray Sources, pp. 157–214. Cambridge University Press, Cambridge (2006)
57. J.E. McClintock, R.A. Remillard, M.P. Rupen et al.: ApJ, **698**, 1398 (2009)
58. I.M. McHardy, E. Körding, C. Knigge et al.: Nature, **444**, 730 (2006)
59. I.M. McHardy, P. Arévalo, P. Uttley et al.: MNRAS, **382**, 985 (2007)
60. M. Méndez, M. van der Klis: ApJ, **479**, 926 (1997)
61. J.M. Miller, R. Wijnands, J. Homan et al.: ApJ, **563**, 928 (2001)
62. J.M. Miller, J. Homan, G. Miniutti: ApJ, **652**, L113 (2006)
63. J.M. Miller, J. Homan, D. Steeghs: ApJ, **653**, 525 (2006)
64. S. Miyamoto, K. Kimura, S. Kitamoto et al.: ApJ, **383**, 784 (1991)
65. S. Miyamoto, S. Kitamoto, S. Iga et al.: ApJ, **391**, L21 (1992)
66. S. Miyamoto, S. Iga, S. Kitamoto et al.: ApJ, **403**, L39 (1993)
67. S. Miyamoto, S. Kitamoto, S. Iga et al.: ApJ, **435**, 398 (1994)
68. S. Miyamoto, S. Kitamoto, K. Hayahida, W. Egoshi: ApJ, **442**, L13 (1995)
69. S. Motta, T.M. Belloni, J. Homan, et al.: MNRAS, in press (arXiv:0908.2451) (2009)
70. E. Nespoli, T. Belloni, J. Homan et al.: A&A, **412**, 235 (2003)
71. P.L. Nolan, D.E. Gruber, J.L. Matteson et al.: ApJ, **246**, 494 (1981)
72. M.A. Nowak: MNRAS, **318**, 361 (2000)
73. M.A. Nowak, J. Wilms, W.A. Heindl et al.: MNRAS, **320**, 316 (2001)
74. K. Pottschmidt, J. Wilms, M.A. Nowak et al.: A&A, **407**, 1039 (2003)
75. D. Psaltis, C. Norman: ApJ, submitted (astro-ph/0001931v1) (2000)
76. P. Reig, T. Belloni, M. van der Klis: A&A, **412**, 229 (2003)
77. R.A. Remillard, J.E. McClintock: ARA&A, **44**, 49 (2006)
78. R.A. Remillard, G.J. Sobczak, M.P. Muno et al.: ApJ, **564**, 962 (2002)
79. M.J. Ricketts, K.A. Pounds, M.J.L. Turner: Nature, **257**, 657 (1975)
80. J. Rodriguez, S. Corbel, J.A. Tomsick: ApJ, **595**, 1032 (2003)
81. J. Rodriguez, S. Corbel, D.C. Hannikainen et al.: ApJ, **615**, 416 (2004)
82. J. Samimi, G.H. Share, K. Wood et al.: Nature, **278**, 434 (1979)
83. I.C. Shih, A.J. Bird, P.A. Charles et al.: MNRAS, **361**, 602 (2005)
84. G.J. Sobczak, J.E. McClintock, R.A. Remillard et al.: ApJ, **520**, 776 (1999)
85. G.J. Sobczak, J.E. McClintock, R.A. Remillard et al.: ApJ, **544**, 993 (2000)
86. P. Soleri, T. Belloni, P. Casella: MNRAS, **383**, 1089 (2007)
87. S.J. Sturner, C.R. Shrader: ApJ, **625**, 923 (2005)
88. M. Takizawa, T. Dotani, K. Mitsuda et al.: ApJ, **489**, 272 (1997)
89. Y. Tanaka, W.H.G. Lewin: Black-hole binaries. In: W.H.G. Lewin, J. van Paradijs, E.P.J. van den Heuvel (eds.) X-Ray Binaries, pp. 126–174. Cambridge University Press, Cambridge (1995)
90. H. Tananbaum, H. Gursky, E. Kellogg et al.: ApJ, **177**, L5 (1972)
91. N.J. Terrell: ApJ, **174**, L35 (1972)
92. L. Titarchuk, N. Shaposhnikov: Ap, **626**, 298 (2007)
93. S.P. Trudolyubov: ApJ, **558**, 276 (2001)
94. J.A. Tomsick, S. Corbel, A. Goldwurm et al.: ApJ, **630**, 413 (2005)
95. M. van der Klis: Rapid X-ray variability. In: W.H.G. Lewin, M. van der Klis (eds.) Compact Stellar X-Ray Sources, pp. 39–112. Cambridge University Press, Cambridge (2006)
96. F. Vignarca, S. Migliari, T. Belloni et al.: A&A, **397**, 729 (2003)
97. B. Warner, P.A. Woudt, M.L. Pretorius: MNRAS, **344**, 1193 (2003)
98. K. Watarai, J. Fukue, M. Takeuchi et al.: PASJ, **52**, 133 (2000)

 99. R. Wijnands, M. van der Klis: ApJ, **514**, 939 (1999)
100. J. Wilms, M.A. Nowak, K. Pottschmidt et al.: MNRAS, **320**, 327 (2001)
101. J. Wilms, M.A. Nowak, K. Pottschmidt et al.: A&A, **447**, 245 (2006)
102. W. Yu, M. Klein-Wolt, R. Fender et al.: ApJ, **589**, L33 (2003)
103. W. Yu, M. van der Klis, R. Fender: ApJ, **611**, L121 (2004)
104. W. Yu, F.K. Lamb, R. Fender, M. van der Klis: ApJ, **663**, 1309 (2007)
105. A.A. Zdziarski, J.E. Grove, J. Poutanen et al.: ApJ, **554**, L45 (2001)
106. A.A. Zdziarski, M. Gierliński, J. Mikołajewska et al.: MNRAS, **351**, 791 (2004)

Chapter 4
Radio Emission and Jets from Microquasars

E. Gallo

Abstract To some extent, all Galactic binary systems hosting a compact object are potential "microquasars", so much as all Galactic nuclei may have been quasars, once upon a time. The necessary ingredients for a compact object of stellar mass to qualify as a microquasar seem to be: accretion, rotation, and magnetic field. The presence of a black hole may help, but is not strictly required, since neutron star X-ray binaries and dwarf novae can be powerful jet sources as well. The above issues are broadly discussed throughout this chapter, with a rather trivial question in mind: Why do we care? In other words: are jets a negligible phenomenon in terms of accretion power, or do they contribute significantly to dissipating gravitational potential energy? How do they influence their surroundings? The latter point is especially relevant in a broader context, as there is mounting evidence that outflows powered by supermassive black holes in external galaxies may play a crucial role in regulating the evolution of cosmic structures. Microquasars can also be thought of as a form of quasars for the impatient: what makes them appealing, despite their low number statistics with respect to quasars, are the fast variability timescales. In the first approximation, the physics of the jet-accretion coupling in the innermost regions should be set by the mass/size of the accretor: stellar mass objects vary by 10^5–10^8 times shorter timescales, making it possible to study variable accretion modes and related ejection phenomena over average Ph.D. timescales. At the same time, allowing for a systematic comparison between different classes of compact objects – black holes, neutron stars, and white dwarfs – microquasars hold the key to identify and characterize properties that may be unique to, e.g., the presence (or the lack) of an event horizon.

E. Gallo (✉)
MIT Kavli Institute for Astrophysics and Space Research, 70 Vassar Street, Bldg 37-685, Cambridge, MA 02139, USA, egallo@mit.edu

Gallo, E.: *Radio Emission and Jets from Microquasars*. Lect. Notes Phys. **794**, 85–113 (2010)
DOI 10.1007/978-3-540-76937-8_4

4.1 Radio Observations of Black Holes

The synchrotron nature of the radio emission from X-ray binaries is generally inferred by the non-thermal spectra and high brightness temperatures. The latter translate into minimum linear sizes for the radio emitting region which often exceed the typical orbital separations, making the plasma uncontainable by any known component of the binary. If coupled to persistent radio flux levels, this implies the presence of a continuously replenished relativistic plasma that is flowing out of the system [58, 95, 30]. Thanks to aggressive campaigns of multi-wavelength observations of X-ray binaries in outbursts over the last decade or so, we have now reached a reasonable understanding of their radio phenomenology in response to global changes in the accretion mode.

For the black holes (BHBs), radiatively inefficient, low/hard X-ray states are associated with flat/slightly inverted radio-to-mm spectra and persistent radio flux [39] (the reader is referred to Chap. 3 of this book for a review of X-ray states of black hole X-ray binaries, as well as [80, 60]). In analogy with compact extra-galactic radio sources, the flat spectra are thought to be due to the superimposition of a number of peaked synchrotron spectra generated along a conical outflow, or jet, with the emitting plasma becoming progressively thinner at lower frequencies as it travels away from the jet base [11, 69]. The jet interpretation has been confirmed by high resolution radio maps of two hard state BHBs: Cygnus X-1 [114] (Fig. 4.1,

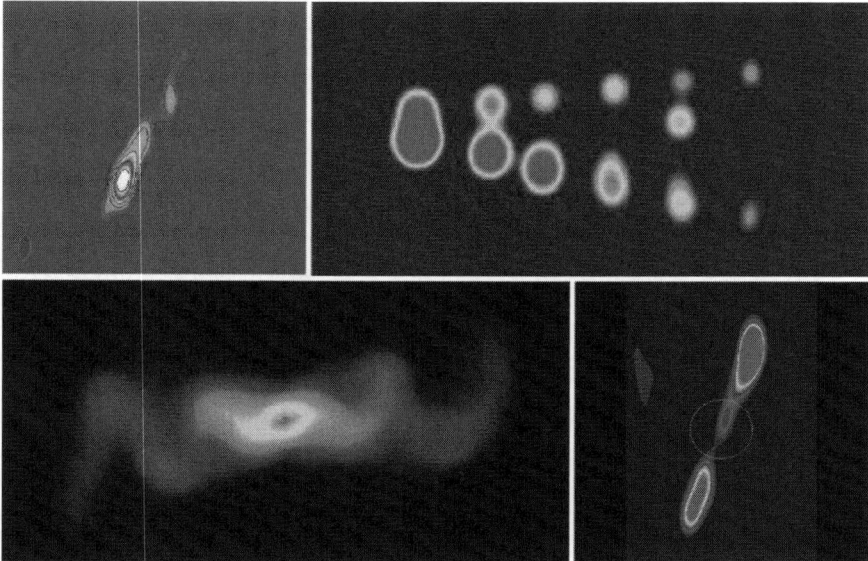

Fig. 4.1 *Top left*: steady, milliarcsec scale jet from the high mass black hole X-ray binary Cyg X-1. From [114]. *Top right*: transient, arcsec-scale radio jets from the superluminal Galactic jet source GRS 1915+105. From [98]. *Bottom left*: arcsec-scale radio jets from the first Galactic source discovered: SS 443. The binary orbit is almost edge-on; the precessing accretion disk of SS 433 causes its jets to trace a "corkscrew" in the sky every 162 days. From [12]. *Bottom right*: fossil, arcmin-scale radio jets around the Galactic center black hole in 1E 140.7-2942. From [99]

top left panel) and GRS 1915+105 [25, 42] are both resolved into elongated radio sources on milliarcsec scales (tens of A.U.) implying collimation angles smaller than a few degrees. Even though no collimated radio jet has been resolved in any BHB emitting X-rays below a few percent of the Eddington limit, L_{Edd}, it is widely accepted, by analogy with the two above-mentioned systems, that the flat radio spectra associated with unresolved radio counterparts of X-ray binaries are originated in conical outflows. Yet, it remains to be proven whether such outflow would maintain highly collimated at very low luminosity levels, in the so-called quiescent regime ($L_X/L_{Edd} \lesssim 10^{-5}$; see Sect. 4.5.1).

Radiatively efficient, high/soft (thermal dominant) X-ray states, on the contrary, are associated with no *flat-spectrum* core radio emission [41]; the core radio fluxes drop by a factor at least 50 with respect to the hard state (e.g., [41, 18]), which is generally interpreted as the physical suppression of the jet taking place over this regime. While a number of sources have been detected in the radio during the soft state [14, 18, 48, 13], the common belief is that these are due to optically thin synchrotron emission – until proven otherwise.

Transient ejections of optically thin radio plasmons moving away from the binary core in opposite directions are often observed as a result of bright radio flares associated with hard-to-soft X-ray state transitions (Fig. 4.1, top right panel). These are surely the most spectacular kind of jets observed from X-ray binaries, in fact, those which have inspired the term "microquasar" [96]. As proven by the case of GRS 1915+105, and more recently by Cygnus X-1 as well [31], the same source can produce either kind of jets, persistent/partially self-absorbed, and transient/optically thin, dependently on the accretion regime.

4.2 Coupling Accretion and Ejection in Black Holes

The question arises whether the flat-spectrum, steady jet, and the optically thin discrete ejections differ fundamentally or they are different manifestations of the same phenomenon. This issue has been tentatively addressed with a phenomenological approach [34]. Broadly speaking, this model aims to put together the various pieces of a big puzzle that were provided to us by years of multiwavelength monitoring of BHBs, and to do so under the guiding notion that the jet phenomenon has to be looked at as an intrinsic part of the accretion process. Reference [34] collected as many information as possible about the very moment when major radio flares occur in BHBs, and proposed a way to "read them" in connection with the X-ray state over which they took place as well as the observed jet properties prior and after the radio flare.

The study makes use of simultaneous X-ray (typically Rossi X-ray Timing Explorer, RXTE) and radio (Australia Telescope Compact Array, ATCA, and/or the Very Large Array, VLA) observations of four outbursting BHB systems: GRS 1915+105, XTE J1550-564, GX 339-4, and XTE J1859+229. X-ray hardness-intensity diagrams (HID) have been constructed for the various outbursts and linked with the evolution of the jet morphology, radio luminosity, total power, Lorentz factor, and so on.

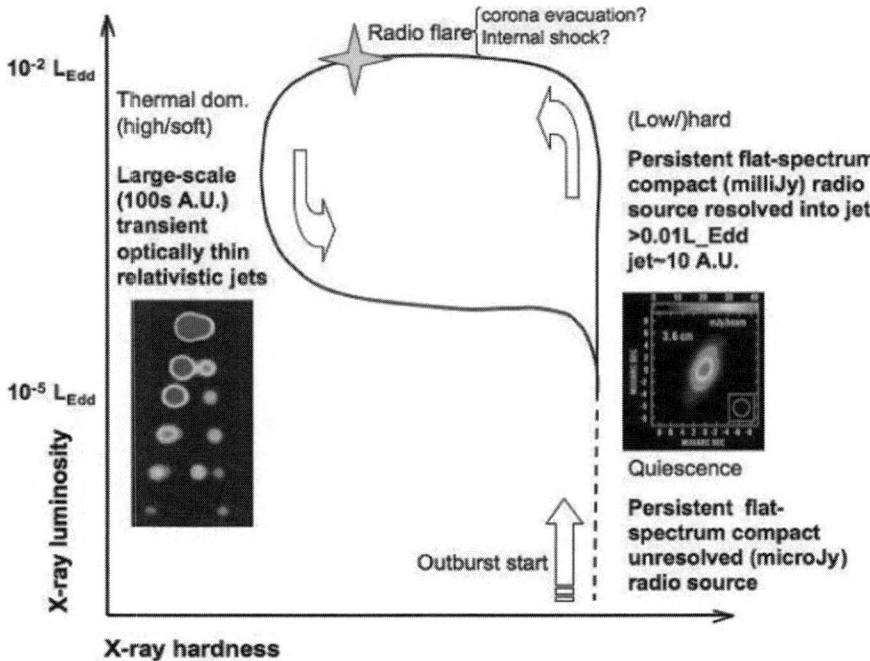

Fig. 4.2 Schematic of radio properties of black hole X-ray binaries over different accretion modes (see Chap. 3 for a description of X-ray states). See http://www.issibern.ch/ teams/proaccretion/Documents.html for a more detailed illustration and [34, 32] for the relative science papers

Figure 4.2 (naively) illustrates our understanding of the so-called jet-accretion coupling in BH X-ray binaries. It represents the HID of a well-behaved outburst, with the time arrow progressing counterclockwise. Starting from the bottom right corner, the system is a low-luminosity, quiescent X-ray state, producing a (supposedly) mildly relativistic, persistent outflow, with flat radio spectrum. Its luminosity starts to increase at all wavelengths, while the X-ray spectrum remains hard. Around a few percent of the Eddington X-ray luminosity, a sudden transition is made (top horizontal branch) during which the global properties of the accretion flow change substantially (hard-to-soft state transition), while a bright radio flare is observed, likely due to a sudden ejection episode [97]. This can be interpreted [34] as the result of the inner radius of a geometrically thin accretion disk moving inward. The Lorentz factor of the ejected material, due to the deeper potential well, exceeds that of the hard state jet, causing an internal shock to propagate through it, and to possibly disrupt it. Once the transition to the high/soft (thermal dominant) state is made, no core radio emission is observed, while large scale (hundreds of A.U.), rapidly fading radio plasmons are often seen moving in an opposite direction with respect to the binary system position, with highly relativistic speed. Toward the end of the outburst, the X-ray spectrum starts to become harder, and the compact, flat-spectrum, core radio source turns on once again. A new cycle begins (with timescales that vary greatly from source to source).

The bright radio flare associated with the state transition could coincide with the very moment in which the hot corona of thermal electrons, responsible for the X-ray power law in the spectra of hard state BHBs, is accelerated and ultimately evacuated. This idea of a sudden evacuation of inner disk material is not entirely new, and in fact dates back to extensive RXTE observations of the rapidly varying GRS 1915+105: despite their complexity, the source spectral changes could be accounted for by the rapid removal of the inner region of an optically thick accretion disk, followed by a slower replenishment, with the timescale for each event set by the extent of the missing part of the disk [7, 8]. Subsequent, multi-wavelength (radio, infrared, and X-ray) monitoring of the same source suggested a connection between the rapid disappearance and follow-up replenishment of the inner disk seen in the X-rays, with the infrared flare starting during the recovery from the X-ray dip, when an X-ray spike was observed [97].

Yet, it remains unclear what drives the transition in the radio properties after the hard X-ray state peak is reached. Specifically, radio observations of GX 339-4, XTE J1550-564, and GRS 1915+105 indicate that in this phase the jet spectral index seems to "oscillate" in an odd fashion, from flat to inverted to optically thin, as if the jet was experiencing some kind of instability as the X-ray spectrum softens. Recent, simultaneous RXTE and INTEGRAL (The INTernational Gamma-Ray Astrophysics Laboratory) observations of GX 339-4 [6] have shown that the high energy cutoff typical of hard state X-ray spectra, either disappears or shifts toward much higher energies within timescales of hours (<8 h) during the transition.

Finally, there are at least a couple of recent results that might challenge some of the premises the unified scheme is based on. The first one is the notion that, for the internal shock scenario to be at work and give rise to the bright radio flare at the state transition, whatever is ejected must have a higher velocity with respect to the pre-existing hard state steady jet. From an observational point of view, this was supported, on one side, by the lower limits on the transient jets' Lorentz factors, typically higher than $\Gamma = 2$ [35], and, on the other hand, by the relative small scatter about the radio/X-ray correlation in hard state BHBs [49] (see Sect. 4.3.1). The latter has been challenged on theoretical grounds [56].

The second premise has to do with the existence of a geometrically thin accretion disk in the low/hard state of BHBs. Deep X-ray observations of hard state BHBs [84, 85, 110, 116] (see also [52]) have shown evidence for a cool disk component extending close to the innermost stable orbit already during the bright phases of the hard state that is prior to the horizontal brunch in the top panel of Fig. 4.2 ($L_X/L_{Edd} \simeq 10^{-3} - 10^{-2}$). This challenges the hypothesis of a sudden deepening of the inner disk potential well as the cause of a high Lorentz factor ejection. Possibly, whether the inner disk radius moves close to the hole prior or during the softening of the X-ray spectrum does not play such a crucial role in terms of jet properties; if so, then the attention should be diverted to a different component, such as the presence/absence, or the size, of a Comptonizing corona [62], which could in fact coincide with the very jet base [76]. It is worth mentioning that a recent paper [67] gives theoretical support to the survival of a thin accretion disk down to low Eddington ratios: within the framework of the disk evaporation model (e.g., [66]), it is found that a weak, condensation-fed, inner disk can be present in the hard state of black hole transient systems for Eddington-scaled luminosities as low as 10^{-3} (depending on the magnitude of the viscosity parameter).

Here I wish to stress that, in addition to solving the above-mentioned issues, much work needs to be done in order to test the consistency of the internal shock scenario as a viable mechanism to account for the observed changes in the radio properties, given the observational and theoretical constraints for a given source (such as emissivities, radio/infrared delays, and cooling times).

Finally, one of the most interesting aspects of the proposed scheme – assuming that is correct in its general principles – is obviously its connection to super-massive BHs in active Galactic nuclei (AGN), and the possibility to mirror different X-ray binary states into different classes of AGN. This is explored in detail in Chap. 5.

4.3 Empirical Luminosity Correlations

4.3.1 Radio/X-Ray

In a first attempt to quantify the relative importance of jet vs. disk emission in BHBs, [49] collected quasi-simultaneous radio and X-ray observations of ten low/hard state sources. This study established the presence of a tight correlation between the X-ray and the radio luminosity, of the form $L_R \propto L_X^{0.7 \pm 0.1}$, first quantified for GX 339-4 [19]. The correlation extends over more than three orders of magnitude in L_X and breaks down around 2%L_{Edd}, above which the sources enter the high/soft (thermal dominant) state, and the core radio emission drops below detectable levels. Given the nonlinearity, the ratio radio-to-X-ray luminosity increases toward quiescence (below a few $10^{-5} L_{Edd}$). This leads to the hypothesis that the total power output of quiescent BHBs could be dominated by a radiatively inefficient outflow, rather than by the local dissipation of gravitational energy in the accretion flow [37, 74].

Even though strictly simultaneous radio/X-ray observation of the nearest quiescent BHB, A 0620-00, seems to confirm that the nonlinear correlation holds down to Eddington ratios as low as 10^{-8} [46], many outliers have been recently been found at higher luminosities [18, 16, 13, 15, 106, 45], casting serious doubts on the universality of this scaling, and the possibility of relying on the best-fitting relation for estimating other quantities, such as distance or black hole mass.

4.3.2 Optical-Infrared/X-Ray

The infrared (IR) spectra of BHBs with a low mass donor star are likely shaped by a number of competing emission mechanisms, most notably: reprocessing of accretion-powered X-ray and ultraviolet photons, either by the donor star surface or by the outer accretion disk, direct thermal emission from the outer disk, and non-thermal synchrotron emission from a relativistic outflow. Reference [109] have collected all the available quasi-simultaneous optical and near-IR data of a large sample of Galactic X-ray binaries over different X-ray states. The optical/near-IR (OIR) luminosity of hard/quiescent BHBs is found to correlate with the X-ray luminosity to the power ~0.6, consistent with the radio/X-ray correlation slope down to $10^{-8}L_{Edd}$ [46]. Combined with the fact that the near-IR emission is largely suppressed in the soft state, this leads to the conclusion that, for the BHBs, the spectral break to the optically thin portion of the jet takes place most likely in the mid-IR (2–$40\,\mu m$). A similar correlation is found in neutron stars (NSs) in the hard state. By comparing the observed relations with those expected from models of a number of emission processes, [109] are able to constrain the dominant contribution to the OIR portion of the spectral energy distribution (SED) for the different classes of X-ray binaries. They conclude that for hard state BHs at high luminosities (above 10^{-3} times the Eddington limit) jets are contributing 90% of the near-IR emission. The optical emission could have a substantial jet contribution; however, the optical spectra show a thermal spectrum indicating X-ray reprocessing in the disk dominates in this regime. In contrast, X-ray reprocessing from the outer accretion disk dominates the OIR spectra of hard state NSs, with possible contributions from the synchrotron emitting jets and the viscously heated disk only at very high luminosities.

4.4 Jet–ISM Interaction

4.4.1 Jet-Driven Nebulae

It is worth stressing that none of the above scaling relations deals with actual measurements of the *total* jet power, which is a function of the observed radio luminosity, corrected for relativistic effects, and of the unknown radiative efficiency. A fruitful method, borrowed from the AGN community, is that to constrain the jet power × lifetime product by looking at its interaction with the surrounding interstellar medium (ISM). Beside the arcmin-scale, fossil jets around the so-called Great Annihilator (Fig. 4.1, bottom right panel; [99]), a well-known case is that of the nebula around the first Galactic jet source discovered: SS 433. The "ears" of W50 (Fig. 4.3, top right panel) act as an effective calorimeter for the jets' mechanical power, which is estimated to be greater than 10^{39} erg s^{-1} [9]. Similarly, the neutron star X-ray binary in Circinus X-1 is embedded in an extended, jet-driven radio nebula (see Fig. 4.3, top left panel). In this particular case, it is likely that we are actually looking toward the central X-ray binary system through the jet-powered radio lobe, making this the only known case of a Galactic *microblazar*. Results from modeling suggest an age for the nebula of $\lesssim 10^5$ yr and a corresponding time-averaged jet power in excess of 10^{35} erg s^{-1}. During flaring episodes, the instantaneous jet power may reach values of similar magnitude to the X-ray luminosity [119].

More recently, a low surface brightness arc of radio/optical emission has been discovered around Cygnus X-1 [47, 108] (Fig. 4.3, bottom panels) and interpreted

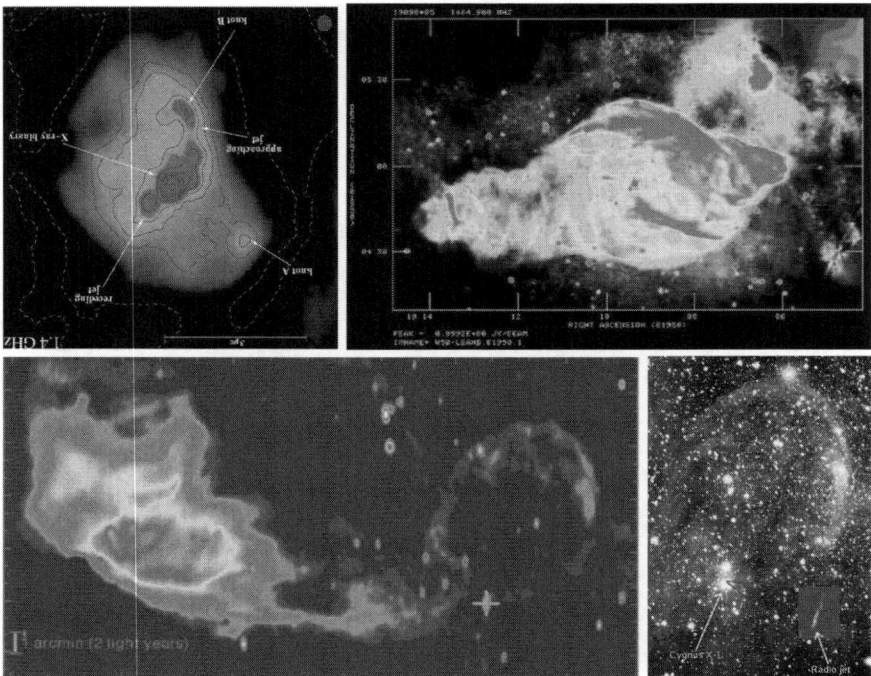

Fig. 4.3 *Top left*: the jet-powered radio nebula of the "microblazar" Circinus X-1 imaged with the Australia Telescope Compact Array; from [119]. *Top right*: W50 nebula surrounding SS 433 (see Fig. 4.1, *bottom left panel* for an image of the arcsec-scale radio jets). The jets of SS 433 are drilling their way into the supernova remnant and give rise to these characteristic "ears" [27]. *Bottom left*: a jet-powered radio nebula around the black hole X-ray binary Cygnus X-1, as seen by the Westerbork Synthesis Radio Telescope at 1.4 GHz. From [47]. *Bottom right*: the optical counterpart to the Cyg X-1 nebula, as observed with the 2.5m Isaac Newton optical telescope. From [108]

in terms of a shocked compressed hollow sphere of free–free emitting gas driven by an under-luminous synchrotron lobe inflated by the jet of Cygnus X-1. The lack of a visible counter arc is ascribed to the lower interstellar matter density in the opposite direction. In fact, there exist to date relatively few cases where jet–ISM interactions have been directly observed [87]. Carrying the analogy of AGN jet–ISM interactions over to microquasars, it has been argued that microquasars are located, dynamically speaking, in much more tenuous atmosphere. As a consequence, compared to AGN, microquasar jets require particularly dense environments to produce visible signs of interaction with the surroundings [54, 57].

4.4.2 X-Ray Jets

It is well known that, in AGN, optical and X-ray jets are also frequently seen. With the exception of the large-scale (tens of pc) diffuse X-ray emission detected from the

X-ray binary SS 433 with the Einstein Observatory [112], X-ray jets were not seen for Galactic systems prior to the launch of the Chandra X-Ray Observatory in 1999. With a large improvement in angular resolution over previous missions, Chandra detected arcsec-scale (∼0.025 pc) X-ray jets for the first time in SS 433 [75]. This was the first of a number of discoveries.

Perhaps the most extreme case in terms of energetics has been the detection of decelerating arcsec-scale X-ray (and radio) jets in the microquasar XTE J1550-564, a few years after the ejection event [21, 117, 68]. The detection of optically thin synchrotron X-ray emission from discrete ejection events implies in situ particle acceleration up to several TeV, possibly due to interaction of the jets with the interstellar medium. More recently, a similar large-scale jet has been reported in H 1743-322 [17] (Fig. 4.4, left panels). As for XTE J1550-564, the spectral energy distribution of the jets during the decay phase is consistent with a classical synchrotron spectrum

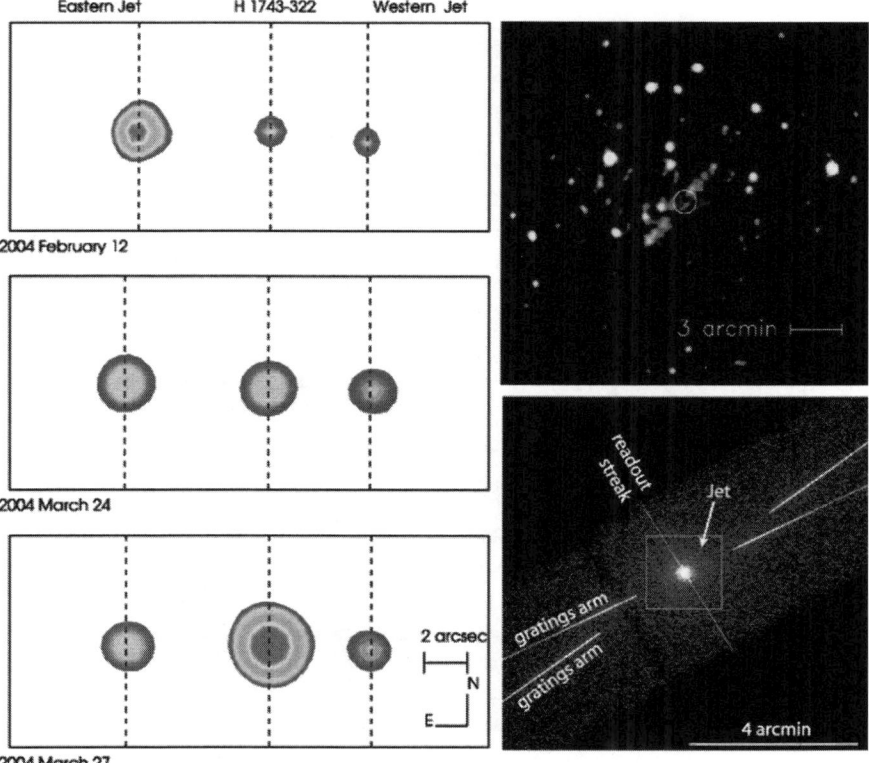

Fig. 4.4 *Left panels*: arcsec-scale, transient X-ray plasmons ejected by the BH candidate H 1743-322 (Chandra ACIS-S). The detection of optically thin synchrotron X-ray emission implies in situ particle acceleration up to several TeV. From [17]. *Top right*: arcmin-scale fossil X-ray jets in the field of 4U 1755-33 [5] (XMM-Newton image). *Bottom right*: arcsec-scale transient X-ray jet from Circinus X-1 seen with the Chandra gratings. From [53]. The X-ray jet direction coincide with that of the ultra-relativistic radio jets [33]

of a single electron distribution from radio up to X-rays, implying the production of very high energy (>10 TeV) particles in those jets.

Another interesting example is that of a fossil arcmin-scale X-ray jet seen by the XMM-Newton telescope in the surroundings of 4U 1755-33 [5] (Fig. 4.4, top right panel). Finally, evidence for a transient X-ray jet has been recently claimed in the neutron star Circinus X-1 during a 50 ks Chandra gratings observation (Fig. 4.4, bottom right panel), taken during a low flux state [53]. The direction of this X-ray feature is consistent with the direction of the northwestern jet seen in the radio [33], suggesting that it originates either in the jet itself or in the shock that the jet is driving into its environment. The inferred jet kinetic power is significantly larger than the minimum power required for the jet to inflate the large-scale radio nebula.

4.5 Quiescence (to Eject or Not to Eject?)

The role of outflows is especially interesting at very low X-ray luminosities, in the so-called "quiescent" regime, i.e., below a few $10^{-5}L_{Edd}$. Persistent steady radio counterparts to BHBs appear to survive down to low quiescent X-ray luminosities (as low as $10^{-8.5}L_{Edd}$ [46]), even though sensitivity limitations on current radio telescopes make it extremely difficult to reach the signal-to-noise ratios required to assess their presence for systems further than 2 kpc or so.

In the context of X-ray binaries, as well as super-massive black holes, the term "jet" is typically used as a synonym for relativistic outflow of plasma and implies a high degree of collimation. As a matter of fact, high spatial resolution radio observations of BHBs in the hard state have resolved highly collimated structures in two systems only: Cyg X-1 [114] and GRS 1915+105 [25, 42] are both resolved into elongated radio sources on milliarcsec scales – that is tens of A.U. – implying collimation angles smaller than a few degrees on much larger scales than the orbital separation. Both systems display a relatively high X-ray (and radio) luminosity, with GRS 1915+105 being persistently close to the Eddington luminosity [32], and Cyg X-1 displaying a bolometric X-ray luminosity around 2%L_{Edd} [26].

This, however, should not be taken as evidence against collimated jets in low luminosity, quiescent systems: because of sensitivity limitations on current high resolution radio arrays, resolving a radio jet at microJy level simply constitutes an observational challenge. In addition, at such low levels, the radio flux could be easily contaminated by synchrotron emission from the donor star.

In principle, the presence of a collimated outflow can also be inferred by its long-term action on the local interstellar medium, as in the case of the hard state BHBs 1E 1740.7-2942 and GRS 1758-258, both associated with arcmin-scale radio lobes [99, 79]. Further indications can come from the stability in the orientation of the electric vector in the radio polarization maps, as observed in the case of GX 339-4 over a 2-year period [22]. This constant position angle, being the same as the sky position angle of the large-scale, optically thin radio jet powered by GX 339-4 after its 2002 outburst [48], clearly indicates a favored ejection axis in the system. However, all three systems emit X-rays at "intermediate" luminosities

$(10^{-3}–10^{-2} L_{Edd})$, and tell us little about outflows from quiescent BHs. On the other hand, failure to image a collimated structure in the hard state of XTE J1118+480 down to a synthesized beam of 0.6×1.0 mas^2 at 8.4 GHz [94] poses a challenge to the collimated jet interpretation, even though XTE J1118+408 was observed at roughly one order of magnitude lower luminosity with respect to, e.g., Cyg X-1 $(10^{-3} L_{Edd})$. Under the (naive) assumption that the jet size scales as the radiated power, one could expect the jet of XTE J1118+408 to be roughly ten times smaller than that of Cyg X-1 (which is 2×6 mas^2 at 9 GHz, at about the same distance), i.e., still point-like at Very Long Based Array (VLBA) scales [94].

In fact, [50] have pointed out that long period ($\gtrsim 1$ day) BHBs undergoing outbursts tend to be associated with spatially resolved optically thin radio ejections, while short period systems would be associated with unresolved, and hence physically smaller, radio ejections. If a common production mechanism is at work in optically thick and optically thin BHB jets, then the above arguments should apply to steady optically thick jets as well, providing an alternative explanation to the unresolved radio emission of XTE J1118+480 (which, with its 4 h orbital period, is one of the shortest known). By analogy, a bright, long period system, like for instance V404 Cyg, might be expected to have a more extended optically thick jets. This is further explored in the next section.

4.5.1 The Brightest: V404 Cyg

In order to eventually resolve the radio counterpart to a quiescent X-ray binary, the black hole V404 Cyg was observed with the High Sensitivity Array (HSA, composed of VLBA plus the Green Bank Telescope, Effelsberg, and the phased VLA) in December, 2007 for 4.25 h [86]. These observations failed to resolve the radio source (Fig. 4.5, left panel), yielding an upper limit of 1.3 milliarcsec on the 8.4 GHz source size – 5.2 A.U. at 4 kpc – and a corresponding lower limit of 7×10^6 K on its brightness temperature (confirming that the radio emission must be non-thermal). Interestingly, the inferred upper limit on the radio source size is already two times smaller than the steady jet resolved in Cyg X-1 with the VLBA (Fig. 4.1, top left panel).

A small flare was detected from V404 Cyg with both the VLA and the VLBA, on a timescale of 1 h (Fig. 4.5, right panel), in which the source flux density rose by a factor of 3. As the brightest black hole X-ray binary in quiescence (few 10^{-5} L_{Edd} in X-rays, 0.4 mJy at GHz frequencies), V404 Cyg is the only source in which such flaring activity has been detected in the quiescent state (the flare is certainly intrinsic to the source and cannot be caused by interstellar scintillation [86]). The question of whether flares are unique to this source, or are common in such systems, has direct implications for the nature of the accretion process at low luminosities. Observations of XTE J1118+480 and GX 339-4 in their hard states have shown evidence for fast variability in the optical and X-ray bands [65, 64, 63], with properties inconsistent with X-ray reprocessing and more indicative of synchrotron variability.

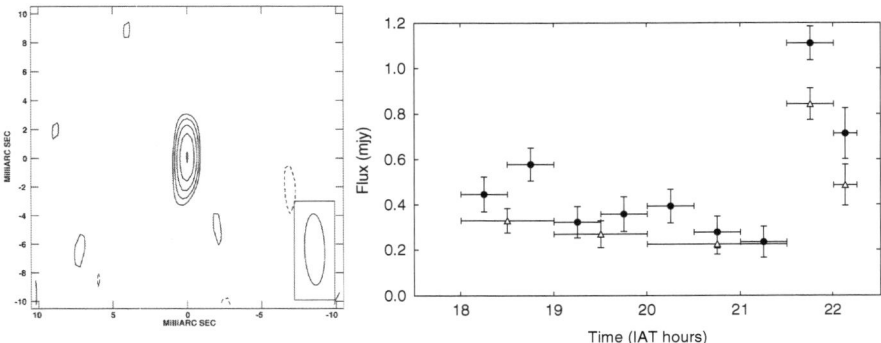

Fig. 4.5 *Left*: the quiescent black hole V404 Cyg was observed with the High Sensitivity Array for 4.25 hr at 8.4 GHz. The radio source is unresolved, yielding an upper limit of 5.2 A.U. to the size of the emitting region. *Right*: the source radio light curve during the observation, showing a flare with 30 min rising time. From [86]

The timescales are shorter than seen in V404 Cyg and there were no high-time resolution radio data for comparison. A better comparison is the quiescent BH in the center of the Milky Way, Sgr A*. It shows radio and infrared flaring activity, which has been explained as an adiabatic expansion of a self-absorbed transient population of relativistic electrons [121]. Whether there is a truly steady underlying jet as assumed by standard jet models [11, 69] or whether the emission is composed of multiple overlapping flares [70] remains to be determined. More sensitive instruments are necessary to probe these short timescale arcs and to determine the nature of the quiescent jet emission (see Sect. 4.11).

4.5.2 The Faintest: A 0620-00

In spite of the large degree of uncertainty on the overall geometry of the accretion flow in this regime, there is general agreement that the X-ray emission in quiescent BHBs comes from high-energy electrons near the BH. The SEDs of quiescent BHBs and low-luminosity AGN are often examined in the context of the advection-dominated accretion flow (ADAF) solution [103, 102], whereby the low X-ray luminosities are due to a highly reduced radiative efficiency, and most of the liberated accretion power disappears into the horizon. Here, due to the low densities, a two-temperature inflow develops, a significant fraction of the viscously dissipated energy remains locked up in the ions as heat, and is advected inward, effectively adding to the BH mass. The ADAF model successfully accounts for the overall shape of the UV-optical-X-ray spectra of quiescent BHBs (see e.g., [81] for an application to the high quality data of XTE J1118+480). Nevertheless, alternative suggestions are worth being considered. Reference [10] elaborated an "adiabatic inflow–outflow solution" (ADIOS), in which the excess energy and angular momentum are lost to an outflow at all radii; the final accretion rate into the hole may be only a tiny fraction of the mass supply at large radii.

Alternatively, building on the work by [28] on AGN jets, a jet model has been proposed for hard state BHBs [78, 77, 76]. Figure 4.6, (top panel) shows a fit to the radio-to-X-ray SED of A 0620-00, the lowest Eddington-ratio BHB with a detected radio counterpart ($L_X/L_{Edd} \simeq 10^{-8}$; $F_{8\,GHz} = 50\,\mu Jy$) with such a "maximally jet-dominated" model [44]. This is the first time that such a complex model was applied in the context of quiescent BHBs, and with the strong constraints on the jet break frequency cutoff provided by the Spitzer Space Telescope data in the mid-IR regime (see Sect. 4.7 for more details). In terms of best-fitting parameters, the major difference with respect to higher luminosity sources for which this model has been tested [78, 77] (see Chap. 6) is in the value of the acceleration parameter f compared to the local cooling rates, which turns out to be two orders of magnitude lower for A 0620-00. This "weak acceleration" scenario is reminiscent of the Galactic Center super-massive BH Sgr A*. Within this framework, the SED of Sgr A* does not require a power law of optically thin synchrotron emission after the break from its flat/inverted radio spectrum. Therefore, if the radiating particles have a power-law distribution, it must be so steep as to be indistinguishable from a Maxwellian in the optically thin regime. In this respect, they must be only weakly accelerated. In the framework of a jet-dominated model for the quiescent regime, it appears that something similar, albeit less extreme, is occurring in the quiescent BHB A 0620-00; either scenario implies that acceleration in the jets is inefficient at 10^{-9}–$10^{-8}\,L_{Edd}$.

4.6 Neutron Stars

The mechanism(s) of jet production, from an *observational* point of view, remains essentially unconstrained. While in the case of super-massive BHs in AGN it is often implicitly assumed that the jets extract their energy from the rotation of the centrally spinning black hole via large-scale magnetic field lines that thread the horizon; in the case of X-ray binaries, the relatively low (lower limit on the) jets' Lorentz factors [35] do not appear to require especially efficient launching mechanisms. While on the "experimental" side, substantial improvements are being made with fully relativistic magneto-hydrodynamic simulations (e.g., [82], and references therein), from the observer perspective it seems that a fruitful – and yet relatively unexplored – path to pursue is that to compare in a systematic fashion the properties of jets in blackhole systems to that of e.g., low magnetic field NSs.

A comprehensive study comparing the radio properties of BHs and NSs [91] has highlighted a number of relevant difference/similarities (see Fig. 4.7): (i) Below a few per cent of the Eddington luminosity (in the hard, radiatively inefficient states) both BHs and NSs produce steady compact jets, while transient jets are associated with variable sources/flaring activity at the highest luminosities. (ii) For a given X-ray luminosity, the NSs are less radio loud, typically by a factor of 30 (Fig. 4.7,

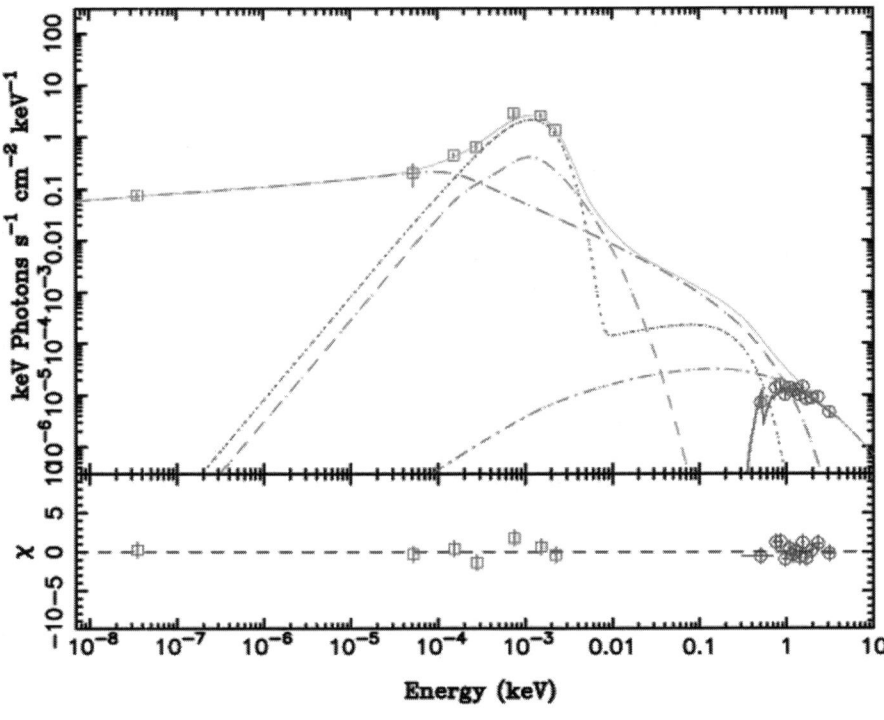

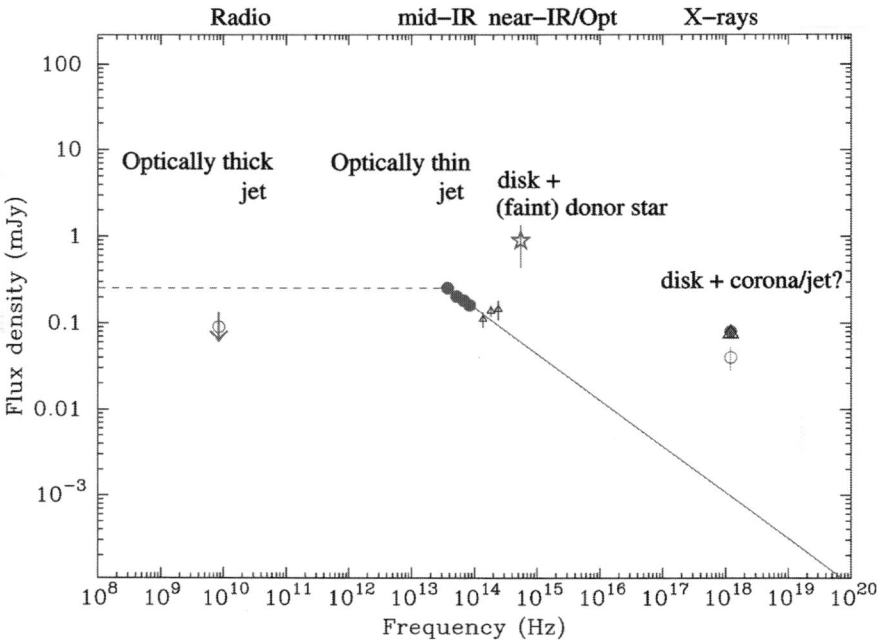

Fig. 4.6 (continued)

top panel). (iii) Unlike BHs, NSs do not show a strong suppression of radio emission in the soft/thermal dominant state. (iv) Hard state NSs seem to exhibit a much steeper correlation between radio and X-ray luminosities.

Highly accreting NSs, called Z-type (Fig. 4.7, bottom left panel), show periodic X-ray state transitions on timescales of a few days. Z sources can be considered the NS counterparts of transient, strongly accreting BHs such as GRS 1915+105. At the same time, they are known to display hard non-thermal tails in the X-ray energy spectra. The physical origin of this non-thermal component is still an area of controversy, with two main competing models: (1) inverse Compton scattering from a non-thermal electron population in a corona [105] and (2) bulk motion Comptonization [115]. Recently, simultaneous VLA/RXTE observations of the Z source GX 17+2 have shown a positive correlation between the radio emission and the hard tail power-law X-ray flux in this system (Fig. 4.7, bottom right panel, [89]). If further confirmed with a larger sample and improved statistics, this relation would point to a common mechanism for the production of the jet and hard X-ray tails.

4.7 Jet Power: The Mid-IR Leverage

Observations of hard state BHBs have established that synchrotron emission from the steady jets extends all the way from the radio to the mm-band [40], above which the break to the optically thin portion of the spectrum is thought to occur. However, even for the highest quality SED, disentangling the relative contributions of inflow vs. outflow to the radiation spectrum and global accretion energy budget can be quite challenging, as exemplified by the emblematic case of XTE J1118+480 [81, 77, 122]. Estimates of the total jet power based on its radiation spectrum depend crucially on the assumed frequency at which the flat, partially self-absorbed spectrum turns and becomes optically thin, as the jet "radiative efficiency" depends ultimately on the location of the high-energy cutoff induced by the higher synchrotron cooling rate of the most energetic particles. From a theoretical point of view, the "break

──────────◄──

Fig. 4.6 *Top*: radio-to-X-ray SED of the quiescent black hole X-ray binary A 0620-00 fit with a "maximally-dominated" jet model. *Solid gray line*: total spectrum; *dot-long-dashed line*: pre-acceleration inner jet synchrotron emission; *dotted line*: post-acceleration outer jet synchrotron; *triple dot-dashed line*: Compton emission from the inner jet, including external disk photons as well as synchrotron self-Compton; *double-dot-dashed line*: thermal multicolor-blackbody disk model plus single blackbody representing the star. The symbols represent the data, while the *solid black (red in the color version) line* is the model fit in detector space. From [44]. *Bottom*: evidence for optically thin synchrotron emission from a jet in the NS 4U 0614+091. By comparison, if the mid-IR excess detected in the BHs A 0620-00 is indeed due to jet emission, this means that the power content of the BH jet is at least 10 times higher than in the NS, where the jet spectrum breaks to optically thin already at mid-IR frequencies. From [90]

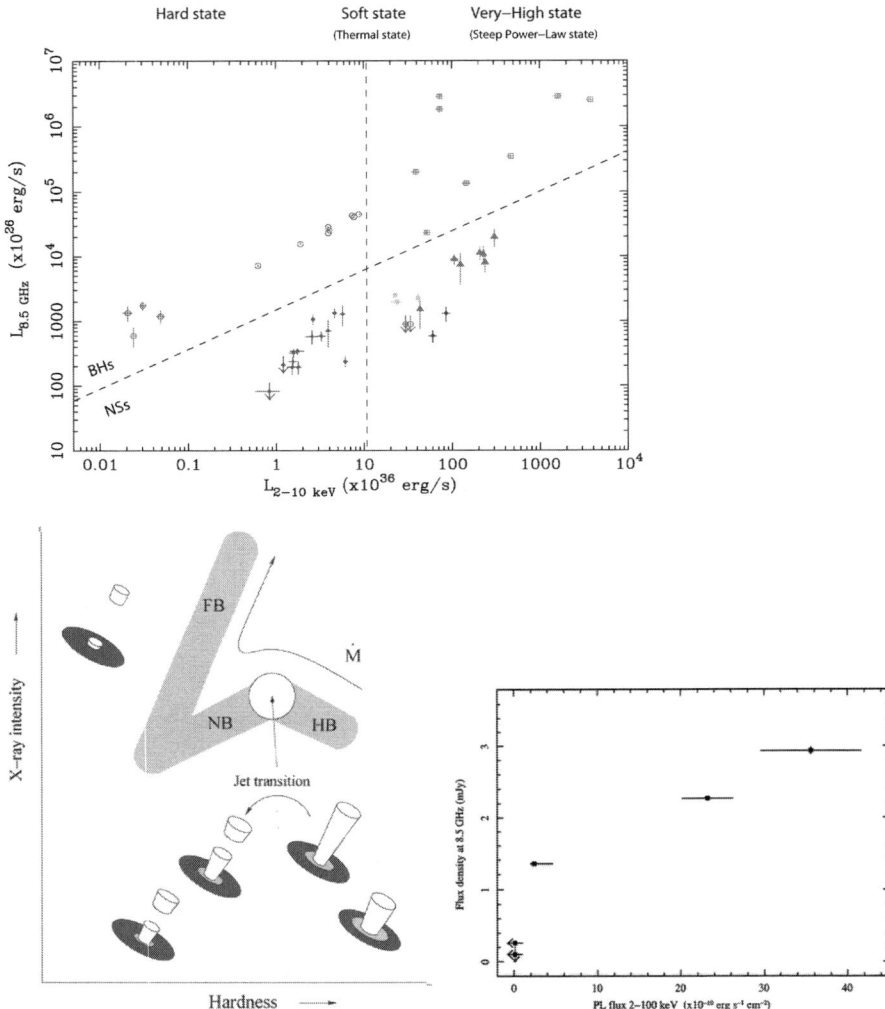

Fig. 4.7 *Top*: radio vs. X-ray luminosity for (two representative) black holes (*top left*) and neutron stars (*bottom right*). On average, neutron stars are 30 times fainter in the radio band. From [91]. *Bottom left*: sketch of the disk-jet coupling in Z-type sources, which are believed to be the NS equivalents of the BH GRS 1915+105, constantly emitting near L_{Edd} and producing powerful jets associated with rapid state transitions. From [91]. *Bottom right*: a correlation between the core radio flux density and the X-ray flux in the hard X-ray in the Z source GX 17+2. From [89]

frequency", here defined as the frequency at which the partially self-absorbed jet becomes optically thin, is inversely proportional to the BH mass: as jet spectral breaks are often observed in the GHz/sub-mm regime in AGN, they are expected to occur in the IR-optical band for 10^{5-7} times lighter objects. We know however from observations of GX 339-4, the only BHB where the optically thin jet spectrum has been perhaps observed [20, 61], that the exact break frequency can vary with the overall luminosity, possibly reflecting changes in the magnetic field energy density, the particle density, and the mass loading at the jet base. Determining the location of the jet break as a function of the bolometric luminosity is also important to assess the synchrotron contribution to the hard X-ray band. As an example, that the optically thin jet IR-emission in GX 339-4 connects smoothly with the hard X-ray power law has led to challenge the "standard" Comptonization scenario for the hard X-ray state [77].

Because of the low flux levels expected from the jets in this regime (tens to hundreds of µJy based on extrapolation from the radio band), combined with the companion star/outer disk contamination at near-IR frequencies, the sensitivity and leverage offered by Spitzer is crucial in order to determine the location of the jet break. In fact, Spitzer observations of three quiescent black hole X-ray binaries, with the Multi-band Imaging Photometer (MIPS), have shown evidence for excess emission with respect to the Rayleigh-Jeans tail of the companion star between 8 and 24 µm. This excess, which has been interpreted as due to thermal emission from cool circumbinary material [100], is also consistent with the extrapolation of the measured radio flux assuming a slightly inverted spectrum, typical of partially self-absorbed synchrotron emission from a conical jet [44]. If so, then the jet synchrotron luminosity exceeds the measured X-ray luminosity by a factor of a few in these systems. Accordingly, the mechanical power stored in the jet exceeds the bolometric X-ray luminosity at least by four orders of magnitude (based on kinetic luminosity function of Galactic X-ray binary jets [55]).

Despite their relative faintness with respect to BHs at low (radio) frequencies, the same multi-wavelength approach can be undertaken in NSs, in the hope to detect excess mid-IR emission. Indeed, coordinated radio (VLA), mid-IT (Spitzer), optical (YALO SMARTS), and X-ray (RXTE) observations have yielded the first spectro-scopical evidence for the presence of a steady jet in a low-luminosity, ultra-compact[1] neutron star X-ray binary (4U 0614+091 [90]). The Spitzer data (Fig. 4.6, bottom panel) show a neat optically *thin* synchrotron spectrum ($F_\nu \propto \nu^{-0.6}$), indicating that the jet break occurs at much lower frequencies in this neutron star system with respect to blackhole X-ray binaries. As a consequence, unlike in the black holes, the jet cannot possibly contribute significantly to the X-ray emission.

The Spitzer results on low-luminosity X-ray binaries (three BHs and one NS; [90, 44]) seem to point toward different energy dissipation channels in different classes of objects, namely: the former seem to be more efficient at powering synchrotron emitting outflows, having partially self-absorbed jets which extend their spectrum up to the near-IR band. However, it is worth reminding that the most relativistic jet discovered in the Galaxy so far is the neutron star X-ray binary Circinus X-1 [33], for which the inferred Lorentz factor exceeds 15 (see Fig. 4.8).

---

[1] That is, with orbital period shorter than 1 h.

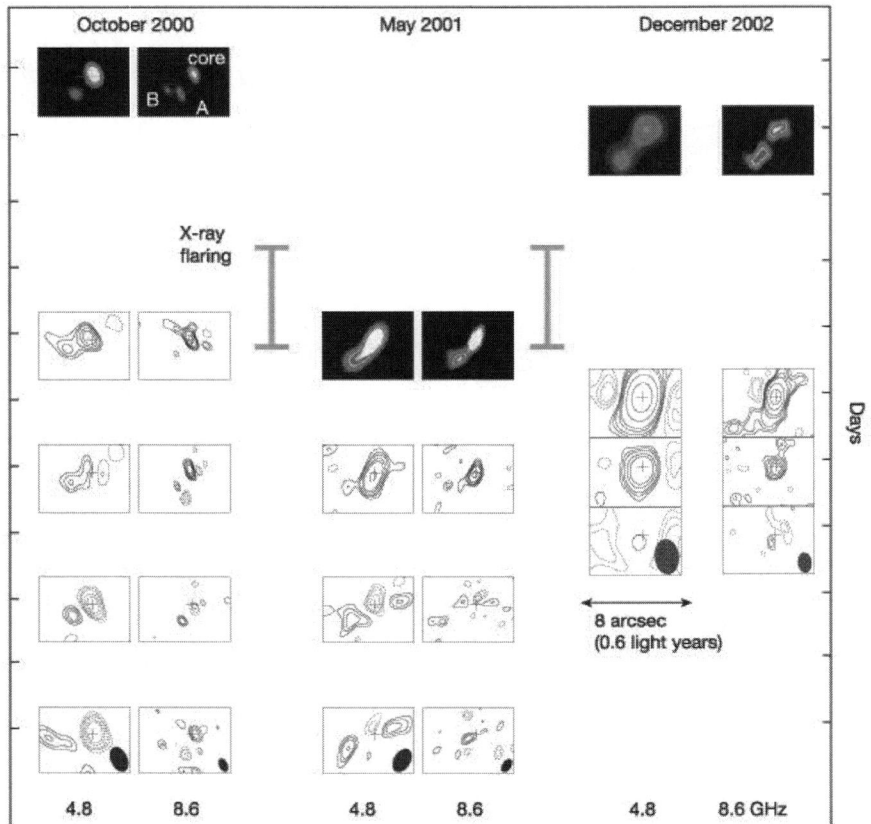

Fig. 4.8 An ultra-relativistic jet powered by the neutron star X-ray binary Circinus X-1. From [33]. The inferred bulk Lorentz factor exceeds $\Gamma = 15$, making this the most relativistic Galactic jet source to date

4.8 Jets, Advection, and Event Horizon

One of the most cited accomplishments of the ADAF model is to naturally account for the relative dimness of quiescent BHBs with respect to quiescent NS X-ray binaries [101], whereas in the case of a BH accretor all the advected energy disappears as it crosses the horizon, adding to the BH mass, the same energy is released and radiated away upon impact in the case of an accretor with solid surface. Under the ADAF working hypothesis, such luminosity difference – which is indeed observed in X-rays [51] – is actually taken as observational evidence for the existence of a horizon in BHBs.

However, while the observed luminosity gap may well be a natural byproduct of advection of energy through the BH horizon, the possibility exists that the very mechanism of jet production differs between the two classes and that the observed

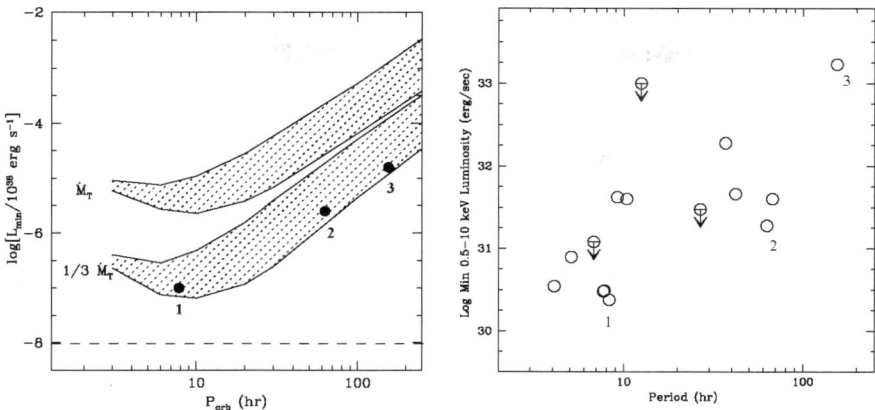

Fig. 4.9 *Left*: in the ADAF scenario, the minimum luminosity around a few 10^{30} erg s^{-1} corresponds to roughly two-thirds of the outer accretion rate lost to a wind/outflow. From [83]. *Right*: quiescent X-ray luminosities/upper limits for 15 BH X-ray binaries, plotted against the systems' orbital periods. From [43]

luminosity difference is due to the different channels for dissipating a common energy reservoir, with the black hole "preferring" jets (as discussed in Sect. 4.6, the NSs are indeed fainter than the BHs in the radio band).

Out of 15 BHB systems with sensitive X-ray observations while in the quiescent regime, 12 have now been detected in X-rays (see Fig. 4.9, right panel; updated from [118] after [18, 46, 59]). For those 12, the quiescent luminosities range between a few 10^{30} and 10^{33} erg s^{-1}. The nearest BH, A 0620-00, has been steadily emitting at $\simeq 2$–3×10^{30} erg s^{-1} at least for the past 5 years [46]; this is approximately the same luminosity level as XTE J1650-500 [43], XTE J1118+480 [81], and GS 2000+25 [51], suggesting that this might be some kind of limiting value.

In fact, for low-mass X-ray binaries one can make use of binary evolution theory, combined with a given accretion flow solution, to predict a relation between the minimum quiescent luminosity and the system orbital period, P_{orb} [83, 66]. Independently of the actual solution for the accretion flow in quiescence, the existence of a minimum luminosity in low-mass X-ray binaries stems directly from the existence of a bifurcation period, P_{bif}, below which the mass transfer rate is driven by gravitational wave radiation (j-driven systems), and above which it is dominated by the nuclear evolution of the secondary star (n-driven systems). As long as the luminosity expected from a given accretion flow model scales with a positive power of the outer accretion rate, systems with orbital periods close to the bifurcation period should display the lowest quiescent luminosity. As an example, the left panel of Fig. 4.9 (from [83]) illustrates how the predicted luminosity of quiescent BHs powered by ADAFs depends on the ratio between the outer mass transfer rate and the ADAF accretion rate. The lower band, for instance, corresponds to $\sim 1/3$ of the outer mass transfer being accreted via the ADAF, implying that the remaining 2/3 is lost to an outflow (effectively making this inflow an ADIOS [10]). Interestingly, this lower band roughly reproduces the observed luminosities of three representative systems spanning the whole range of detected systems ((1) A 0620-00, (2) GRO J1655-40, and (3) V404 Cyg – marked in both panels).

4.9 White Dwarfs

Accreting WDs come in three main classes: cataclysmic variables (CVs), super-soft X-ray binaries, and symbiotic stars. Perhaps the best understood of all accretion disks are those in cataclysmic variables (CVs). Although outflows have been observed from an increasing number of accreting WDs in symbiotic stars and super-soft X-ray sources [113, 71], these are generally not spatially well resolved, and questions on the jet-accretion coupling remain open. Most WD ejecta have velocities of hundreds to thousands of km s^{-1}, and even in those cases when the emission can be spatially separated from the binary core, it usually appears to be due to free–free emission from shocks. However, a handful of observations suggest that the out-flowing particles can be accelerated to high enough energies that non-thermal radio and/or X-ray emission can also be produced [23, 104].

Two recent works [107, 72] have provided us with the best observational evidence for synchrotron emission from a WD jet. As illustrated in Fig. 4.10, VLBA observations of the recurrent nova RS Ophiuchi, taken about 21 and 27 days after the outburst peak, have imaged a resolved synchrotron component well to the east of the shell-like shock feature.[2] Interestingly, the inferred jet velocity is comparable to the escape velocity from a WD.

A second example is the detection of a transient radio jet erupting from the dwarf nova SS Cyg [72]. Radio observations of this system, conducted during its April 2007 outburst with the VLA, revealed a variable, flat-spectrum source with high brightness temperature, most likely due to partially self-absorbed synchrotron emission from a jet [72]. As apparent from Fig. 4.11, the behavior of SS Cyg (right panel) during the outburst closely resembles that of X-ray binaries hosting relativistic compact objects. Plotted here, from left to right, are the "disk-fraction vs. luminosity diagrams" (which can be seen as the generalization of HIDs generally adopted for X-ray binaries; see [73]) for three different accreting objects: a BH, NS, and the WD SS Cyg. Unlike for BHs and NSs, where the x-axis is the X-ray spectral hardness, for the WD case this variable has been replaced by the power-law fraction, which quantifies the prominence of the power-law component over the boundary layer/thermal disk component of the energy spectrum, as measured in the optical/UV band. Evidently, the three systems occupy the same region of the parameter space during a typical outburst (see Fig. 4.2 for a schematic of typical BH outburst HID, and Fig. 4.7, bottom left panel, for Z-type NSs), the most notable

---

[2] Typically, post-outburst radio emission from RS Ophiuchi originates from dense ISM that is swept up, compressed, and shocked by the relatively rarefied shell that has been ejected [111].

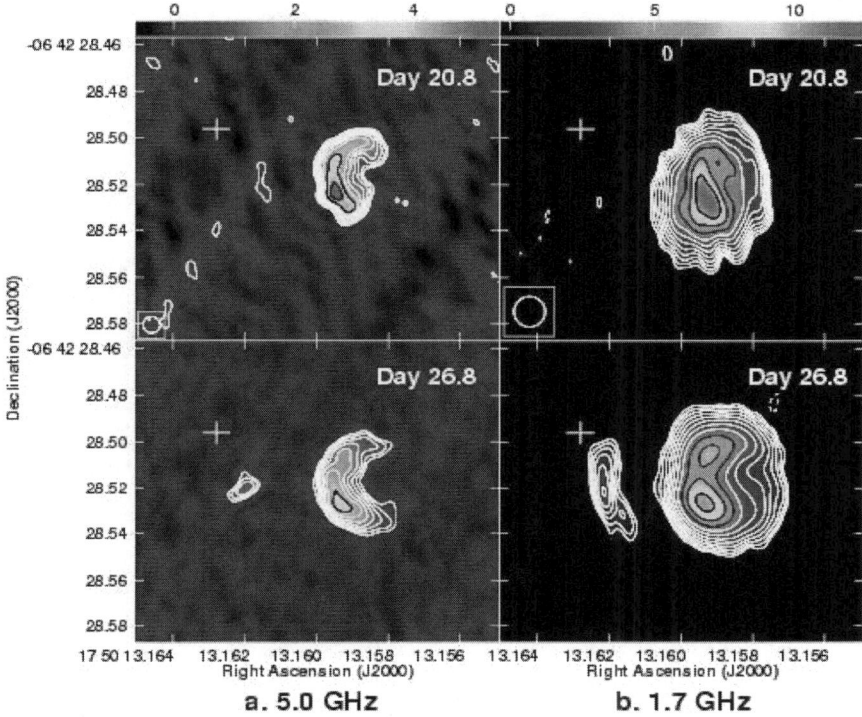

Fig. 4.10 Two epoch, post-outburst, radio observations of the recurrent nova RS Ophiuchi (VLBA). The second epoch observations (*bottom*) show evidence for an additional resolved component to the east of the shell-like feature visible already in the first epoch (*top*). From [107]

difference being that – for the BHs – the core, flat-spectrum radio emission drops below detectable levels when in the soft/thermal dominant state, while it seems to be only somewhat weakened in NSs and WDs.

These results provide further support to the hypothesis that the mere existence of accretion coupled to magnetic fields may be sufficient ingredient for a jet to form independently of the presence/absence of an horizon. However, the latter (or, put it another way, the presence of a physical surface) could play a role in (i) making the jets more radio loud and (ii) enhancing the magnitude of the jet suppression mechanism during bright soft states.

4.10 Jet Production, Collimation, Matter Content

It is often assumed that the velocity of the steady jet is only mildly relativistic, with $\Gamma \simeq 2$ [49]. This comes from the relative spread about the radio/X-ray correlation, interpreted as evidence for a low average Lorentz factor. However, this argument has been confuted on theoretical grounds [56]. As far as the transient jets are concerned,

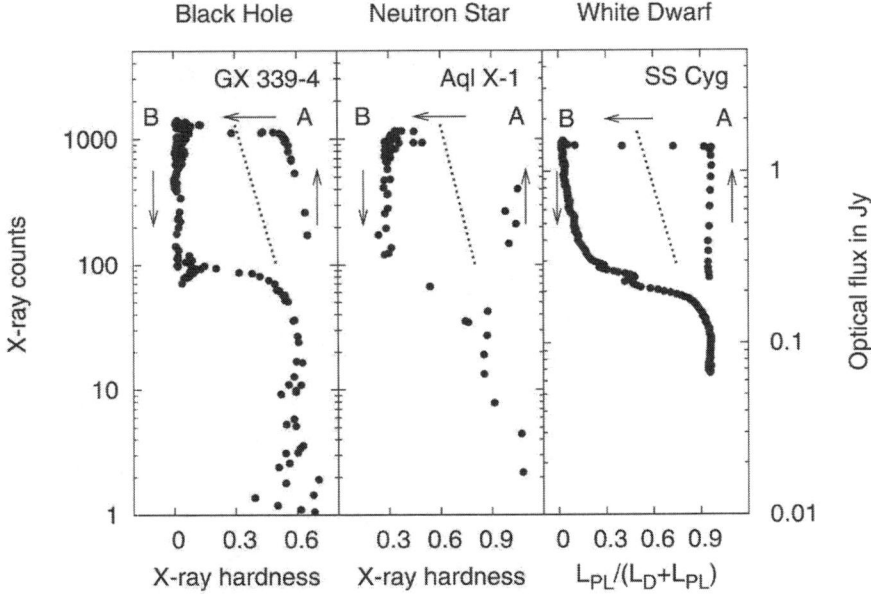

Fig. 4.11 Disk fraction vs. luminosity diagrams for Galactic binary stellar systems and microquasar sources hosting different compact objects: a BH, NS, and WD. From [72]. This study provides the strongest observational evidence to date for a synchrotron-emitting outflow from a dwarf nova system and strengthens the similarities between jet sources across the whole mass spectrum

there is a high degree of uncertainty in estimating their Lorentz factors, mainly because of distance uncertainties [35].

A recent work [88] made a substantial step forward in constraining the Lorentz factor of microquasars by means of the observational upper limits on the jets' opening angles. This method relies on the fact that, while the jets could undergo transverse expansion at a significant fraction of the speed of light, time dilation effects associated with the bulk motion will reduce their apparent opening angles. Reference [88] have calculated the Lorentz factors required to reproduce the small opening angles that are observed in most X-ray binaries, with very few exceptions, under the crucial assumption of no confinement. The derived values, mostly lower limits, are larger than typically assumed, with a mean $\Gamma_m > 10$. No systematic difference appears to emerge between hard state steady jets and transient plasmons (Fig. 4.12). If indeed the transient jets were as relativistic as the steady jets, as already mentioned, this would challenge the hypothesis of internal shocks at work during hard-to-thermal state transitions in BHBs. In order for that scenario to be viable, the transient jets must have higher Lorentz factor; in other words, steady jets ought to be laterally confined. The issue of the jets' matter content remains highly debated. Perhaps with the exception of SS 433 [120, 92], where atomic lines have been detected as optical at X-ray wavelengths along the jets (see Fig. 4.13), various

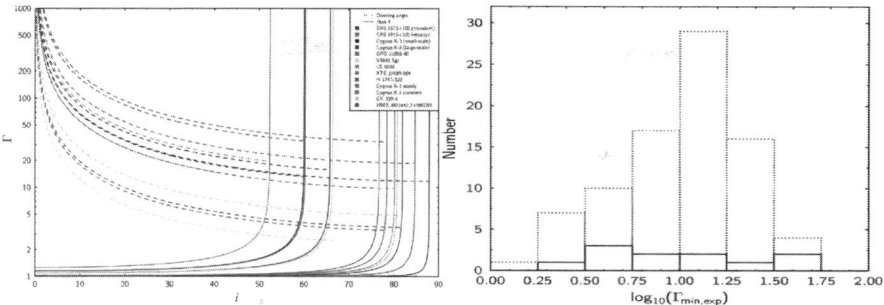

Fig. 4.12 *Left*: Lorentz factors derived from opening angles (Eq. 4 in [88], *dashed lines*) and from the inferred $\beta \times \cos i$ products for a number of microquasars. Since the Lorentz factors are derived from upper limits on the opening angles they are in fact lower limits (assuming freely-expanding jets with an expansion speed c). *Right*: mean Lorentz factors for X-ray binaries from opening angles' constraints (*solid*), compared to AGN from proper motions' (*dashed*). From [88]

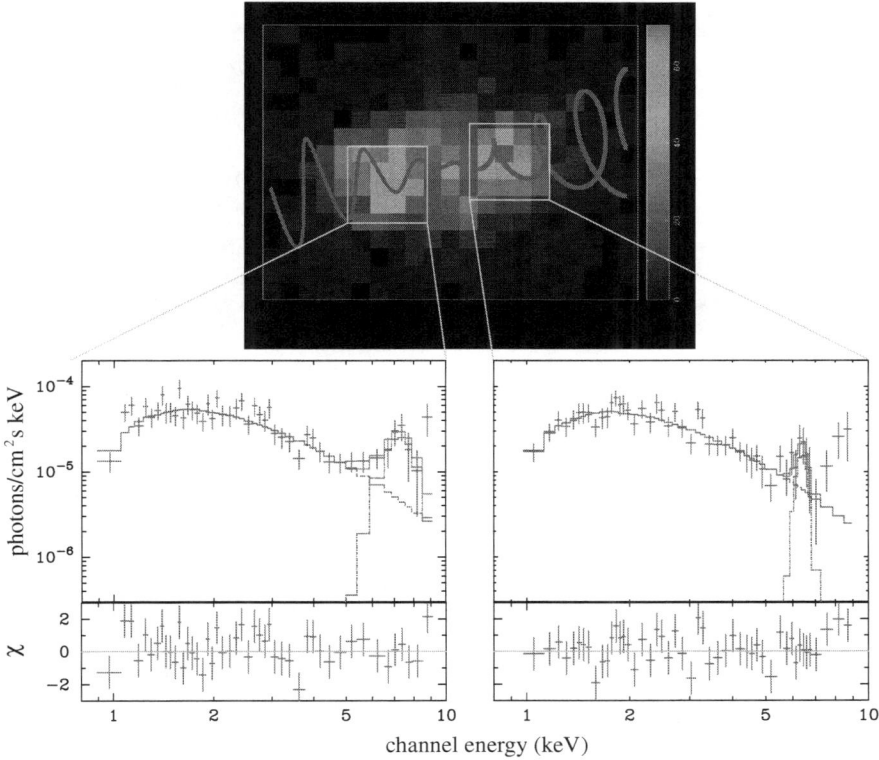

Fig. 4.13 Chandra image of SS 433 with the projected precession cycle of the jets superimposed on it. These observations reveal evidence for a hot continuum and Doppler-shifted iron emission lines from spatially resolved regions. From [92]

studies come to different conclusions for different sources. For example, circular polarization, which in principle can provide an excellent tool for investigating the baryonic content of the jets, is only detected in a handful of sources [38], where no strong conclusion could be placed yet. Entirely different studies (e.g., based on modeling large-scale jet–ISM interaction structures by means of self-similar jet fluid models) also draw different conclusions: in the case of Cygnus X-1 for instance, some authors [47, 54] argue for cold baryons in the flow, while an electron/positron jet seems to be favored in GRS 1915+105 based on energetics arguments [36].

4.11 Future Prospects

In this last section, I choose to briefly introduce a number of observational capabilities that are about to become available to the community – or have just started to – that I believe will literally revolutionize our understanding of microquasars (and not only).

4.11.1 Next Generation Radio Interferometers

With an imaging resolution of 50 milliarcsec at 40 GHz, the imaging capability of the VLA is comparable with the highest resolution of Advanced Camera for Surveys onboard the Hubble Space Telescope. However, the fundamental data-processing capabilities of the VLA remained essentially unchanged since the 1970s. The Expanded VLA (EVLA[3]) will dramatically improve its ability to make high-sensitivity and high-resolution images. The EVLA will attain unprecedented image quality with 10 times the sensitivity and 1000 times the spectroscopic capability of the existing array. Finally, the addition of eight new antennas will provide an order of magnitude increase in angular resolution. At the time of writing, the number of EVLA antennas continues to increase at a rate of one every 2 months. Each upgraded EVLA antenna produces 100 times more data than an original VLA antenna. When completed in 2012, the EVLA will be the most powerful centimeter-wavelength radio telescope in the world. The technology developed for the EVLA will enable progress on the next generation radio telescope: the Square Kilometer Array.

With baselines of up to 217 km, the Multi-Element Radio Linked Interferometer Network (MERLIN) provides cm-wavelength imaging at 10–150 milliarcsec resolution, effectively covering the gap between arrays such as the Westerbork Synthesis Radio Telescope (WSRT) and the VLA, and Very Long Based Interferometry (VLBI) arrays such as the VLBA and the European VLBI Network (EVN). e-MERLIN[4] is a major UK project aimed at increasing the bandwidth and thus the sensitivity of MERLIN by about an order of magnitude. This increased sensitivity, together with the high resolution provided by the long baselines, will enable a wide

---

[3] http://www.aoc.nrao.edu/evla/
[4] http://www.jb.man.ac.uk/research/rflabs/eMERLIN.html

range of new astronomical observations, including of course Galactic microquasars. One other major development which is part of the upgrade is frequency flexibility, as e-MERLIN will be able to switch rapidly between 1.4, 5, 6, and 22 GHz. The e-MERLIN telescope array is now nearing completion and will soon start to acquire data for its approved legacy programs.

In terms of microquasar studies, the EVLA and e-MERLIN will be simply a revolution. They will allow for the detection of faint quiescent systems in short exposures, enabling us to test whether collimated jets survive in quiescent blackholes (such as V404 Cyg). It will be possible to perform systematic searches for radio counterparts to NSs. Radio outbursts will hopefully receive the same daily coverage as X-rays outbursts do, and with amazingly fast frequency switching capabilities. Deep searches for low surface brightness jet-powered nebulae will be carried out at relatively (comparatively) limited expenses in terms of telescope time. Finally, we will be able to perform spectroscopic studies of microquasars jet, search for lines, possibly constrain the jet baryon content *directly*.

LOFAR[5] (Low Frequency Array) is a next-generation radio telescope currently under construction in The Netherlands, with baseline stations under development over a number of EU countries. The array will operate in the 30–80 and 120–240 MHz bands, thus representing the largest of the pathfinders for the lowest frequency component of the Square Kilometer Array (SKA) project (see [29]). Core Station One of LOFAR is currently operating; according to the plan, 36 stations will be deployed by the end of 2009.

LOFAR will literally revolutionize the study of bursting and transient radio phenomena at low radio frequencies. So far, the primary instruments for detecting extragalactic Gamma-ray bursts and Galactic microquasars have been orbiting satellites, such as BeppoSAX, INTEGRAL, RXTE, Swift, and others. One of the breakthroughs in this field was the localization of Gamma-ray burst and their subsequent identification with supernovae/galaxies at high red shifts. From the empirical relation between radio and X-ray emission for these systems (see Chap. 5) it is apparent that the all-sky monitoring with LOFAR will be a factor of 5–10 more effective in discovering such events than previous all-sky-monitors [29]. Furthermore, LOFAR will allow much more accurate localization of these events, enabling fast response science and follow-ups at other wavelengths.

4.11.2 γ-Ray Binaries

A new observational window has just been opened by ground-based Cherenkov telescopes, HESS[6] (High Energy Stereoscopic System) and MAGIC[7] (Major Atmospheric Gamma-ray Imaging Telescope), that survey the sky above 100 GeV. Because

---

[5] http://www.lofar.org/

[6] http://www.mpi-hd.mpg.de/hfm/HESS/HESS.html

[7] http://wwwmagic.mppmu.mpg.de/

of their high sensitivity, and high angular and energy resolution, these telescopes are revealing and identifying a plethora of new extragalactic and Galactic sources of very high energy (>100 GeV) radiation. The Galactic Center, supernovae remnants, pulsar-wind nebulae, and some "γ-ray binaries" have all been identified as very high γ-ray sources in the Galaxy [93]. The HESS collaboration reported the detection of TeV γ-ray emission from the Be-type binary system PSR B1259-63, close to the periastron passage [1]. Also TeV γ-rays have been detected from LS 5039 [2] and LS I +61 303 [4]. The idea of Galactic jet sources as capable of accelerating particles up to very high energies has been strengthened by the direct detection of large-scale X-ray jets resulting from in situ shocks where electrons are accelerated up to TeV energies [21, 17]. The synergy between very high energies and lower frequency observations is exemplified by the recent detection of a γ-ray flare from the BH X-ray binary Cygnus X-1, reported by the MAGIC collaboration. The flare, seen simultaneously with RXTE, Swift, and INTEGRAL, was compatible with a point-like source at a position consistent with the binary system, thus ruling out its arcmin-scale jet-driven radio nebula [3]. Alternatively, relativistic particles can be injected in to the surrounding medium by the wind from a young pulsar (see, e.g., the case of LS I +61 303 [24]). Coordinated multi-wavelength monitoring can discriminate between competing models, as they predict different emission regions for the radio and γ-ray radiation (core vs. arc-minute scale γ emission can now be resolved by high resolution γ-ray facilities), as well as different flux variations with the orbital phase.

Finally, scheduled to launch in the mid-2008, the Gamma-ray Large Area Space Telescope[8] (GLAST) is an international and multi-agency space mission which will study the cosmos in the energy range 10 keV–300 GeV, complementing ground-based Cherenkov telescopes with wider field. In the 90s, EGRET (the Energetic Gamma Ray Experiment Telescope) made the first complete survey of the sky in the 30 MeV–10 GeV range, showing the γ-ray sky to be surprisingly dynamic and diverse. *Most* of the EGRET sources remain unidentified (170 over 271): this outlines the importance and potential for new discoveries of the GLAST mission, whose Large Area Telescope (LAT) has a field of view about twice as wide (more than 2.5 steradians) and sensitivity about 50 times at 100 MeV.

The future is bright.

Acknowledgments The author is supported by NASA through a Hubble Fellowship grant HST-HF-01218 issued from the Space Telescope Science Institute, which is operated by the Association of Universities for Research in Astronomy, Incorporated, under NASA contract NAS5-26555. I wish to thank Rob Fender, Elmar Körding, James Miller-Jones, and Valeriu Tudose for providing me with original figures, in part still unpublished.

---

[8] http://glast.gsfc.nasa.gov/

References

1. F. Aharonian, A.G. Akhperjanian, K.-M. Aye et al.: A&A, **442**, 1 (2005a)
2. F. Aharonian, A.G. Akhperjanian, K.-M. Aye et al.: Science, **309**, 746 (2005b)
3. J. Albert, E. Aliu, H. Anderhub et al.: ApJ, **665**, L51 (2007)
4. J. Albert, E. Aliu, H. Anderhub et al.: Science, **312**, 1771 (2006)
5. L. Angelini, N.E. White: ApJ, **586**, L71 (2003)
6. T.M. Belloni, I. Parolin, M. Del Santo et al.: MNRAS, **367**, 1113 (2006)
7. T.M. Belloni, M. Méndez, A.R. King et al.: ApJ, **488**, L109 (1997)
8. T.M. Belloni, M. Méndez, A.R. King et al.: ApJ, **479**, L145 (1997)
9. M.C. Begelman, A.R. King, J.E. Pringle: MNRAS, **370**, 399 (1980)
10. R.D. Blandford, M.C. Begelman: MNRAS, **303**, L1 (1999)
11. R.D. Blandford, A. Königl: ApJ, **232**, 34 (1979)
12. K. Blundell, M.G. Bowler: ApJ, **616** L159 (2004)
13. C. Brocksopp, S. Corbel, R.P. Fender et al.: MNRAS, **356**, 125 (2005)
14. C. Brocksopp, R.P. Fender, M. McCollough et al.: MNRAS, **331**, 765 (2002)
15. M. Cadolle-Bel, M. Ribó, J. Rodriguez et al.: ApJ, **659**, 549 (2007)
16. S. Chaty: Multi-wavelength observations of the microquasar XTE J1720-318: a transition from high-soft to low-hard state. In: T. Belloni (ed.) Proceedings of the VI Microquasar Workshop: Microquasars and Beyond (Proceedings of Science) p. 141 (2007)
17. S. Corbel, P. Kaaret, R.P. Fender et al.: ApJ, **632**, 504 (2005)
18. S. Corbel, R.P. Fender, J.A. Tomsick et al.: ApJ, **617**, 1272 (2004)
19. S. Corbel, M. Nowak, R.P. Fender et al.: A&A, **400**, 1007 (2003)
20. S. Corbel, R.P. Fender: ApJ, **573**, L35 (2002)
21. S. Corbel, R.P. Fender, A.K. Tzioumis et al.: Science, **298**, 196 (2002)
22. S. Corbel, R.P. Fender, A.K. Tzioumis et al.: A&A, **359**, 251 (2000)
23. M.M. Crocker, R.J. Davis, S.P.S. Eyres et al.: MNRAS, **326**, 781 (2001)
24. V. Dhawan, A. Mioduszewski, M. Rupen: LS I +61 303 is a Be-Pulsar binary, not a Microquasar. In: T. Belloni (ed.) Proceedings of the VI Microquasar Workshop: Microquasars and beyond (Proceedings of Science) p. 52 (2007)
25. V. Dhawan, I.F. Mirabel, L.F. Rodríguez: ApJ, **543**, 373 (2000)
26. T. Di Salvo, C. Done, P.T. Zycki et al.: ApJ, **547**, 1024 (2001)
27. G.M. Dubner, M. Holdaway, W.M. Goss et al.: Astron. J., **116**, 1842 (1998)
28. H. Falcke, P.L. Biermann: A&A, **293**, 665 (1995)
29. R.P. Fender: LOFAR Transients and the Radio Sky Monitor. In: A. Tzioumis, T.J. Lazio, R. Fender (eds.) Bursts, Pulses and Flickering: Wide-field Monitoring of the Dynamic Radio Sky (Proceedings of Science) (2008) [arXiv0805.4349]
30. R.P. Fender: Radio emission and jets from X-ray binaries. In: W.H.G. Lewin, M. van der Klis (eds.) Compact Stellar X-Ray Sources Cambridge University Press, Cambridge (2006)
31. R.P. Fender, A.M. Stirling, R.E. Spencer et al.: MNRAS, **369**, 603 (2006)
32. R.P. Fender, T.M. Belloni: ARA&A, **42**, 317 (2004)
33. R.P. Fender, K. Wu, H. Johnston et al.: Nature, **427**, 222 (2004)
34. R.P. Fender, T.M. Belloni, E. Gallo: MNRAS, **355**, 1105 (2004)
35. R.P. Fender: MNRAS, **340**, 1353 (2003)
36. R.P. Fender: ARA&A, **288**, 79 (2003)
37. R.P. Fender, E. Gallo, P.G. Jonker: MNRAS, **343**, L99 (2003)
38. R.P. Fender, D. Rayner, S.A. Trushkin et al.: MNRAS, **330**, 212 (2002)
39. R.P. Fender: MNRAS, **322**, 31 (2001)
40. R.P. Fender, R.M. Hjellming, R.P.J. Tilanus: MNRAS, **322**, L23 (2001)
41. R.P. Fender, S. Corbel, A.K. Tzioumis et al.: ApJ, **519**, L165 (1999)
42. Y. Fuchs, J. Rodriguez, I.F. Mirabel et al.: A&A, **409**, L35 (2003)
43. E. Gallo, J. Homan, P.G. Jonker et al.: (2008) ApJ, **683**, L51 (2008)
44. E. Gallo, S. Migliari, S. Markoff et al.: MNRAS, **670**, 600 (2007)

45. E. Gallo: Astrophys. Space Sci, **311**, 161 (2007)
46. E. Gallo, R.P. Fender, J.C.A. Miller-Jones et al.: MNRAS, **370**, 1351 (2006)
47. E. Gallo, R.P. Fender, C. Kaiser et al.: Nature, **436**, 819 (2005)
48. E. Gallo, S. Corbel, R.P. Fender et al.: MNRAS, **347**, L52 (2004)
49. E. Gallo, R.P. Fender, G.G. Pooley: MNRAS, **344**, 60 (2003)
50. M.R. Garcia, J.M. Miller, J.E. McClintock et al.: ApJ, **591**, 388 (2003)
51. M.R. Garcia, J.E. McClintock, R. Narayan et al.: ApJ, **553**, L47 (2001)
52. B. Hiemstra, P. Soleri, M. Mendez et al.: MNRAS, **394**, 2080 (2009)
53. S. Heinz, N.S. Schulz, W.N. Brandt et al.: ApJ, **663**, L93 (2007)
54. S. Heinz: ApJ, **636**, 316 (2006)
55. S. Heinz, H.J. Grimm: MNRAS, **633**, 384 (2005)
56. S. Heinz, A. Merloni: MNRAS, **355**, L1 (2004)
57. S. Heinz: New Astron. Rev., **47**, 565, (2003)
58. R.M. Hjellming, K.J. Johnston: ApJ, **328**, 600 (1988)
59. J. Homan, R. Wijnands, A. Kong et al.: MNRAS, **366**, 235 (2006)
60. J. Homan, T.M. Belloni: Astrophys. Space Sci, **300**, 107 (2005)
61. J. Homan, M. Buxton, S. Markoff et al.: ApJ, **624**, 295 (2005)
62. J. Homan, R. Wijnands, M. van der Klis et al.: ApJ, **132**, 377 (2001)
63. R.I. Hynes, P.A. Charles, R.M. Garcia et al.: ApJ, **611**, L125 (2004)
64. R.I. Hynes, C.A. Haswell, W. Cui et al.: MNRAS, **345**, 292 (2003)
65. R.I. Hynes, C.A. Haswell, S. Chaty et al.: ApJ, **539**, L37 (2000)
66. J.-P. Lasota: Comptes Rendus Physique, **8**, 45 (2007)
67. B.F. Liu, R.E. Taam, E. Meyer-Hofmeister et al.: ApJ, **671**, 695 (2007)
68. P. Kaaret, S. Corbel, J.A. Tomsick et al.: ApJ, **582**, 945 (2003)
69. C.R. Kaiser: MNRAS, **367**, 1083 (2006)
70. C.R. Kaiser, R. Sunyaev, H.C. Spruit: A&A, **356**, 975 (2001)
71. K.S. Kawabata, Y. Ohyama, N. Ebizuka et al.: AJ, **132**, 433 (2006)
72. E. Körding, M. Rupen, C. Knigge et al.: Science, **320**, 1318 (2008)
73. E. Körding, S. Jester, R. Fender: Mon. Not. R. Astron. Soc., **372**, 1366 (2006).
74. E. Körding, R.P. Fender, S. Migliari: MNRAS, **369**, 1451 (2006)
75. H.L. Marshall, C.R. Canizares, N.S. Schulz: ApJ, **564**, 941 (2002)
76. S. Markoff, M.A. Nowak, J. Wilms: ApJ, **635**, 1203 (2005)
77. S. Markoff, H. Falcke, R.P. Fender: A&A, **397**, 645 (2003)
78. S. Markoff, H. Falcke, R.P. Fender: A&A, **372**, L25 (2001)
79. J. Martí, S. Mereghetti, S. Chaty et al.: A&A, **338**, L95 (1998)
80. J.E. McClintock, R.A. Remillard: Black hole binaries. In: W.H.G. Lewin, M. van der Klis (eds.) Compact Stellar X-Ray Sources Cambridge University Press, Cambridge (2006)
81. J.E. McClintock, R. Narayan, M.R. Garcia et al.: ApJ, **593**, 435 (2003)
82. J.C. McKinney: MNRAS, **368**, 1561 (2006)
83. K. Menou, A.A. Esin, R. Narayan et al.: ApJ, **520**, 276 (1999)
84. J.M. Miller, J. Homan, G. Miniutti: ApJ, **652**, L113 (2006)
85. J.M. Miller J. Homan, D. Steeghs et al.: ApJ, **653**, 525 (2006)
86. J.C.A. Miller-Jones, E. Gallo, M. Rupen et al.: MNRAS, **388**, 1751 (2008)
87. J.C.A. Miller-Jones, C.R. Kaiser, T.J. Maccarone et al.: Searching for the signatures of jet-ISM interactions in X-ray binaries. In: M. Bandyopadhyay, S. Wachter, D. Gelino, C.R. Gelino (eds.) Proceedings of A Population Explosion: The Nature and Evolution of X-ray Binaries in Diverse Environments (2008) [arXiv:0802.3446]
88. J.C.A. Miller-Jones, R.P. Fender, E. Nakar: MNRAS, **367**, 1432 (2006)
89. S. Migliari, J.C.A. Miller-Jones, R.P. Fender et al.: ApJ, **671**, 706 (2007)
90. S. Migliari, J.A. Tomsick, T.J. Maccarone et al.: ApJ, **643**, L41 (2006)
91. S. Migliari, R.P. Fender: MNRAS, **366**, 79 (2006)
92. S. Migliari, R.P. Fender, M. Méndez: Science, **297**, 1673 (2002)
93. I.F. Mirabel: Science, **312**, 1759 (2006)

94. I.F. Mirabel, V. Dhawan, R.P. Mignani et al.: Nature, **413**, 139 (2001)
95. I.F. Mirabel, L.F. Rodríguez: ARA&A, **37**, 409 (1999)
96. I.F. Mirabel, L.F. Rodríguez: Nature, **392**, 673 (1998)
97. I.F. Mirabel, V. Dhawan, S. Chaty et al.: A&A, **330**, L9 (1998)
98. I.F. Mirabel, L.F. Rodríguez: Nature, **371**, 46 (1994)
99. I.F. Mirabel, L.F. Rodríguez, B. Cordier et al.: Nature, **358**, 215 (1992)
100. M. Muno, J. Mauerhan: ApJ, **648**, L135 (2006)
101. R. Narayan, M.R. Garcia, J.E. McClintock: ApJ, **478**, L79 (1997)
102. R. Narayan, I. Yi: ApJ, **444**, 231 (1995)
103. R. Narayan, I. Yi: ApJ, **428**, L13 (1994)
104. J. Nichols et al.: ApJ, **660**, 651 (2007)
105. J. Poutanen, P. Coppi: Physica Scripta, **77**, 57 (1998)
106. J. Rodriguez et al.: ApJ, **655**, L97 (2007)
107. M.P. Rupen, A. Mioduszewski, J.L. Sokoloski: ApJ, **688**, 559 (2008)
108. D.M. Russell, R.P. Fender, E. Gallo et al.: MNRAS, **376**, 1341
109. D.M. Russell, R.P. Fender, R.I. Hynes et al.: MNRAS, **371**, 1334 (2006)
110. E.S. Rykoff, J.M. Miller, D. Steeghs et al.: ApJ, **666**, 1129 (2007)
111. E.R. Seaquist: Radio emission from novae. In: M.F. Bode, A. Evans (eds.) Classical Novae, p. 143. Wiley, Chichester (1989)
112. F. Seward, J. Grindlay, E. Seaquist et al.: Nature, **287**, 806 (1980)
113. J.L. Sokoloski, S.J. Kenyon, C. Brocksopp et al.: Jets from Accreting White Dwarfs. In: G. Tovmassian, E. Sion (eds.) Revista Mexicana de Astronomia y Astrofisica Conference Series, vol. 20, pp. 35–36 (2004)
114. A.M. Stirling, R.E. Spencer, C.J. de la Force et al.: MNRAS, **327**, 1273 (2001)
115. L. Titarchuk, T. Zannias: ApJ, **493**, 863 (1998)
116. J.A. Tomsick, E. Kalemci, P. Kaaret et al.: ApJ, **680**, 593 (2008)
117. J.A. Tomsick, S. Corbel, R.P. Fender et al.: ApJ, **582**, 933 (2003)
118. J.A. Tomsick, S. Corbel, R.P. Fender et al.: ApJ, **597**, L133 (2003)
119. V. Tudose, R.P. Fender, C.R. Kaiser et al.: MNRAS, **372**, 417 (2006)
120. M. Watson, G. Stewart, A. King et al.: MNRAS, **222**, 261 (1986)
121. F. Yuan, Z.Q. Shen, L. Huang: ApJ, **642**, L45 (2006)
122. F. Yuan, W. Cui, R. Narayan: ApJ, **620**, 905 (2005)

Chapter 5
'Disc–Jet' Coupling in Black Hole X-Ray Binaries and Active Galactic Nuclei

R. Fender

Abstract In this chapter I will review the status of our phenomenological under-standing of the relation between accretion and outflows in accreting black hole systems. This understanding arises primarily from observing the relation between X-ray and longer wavelength (infrared, radio) emission. The view is necessarily a biased one, beginning with observations of X-ray binary systems, and attempting to see if they match with the general observational properties of active galactic nuclei.

Black holes are amongst the most esoteric objects conceived of by man. They do not however lurk only in our imaginations or at the fringes of reality but appear, as far as we can tell from the objective interpretation of astrophysical observations, to play a major role in the history of the entire Universe.

In particular, feedback from accreting black holes, in the form of both radiation and kinetic energy (i.e., jets and winds) has been a key element throughout cosmic evolution. This is summarized in Fig. 5.1 and some key phases can be identified as follows:

- *Reionization:* By redshift of $z \sim 10$, less than a billion years after the Big Bang, UV and X-ray radiations from the first accreting black holes (as well as the first stars) undoubtedly played a key role in ending the so-called "Dark Ages" and reionizing the Universe.
- *The epoch of Active Galactic Nuclei:* Following reionization, supermassive black holes at the centers of active galactic nuclei (AGN) grew rapidly, probably fed by galactic mergers. The peak of AGN activity occurred around a redshift of $z \sim 2$, around 4 billion years after the big bang. Radiation from the accretion flows around these AGN formed what we now observe as the cosmic X-ray background. Kinetic feedback from AGN also acted to stall and reheat cooling flows at the centers of galaxies and to regulate the growth of galaxies.

R. Fender (✉)

School of Physics and Astronomy, University of Southampton, Southampton So17 1BJ, UK,
rpf@phys.soton.ac.uk

Fender, R.: *'Disc–Jet' Coupling in Black Hole X-Ray Binaries and Active Galactic Nuclei.* Lect. Notes Phys. **794**, 115–142 (2010)
DOI 10.1007/978-3-540-76937-8_5

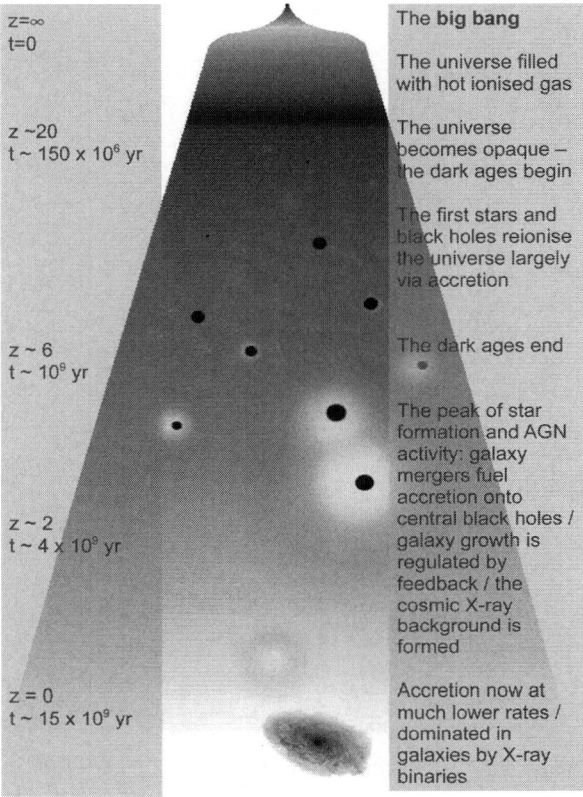

Fig. 5.1 The role of feedback from black hole accretion, in the form of both radiation and kinetic energy, has played a key role in the evolution of our Universe

- *The epoch of stellar-mass black holes:* In the nearby Universe, feeding of the supermassive black holes has declined and their feedback to the ambient medium is dominated by kinetic power. The X-ray luminosity of galaxies is now dominated by accretion onto black holes of mass $M_{BH} \sim 10_\odot$ in binary systems (as well as neutron stars).

Black holes, it seems, despite their oddness, are a key formative component of the Universe. It is essential for our understanding not only of the history of the Universe to this point, but also of the future evolution of the Universe that we understand how these objects behave. In fact, they do one thing, and one thing only, of major significance[1]: they convert gravitational potential energy to radiation and kinetic energy, which feeds back into the Universe.

---

[1] Not counting information trapping on their surfaces (e.g. [87])

This accretion process is in principle quite simple, as outlined in the next section. However, observationally we find that it has its subtleties and nuances, which manifest themselves most clearly in how the black hole distributes its feedback between radiation and kinetic power. The rest of this chapter explores what we can learn about these idiosyncrasies of black hole accretion by studying low-mass ($M < 20M_\odot$), rapidly varying, black holes in binary systems in our galaxy, and how we might apply that to understanding how supermassive black holes have helped shape the observable Universe.

In addition to the flow of energy and feedback, which are the foci of this review, the physics associated with jet formation and the associated particle acceleration, as well as potential tests of general relativity associated with studying black hole accretion are all extremely interesting astrophysical topics in their own right. See suggestions for further reading at the end of the chapter.

5.1 Simple Physical Theory

The maximum energy release associated with the accretion of matter from infinity to a body of mass M and radius R is given by GM/R. It is the ratio M/R which determines the efficiency of the accretion process.

A non-rotating or Schwarzschild black hole has an event horizon at

$$r_s = 2GM/c^2 = 2r_g,$$

where r_g is referred to as the gravitational radius.

This linear dependence on black hole mass means that all non-rotating black holes have the same maximum potential accretion efficiency, regardless of their mass. There is an innermost stable circular orbit, or ISCO, around such black holes, within which matter will plummet rapidly across the event horizon, and this is at $3r_S$, also independent of mass.

Spinning black holes have smaller event horizons and ISCOs than non-rotating black holes: a maximally rotating or Maximal Kerr black hole has an event horizon and ISCO both at $r_g = \frac{1}{2}r_s$. This means that for a black hole of any mass, the ratio M/R is the same to within a factor of 6. This ratio is dwarfed by the enormous range of masses of black holes we have observed in the Universe, from $\leq 10M_\odot$ to $\geq 10^9 M_\odot$. Beyond mass and spin, black holes possess only one more property that of an electric charge. Given the lack of observation of large-scale charged objects in the Universe, it is assumed that black holes are electrically neutral. Figure 5.2 summarizes the relative sizes of neutron stars, non-rotating, and maximally rotating black holes.

This amazing scale invariance of the theoretical accretion efficiency coupled with the inability of a black hole to possess any other distinguishing characteristics drives us to speculate that the process of accretion onto black holes of all scales should be very similar.

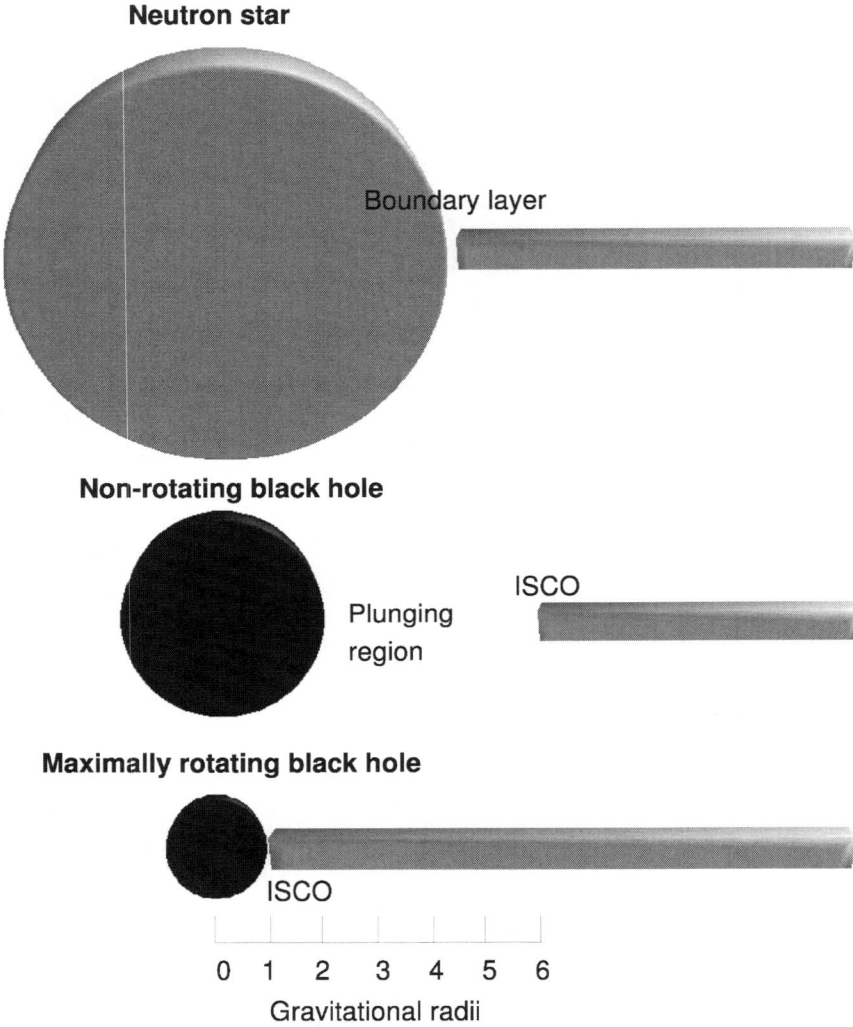

Fig. 5.2 Relative sizes of a neutron star, non-rotating, and maximally rotating black hole in units of gravitational radii $R_G = GM/c^2$. The figure is correct for black holes of any mass. A neutron star of 1.4 $M_\odot$ mass and 10 km radius has a surface at about 4.8 R_G. The accretion disc may extend all the way to the surface where there will be a boundary layer between the inner disc edge and the surface, which will be rotating more slowly. For a non-rotating black hole the event horizon will lie at $2R_G$ and there will be an innermost stable circular orbit (ISCO) at $6R_G$. Inside the ISCO matter will plunge across the event horizon. For a maximally rotating black hole the event horizon and ISCO lie at $1R_G$. Most black holes will be somewhere between these two extremes. If we naively assume radiation only from a disc-like radiatively efficient accretion flow, we can see that accretion onto a neutron star can be more efficient than that onto a non-rotating black hole, but that accretion onto a black hole with significant spin can be the most efficient process

However, the physical conditions in the innermost regions of the accretion flow are not expected to be identical. The fixed M/R ratio necessarily implies that the density of the inner accretion flow (and in fact of the black hole itself) must decrease with increasing mass. Put very simplistically, unless there are strong changes in accretion geometry with mass, the surface area through which matter with significant angular momentum will accrete will scale as M^2. So for two black holes, one "stellar mass" ($\sim 10 M_\odot$) and one "supermassive" ($\sim 10^9 M_\odot$), accreting at the same Eddington ratio, which scales linearly with mass, the density of the accreting matter should scale as $\rho \propto \dot{m}/M^2 \propto M^{-1}$, i.e., a factor of 10^7 more dense in the case of the stellar-mass black hole. Similarly, the effective temperature of the accretion disc varies with mass. This can be illustrated simply as follows: for the same Eddington ratio the luminosity is released in radiatively efficient accretion $L \propto M$ and is emitted over a disc area which scales as M^2. For black body radiation $L \propto AT^4$ (where A is the emitting area and T the temperature), and so we find $T \propto M^{-1/4}$. Therefore, accretion discs in the most luminous AGN should be 100 times cooler than those in the most luminous X-ray binaries [83]. In Fig. 5.3 we present the expected simple scaling of size (both of the black hole and inner accretion disc), luminosity, disc temperature, disc density (also mean density of the matter within

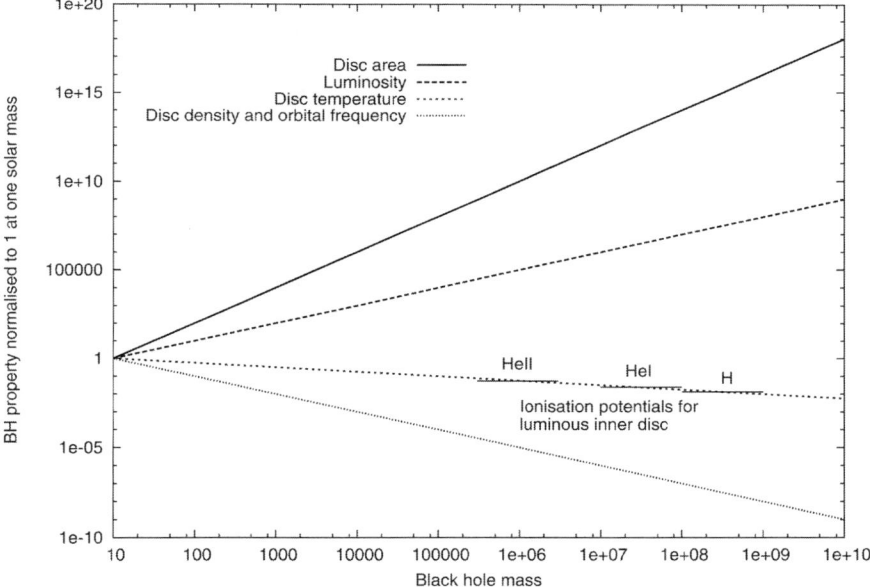

Fig. 5.3 The simple scaling of area (both of event horizon and inner disc), luminosity, disc temperature, disc orbital frequency, and disc density (also mean density within event horizon) with black hole mass for black holes accreting radiatively efficiently at the same Eddington ratio (the combination of which requirements probably means an Eddington ratio ≥ 0.01) and with the same spin. Also indicated are the ionization potentials for H and both HeI and HeII based on a temperature of 1 keV for an Eddington rate $10_\odot$ black hole (i.e., the temperature scale is the actual temperature in keV)

the event horizon), and also orbital frequency as a function of mass for black holes accreting radiatively efficiently at the same Eddington ratio and with the same spin. As we shall see later in this chapter, the requirement for radiative efficiency probably implies an Eddington ratio ≥ 0.01.

How the energy density and organization of the magnetic field in the accretion flow, which are likely to be important both for effective viscosity and jet formation, vary with black hole mass is less well understood.

5.2 Observations of Black Hole X-Ray Binaries

More than 20 objects consistent with black holes of mass $\sim 10 M_\odot$ have been identified in X-ray binary systems within our galaxy (e.g., [52]). This population may be the tip of an iceberg of $\sim 10^8$ stellar-mass black holes within our galaxy (i.e., a far larger total black hole mass than that associated with the $\sim 10^6 M_\odot$ black hole, Sgr A*, at our galactic center). These binary systems show semi-regular outbursts in which they temporarily brighten across the electromagnetic spectrum by many orders of magnitude. The origin of these outbursts is believed to be a disc instability driven by the ionization of hydrogen above a given temperature (e.g., [19]). However while describing general outburst trends, there are many difficulties in explaining all the observational characteristics. When comparing to AGN, we imagine that X-ray binaries near the peaks of their outbursts correspond to the high Eddington ratio systems such as Quasars and that when in quiescence (the most common phase for most systems) they correspond to low-luminosity AGN (LLAGN) such as Sgr A* at the center of our own galaxy.

The evolution of two outbursts from the binary GX 339-4 is illustrated in Fig. 5.4. In very brief summary, most black hole X-ray binaries spend most of their time in a "quiescent" state with X-ray luminosities as low as $\sim 10^{30}$ erg s^{-1} ($\leq 10^{-9} L_{\text{Edd}}$; e.g. [26]). The mass transfer from the companion star, usually via Roche-lobe overflow, progresses at a higher rate than that at which matter is centrally accreted onto the black hole (or neutron star) and so the mass (and temperature) of the disc increase. At some point the effective viscosity of the disc increases, perhaps due to the hydrogen ionization instability mentioned above, and the matter in the disc rapidly drains toward the central accretor. During this phase the central accretion rate is much higher than the time-averaged mass accretion rate from the companion star, and the source becomes very luminous, often approaching the Eddington luminosity.[2]

At an early stage in the outburst the source X-ray spectrum begins to soften (motion to the left in the HIDs in Fig. 5.3). Within a few weeks or months, when some significant fraction of the disc mass has been accreted, the disc cools, and the

---

[2] The *E*ddington luminosity corresponds to the luminosity at which the outward radiation force from accretion balances the inward force of gravity. For spherical accretion of hydrogen it is approximately $L_{\text{Edd}} \sim 1.4 \times 10^{38} (M/M_\odot) \text{erg s}^{-1}$.

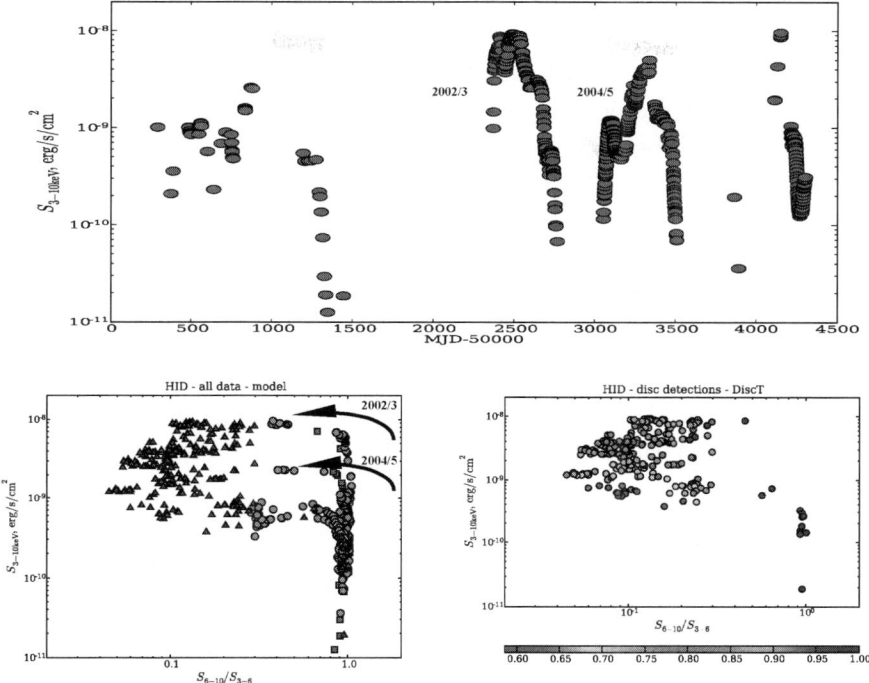

Fig. 5.4 Outbursts of the black hole binary GX 339-4. *Top panel:* the X-ray light curve of GX 339-4 as measured with the PCA instrument onboard the RXTE satellite. Gaps in the data coverage between major outbursts usually indicate low/quiescent flux levels. The outbursts in 2002/2003 and 2004/2005 are covered in the greatest detail and are represented in hardness–intensity diagrams (HIDs) in the lower panels. In these HIDs the abscissae (x-axes) indicate X-ray "color" or hardness ratio, whereas the ordinates (y-axes) indicate luminosity. In the *left panel* the time evolution of the two separate outbursts are indicated; *triangles* indicate those spectra where a strong accretion disc (black body-like) component was required in the spectral fit. Two different outbursts are overplotted – both show hysteretical patterns of behavior, in that the transition from hard → soft states occurs at higher luminosities than the soft → hard transition, but they also clearly differ in that the earlier outburst reached a higher luminosity in the initial stages. The *right panel* indicates the variation of fitted accretion disc temperature (gray scale) in the soft X-ray state, decreasing with luminosity. Adapted from [9]

source returns toward quiescence.[3] In doing so it makes the X-ray spectral transition away from the soft state at a lower luminosity than that at which it entered the state. This results in a hysteretical track in the HID. For more details on black hole outbursts see Chaps. 3, 4 and 6 in this volume.

---

[3] It is worth bearing in mind that two of the sources we use most in our studies of black hole binaries do not really fit this pattern: GX 339-4 does not ever really settle into extended quiescent periods (see Fig. 5.4) and Cyg X-1 is not accreting via simple Roche-lobe overflow.

5.2.1 Relations Between Accretion "State" and Radio Emission

Most of our insights into the relation between modes/rates of accretion and the type and power of any associated feedback (also known as the "disc-jet" coupling[4]) come from (near-)simultaneous radio (also sometimes infrared) and X-ray observations of rapidly varying systems. The radio (and, often, infrared) emission is assumed to arise via synchrotron emission in a jet-like outflow, and the X-ray emission to be a tracer of the accretion flow (rate, geometry, temperature, even composition). It has also been suggested that a significant fraction of the X-ray emission in some states may arise in the jet (see Chap. 6).

A good example of coupling between X-ray state and radio emission is given in Fig. 5.5. In this figure several hours of overlapping radio and X-ray observations of

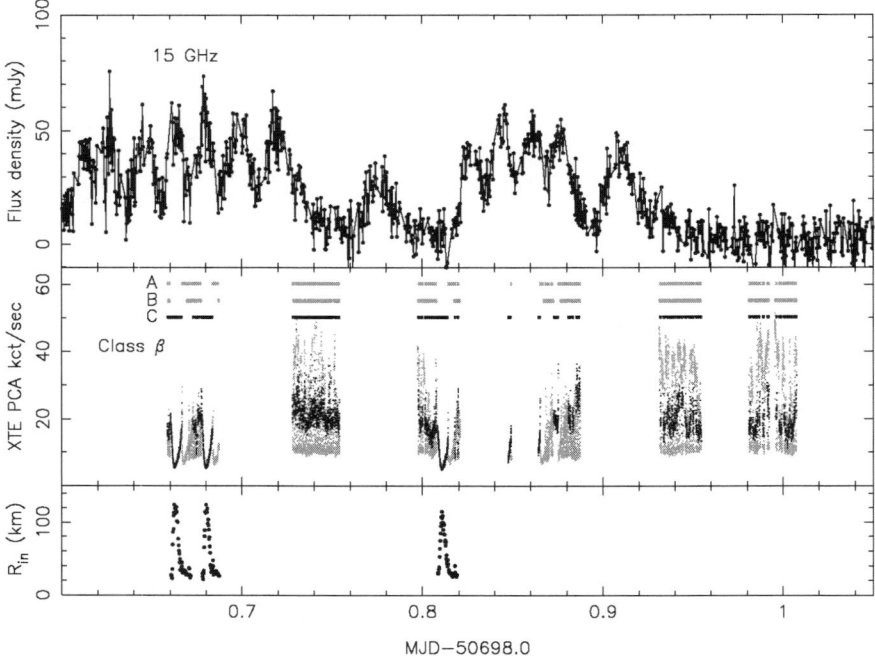

Fig. 5.5 Radio–X-ray coupling in the binary black hole system GRS 1915+105. The *top panel* shows radio monitoring at 15 GHz from the Ryle Telescope, the *middle panel* shows the X-ray flux as measured by RXTE, with the light curve subdivided into "soft" (A and B) and "hard" (C) X-ray states. The radio oscillation events, which we think are associated with individual relativistic ejection events, only occur when long hard (state C) dips are followed by rapid transitions to soft states (e.g., the first and third phases of X-ray coverage) and not when more rapid fluctuations occur, albeit at the same luminosity. While GRS 1915+105 is a complex case, the general pattern of behavior – major ejection events during state transitions – is consistent with other accreting black hole binaries. From [40]

---

[4] This expression now has confusing connotations because people sometimes take the word "disc" in this context to mean only the geometrically thin, optically thick component of the accretion flow.

the powerful jet source and black hole binary GRS 1915+105 are presented (from [40]). The radio emission shows clear phases of oscillatory behavior during periods of strong dips in the hard X-ray light curve, and far less activity when the X-rays, although at the same luminosity, are not showing the long hard-state dips. Similar, maybe identical, patterns are observed at widely separated epochs, demonstrating a clear and repeating pattern of accretion: outflow coupling.

5.2.2 Toward Unified Models for Accretion–Ejection Coupling

Based on more than a decade of X-ray and radio observations, attempts were made in the past few years to find simple and unified patterns for the accretion–outflow ("disc-jet") coupling in black hole X-ray binaries. Our analysis is heavily based on the assumption that radio emission is associated with jet-like outflows, something argued in more detail in e.g. [30] and by many other authors. In terms of relations to X-ray states, [13] demonstrated that during the high/soft X-ray state of the binary GX 339-4 the radio emission was suppressed with respect to the hard state at comparable luminosities. In fact this phenomenon had been observed more than two decades earlier by [86] in the case of Cygnus X-1. Around the same time, [28, 7, 8] demonstrated that while in the hard X-ray state (right-hand side of the the HID) the same binary, GX 339-4, repeatedly displayed a correlation of the following form:

$$L_{\text{GHzradio}} \propto L_{\text{softX-ray}}^b,$$

where $b = 0.7 \pm 0.1$. Reference [14] demonstrated that steady, flat-spectrum radio emission (spectral index $\alpha = \Delta \log S_\nu / \Delta \log \nu \sim 0$) such as observed from GX 339-4 in the hard X-ray state was observed from all hard-state black hole binaries (this characteristic post-outburst radio behavior had been noted by e.g. [31] but not clearly associated with X-ray state).

In a wider-ranging study, [22] found that another binary V404 Cyg (GS 2023+338) displayed the same correlation ($b \sim 0.7$) and the same normalization, within uncertainties. Radio and X-ray measurements for several other black hole binaries in the hard state were also consistent with the same "universal" relation, and repeated suppression of the radio emission in softer X-ray states of Cygnus X-1 was reconfirmed. More recently however, several hard-state black hole binaries have been discovered which are underluminous in the radio band (see [20], and Chap. 4 in this volume for more discussion).

These hard-state, flat-spectrum jets are in fact very powerful, and not just an interesting sideshow to the main event of X-ray production [14]. In fact the combination of two non-linear couplings, $L_{\text{radio}} \propto L_X^{0.7}$ (observed) and $L_{\text{radio}} \propto P_{\text{jet}}^{1.4}$ (theoretical), where P_{jet} is total jet power, but also observed in [43], implies that $P_{\text{jet}} \propto L_X^{0.5}$, and that as the X-ray luminosity of a source declines the jet may come to dominate over radiation in terms of feedback from the accretion process [15]. The key goal then was to determine the normalization for the jet power, which a variety of methods have established as being comparable to the X-rays luminosity at high Eddington ratios, dominating at lower Eddington ratios (e.g. [24, 29]). The current consensus is therefore that in the hard X-ray state black hole binaries produce powerful, flat-spectrum relatively steady jets, whose strength correlates in a (near-)universal way with the X-ray luminosity.

However, really spectacular radio ejection events, during which relativistic, sometimes apparently superluminal radio components were observed to propagate away from the binary, were also known (e.g. [74, 32]). These jets also carry away a large amount of power (see [16] and references therein). How did these relate to the steady radio emission in the hard state, and the apparently suppressed radio emission in the soft state?

Careful examination of X-ray data compared to radio monitoring and imaging observations revealed the answer: such bright ejection events occurred during the transition from hard to soft X-ray states. Combining our knowledge of the hysteretical outburst behavior of black hole binaries (see Chap. 3) with these insights into the coupling to radio emission allowed a first attempt at a "unified model" for the disc-jet coupling, which was put forward by [16], see Fig. 5.6.

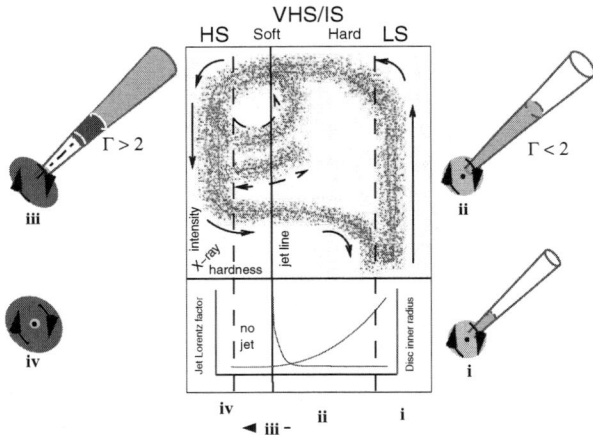

Fig. 5.6 A schematic of the simplified model for the jet–disc coupling in black hole binaries presented by [16]. The *central box panel* represents an X-ray hardness–intensity diagram (HID); "HS" indicates the "high/soft state", "VHS/IS" indicates the "very high/intermediate state," and "LS" the "low/hard state". In this diagram, X-ray hardness increases to the right and intensity upward. The *lower panel* indicates the variation of the bulk Lorentz factor of the outflow with hardness – in the LS and hard-VHS/IS the jet is steady with an almost constant bulk Lorentz factor $\Gamma < 2$, progressing from state (i) to state (ii) as the luminosity increases. At some point – usually corresponding to the peak of the VHS/IS – Γ increases rapidly producing an internal shock in the outflow (iii) followed in general by cessation of jet production in a disc-dominated HS (iv). At this stage fading optically thin radio emission is only associated with a jet/shock which is now physically decoupled from the central engine. As a result the *solid arrows* indicate the track of a simple X-ray transient outburst with a single optically thin jet production episode. The *dashed loop* and *dotted track* indicate the paths that GRS 1915+105 and some other transients take in repeatedly hardening and then crossing zone (iii) – the "jet line" – from *left* to *right*, producing further optically thin radio outbursts. Sketches around the outside illustrate our concept of the relative contributions of jet, "corona" and accretion disc at these different stages

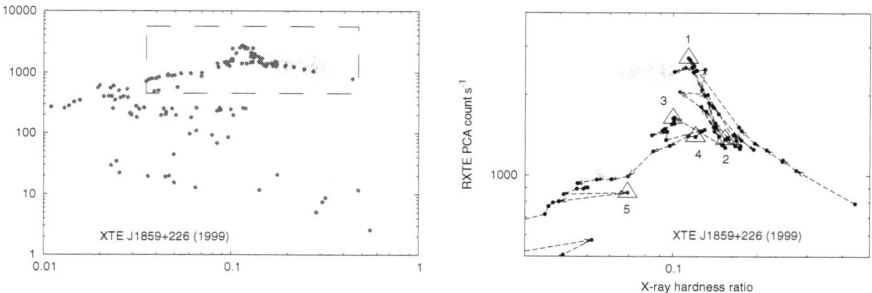

Fig. 5.7 Moments of five radio flare ejection events from the black hole binary XTE J1859+226 (from [5]) as a function of position in the HID. The *left panel* indicates the overall HID for the source which traveled in a generally anti-clockwise direction in the figure, typical for such outbursts. In the *right panel* we focus on the region (indicated by a *box* in the *left panel*) around the time of the radio ejections. *Triangles* indicate the estimated moments of ejection, and *arrows* indicate the temporal evolution in the HID. It is clear that all five radio ejection events took place at approximately the same hardness. Most notable is the last event, number 5, which is associated with a brief excursion back to harder states even as the source is in a softening trend. These data support the idea of a near-constant jet line on the upper branch of the HID, at least within one outburst loop. From [17]

Since this model was first proposed, nearly 5 years ago, we have repeatedly attempted to test whether or not any of the basic empirical relations were wrong. In particular, we have sought to confirm that the major radio outbursts always occur during the hard → soft transitions (i.e., left to right motion at the top of the HID). In a study of over 16 black hole outbursts (compared to four in [16]) we [17] have found no clear exceptions to this. In fact in revisiting the case of the black hole transient XTE J1859+226 we find strong evidence that all of the five recorded radio flare events (which we associate with an ejection) occurred at ~ same X-ray hardness (Fig. 5.7).

While [16] attempted to provide a simple physical interpretation of the empirical relations, several more detailed theoretical models have arisen based on the emerging phenomenology (e.g. [23, 62]; see also Chap. 9). A very interesting combined observational/theoretical result was also presented by [63] who analyzed optical and X-ray observations of the hard-state black hole XTE J1118+480. What they found (see also [34]) was that the very rapid correlated optical and X-ray variability in this source could be explained by a strong synchrotron component in the optical, which arose in a jet/coronal outflow which dominated the feedback or accretion energy in this state – the same conclusion as reached from radio studies (see above).

Reference [16] only really connected the X-ray spectral properties and long-term (i.e., hours and longer) evolution to the state of the jet. Since then several groups have been trying to see if the moment of ejection may be better determined by examining the timing properties, as measured by X-ray power spectra (see also Chap. 3). Certainly the hard to soft transition at the top of the HID is both the time when the major ejections occur and the region when dramatic changes in the power spectra, including the strongest QPOs, are observed (see [39]). In Fig. 5.8 we plot

Fig. 5.8 Location of observed radio flare events (probably connected with relativistic ejections) in the hardness–r.m.s. diagram (which does not show the same hysteresis as the HID). Dramatic drops in X-ray r.m.s. are associated in many cases with the sharpest changes in X-ray power spectra during the overall hard → soft state transition and have been speculatively linked with the relativistic ejections. In the case of XTE J1550-564 (*top panel*) the ejection appears to occur at the beginning of the drop to the low-r.m.s. "zone," and in XTE J1859+226 (*middle panel*) the sequence of five ejection events all happen close to the "zone." However in the case of GX 339-4 (*bottom panel*), a major radio flare event was observed to occur several days before the r.m.s. drop, casting into doubt any direct causal connection between them. From [17]

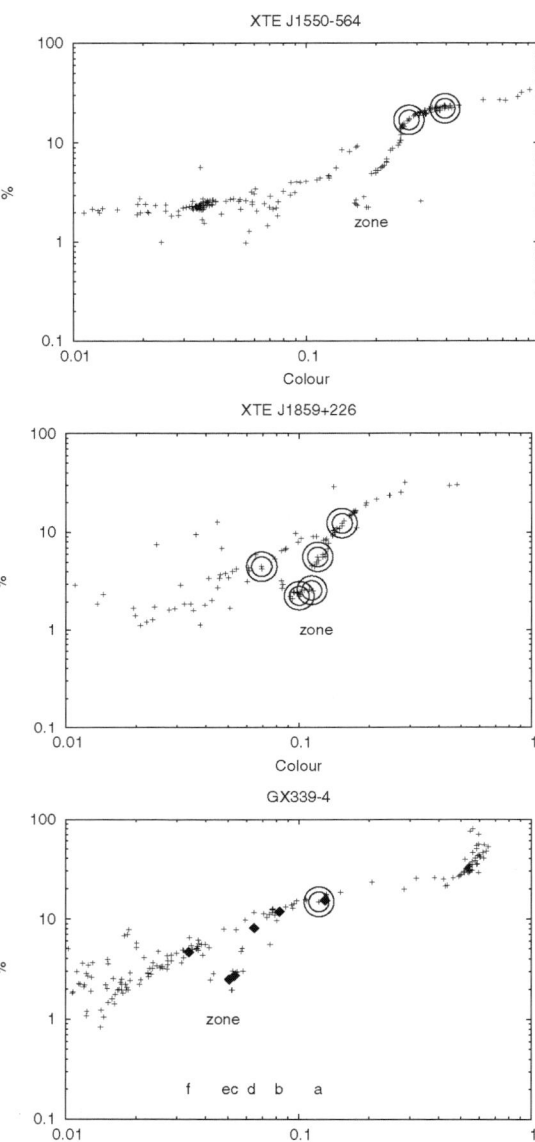

X-ray r.m.s. variability as a function of X-ray color for three sources in outburst. We indicate the moments of bright radio flares (probably associated with relativistic ejection events). Dips in the r.m.s. – color relation, roughly indicated in the figures, have been associated with the sharpest changes in the X-ray timing properties, so they are related directly to the ejections? The result of the comparison is tantalizing: in the case of XTE J1550-564 the bright radio flare occurred between the low-r.m.s. "zone" and an earlier very sharp dip, and in the case of XTE J1859+226 the five radio flares all occurred around the "zone". However, in GX 339-4, the radio flare reported by [23] occurred days before the "zone". Currently therefore the relation of major ejections to source timing properties remains unclear.

5.3 Connections to Active Galactic Nuclei

As the wealth of data on the coupling between accretion and ejection in black hole X-ray binaries grew, more serious attempts were made to scale physical properties up to the AGN. The binary studies had made it clear that there was a strong dependence on accretion rate for a given black hole (not surprisingly), and of course a strong dependence of a variety of properties with black hole mass was also expected (e.g. Fig. 5.2). Therefore some of the first attempts to quantitatively scale properties between black hole binaries and AGN involved black hole mass M, accretion rate $\dot{m}$ (or some proxy for it), and some other property (namely radio luminosity or power spectral break frequency).

5.3.1 Luminosity Scalings

As noted above, Corbel et al. and Gallo et al. reported a non-linear correlation between radio and X-ray luminosities in a number of hard-state black hole binaries. Shortly afterward two groups [67, 12] independently established the existence of a plane linking these binaries with a large population of active galactic nuclei (AGN) hosting supermassive black holes. The plane is based on the relation of the radio luminosity L_{radio}, the X-ray luminosity L_X, and the black hole mass. This should be considered as one of the major steps in the unification of black hole accretion on all mass scales.

In the Merloni et al. formalism, the fundamental plane can be represented as

$$L_{radio} \propto L_X^{0.6} M^{0.8},$$

where the power-law indices are fitted values to a large sample of XRBs and AGN.

The most recent refinements of the plane are presented in [25, 42]. Criticisms of the plane have been rebuked by a consortium of all the original discovery authors, in [68].

Of the three parameters of the fundamental plane, one is genuinely fundamental (M), and one is a good indication of the total radiative output of the system and is therefore pretty fundamental (L_X). However, the third parameter, L_{radio} is merely a tiny tracer of the enormous power carried by the jets from these systems. The fact that it seems to correlate so perfectly with L_X is itself quite amazing and indicates a remarkable stability and regularity in the jet formation process. For example, for a X-ray binary in a bright hard state, the radio emission can be estimated to constitute about 10^{-7} of the total jet power, and yet over long timescales, including phase of jet disruption and reformation, the correlation between radio and X-ray luminosities holds very well (albeit not perfectly – S. Corbel, private communication).

Returning to the plane, it might be a more useful indicator of physical quantities and the flow of matter and power around the black hole if L_{radio} could be replaced with, say, the total jet power L_J or the mass accretion rate $\dot{m}$. In fact we can do both based on the relations established in [43] (hereafter KFM03; see also, e.g., [29, 30]. In Fig. 5.9 we use the relations from this chapter to "calibrate" the plane, by replacing the axes of Merloni et al. with the physical quantities of jet power (left) and accretion rate (right).

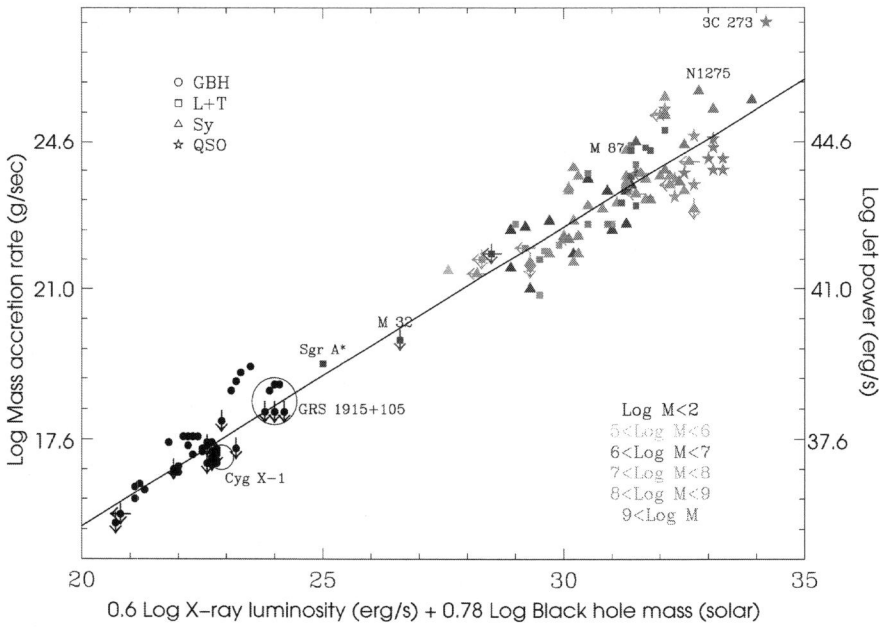

Fig. 5.9 The "fundamental plane" of black hole activity as presented originally in [67], in which their ordinate (y-axis) of radio luminosity has been replaced by the jet power and mass accretion rate (which scale linearly with each other) as estimated by [46]

5.3.2 Variability Scaling

AGN X-ray timing studies started in the 1980s with the launch of EXOSAT, which probed AGN X-ray variability on timescales up to a few days. It was soon established that on these timescales AGN showed red-noise (i.e., "1/f") variability with steep unbroken PSDs [51, 55]. However, McHardy (1988) noted using sparse long-term archival data that the PSDs appeared to flatten or break on longer timescales, similar to the PSD shapes seen in BH XRBs. The launch of RXTE in 1995 allowed long-term light curves to be obtained with extremely good sampling and categorically proved the existence of PSD breaks on timescales close to those expected by scaling the BH XRB break timescales up by the AGN BH mass (e.g., [90, 66, 56]), although with some considerable scatter in the mass–timescale relation.

Recently, [71] established that for a small sample of X-ray binaries there was a positive correlation between radio luminosity and the frequencies of timing features. This was to be expected, since we are confident that on the whole both timing frequencies and radio luminosity are increasing functions of accretion rate (although we also know there are other, state-related, dependencies at high accretion rates).

Reference [57] have now fitted a plane which relates mass and accretion rate to the break frequency in X-ray power spectra, such that

$$T_{\mathrm{break}} \propto M^{-2.1} L_{\mathrm{bol}}^{-1},$$

where T_{break} is the break timescale, reciprocal of the break frequency, M is the black hole mass, and L_{bol} is the bolometric luminosity (Fig. 5.10). All of the sources used

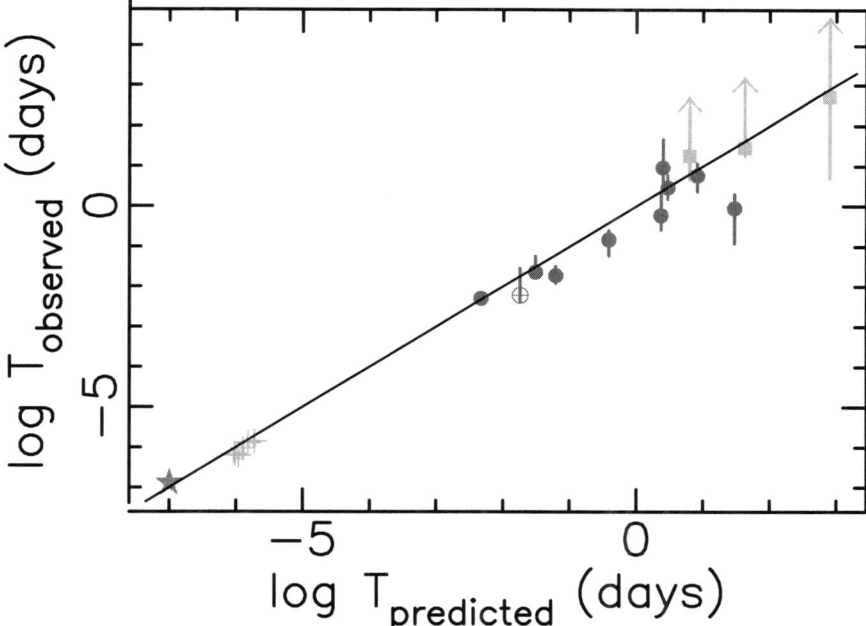

Fig. 5.10 A second fundamental plane, essentially relating characteristic timescales to mass and accretion rate, from [57]. More recently [46] have shown this to extend "hard-state" black hole X-ray binaries and neutron star systems

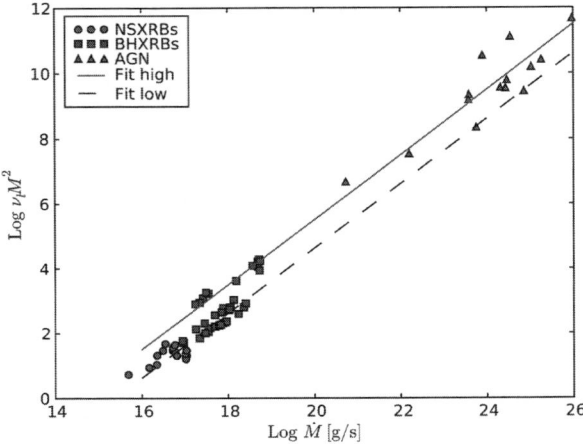

Fig. 5.11 The black hole timing plane extended to lower luminosity black holes and neutron stars. From [46]. The two lines hint at a further parameter, related to X-ray state, which may be required to fully fit all sources

in the correlation are believed to be in radiatively efficient states and so L_{bol} is used as a proxy for accretion rate. Using this substitution converted to accretion rate and integer power-law indices (i.e., 2–2.1), we arrive at

$$T_{break} \propto M/(\dot{m}/\dot{m}_{Edd})$$

revealing the expected linear correlation of break timescales with black hole mass, albeit for a fixed Eddington ratio of accretion rate.

Reference [46] have extended and refined this variability plane to include lower luminosity black hole binaries and neutron star binaries (and even considers white dwarf accretors in cataclysmic variable binaries). Figure 5.11 plots the extended plane; note that the two lines indicate that a further parameter related to X-ray state should be required to fit all sources (as we see in X-ray binaries that timing frequencies can vary at the same X-ray luminosity, if the spectral state is changing).

5.3.3 Further Similarities

As noted earlier, black hole X-ray binaries seem to (nearly) always follow a pattern of behavior in outburst similar to that sketched in [16, 33] (see Fig. 5.6). However, it is clear that between different outbursts of the same source, or outbursts of different sources, the luminosities at which the hard → soft and soft → hard-state transitions may occur can vary quite significantly (e.g. Fig. 5.4; [2, 39]). As a result, an ensemble of X-ray binaries would present a pattern in the hardness:luminosity diagram with a long handle and a filled-in head. Such an ensemble is obviously what we are going to deal with if we want to compare patterns of disc: jet coupling in XRBs with those in AGN.

In [44] we have attempted to do this. First we constructed the disc fraction – luminosity diagram, in which hardness is replaced by the ratio of power law to total luminosity, a number which approaches zero for disc-dominated soft states and unity for hard states. This is necessary for a physical comparison, since the accretion discs temperature is a decreasing function of black hole mass, and for AGN does not contribute significantly in the X-ray band. We then simulated an ensemble of BHXRBs based on [16] and the slight refinement (suggested in [2]) that the "jet line" might be diagonal in such a diagram. This was then compared to a sample of AGN from the SDSS DR5 for which there were X-ray detections and either radio detections or limits, plus a sample of low-luminosity AGN (LLAGN). The similarity was striking (see Fig. 5.12) and suggests that the radio loudness is determined by the combination of "state" and luminosity in a similar way for accreting black holes of all masses. Note that, while it is tempting to consider, the diagram does not indicate that AGN necessarily follow the same anti-clockwise loop in the diagram as XRBs: the motion

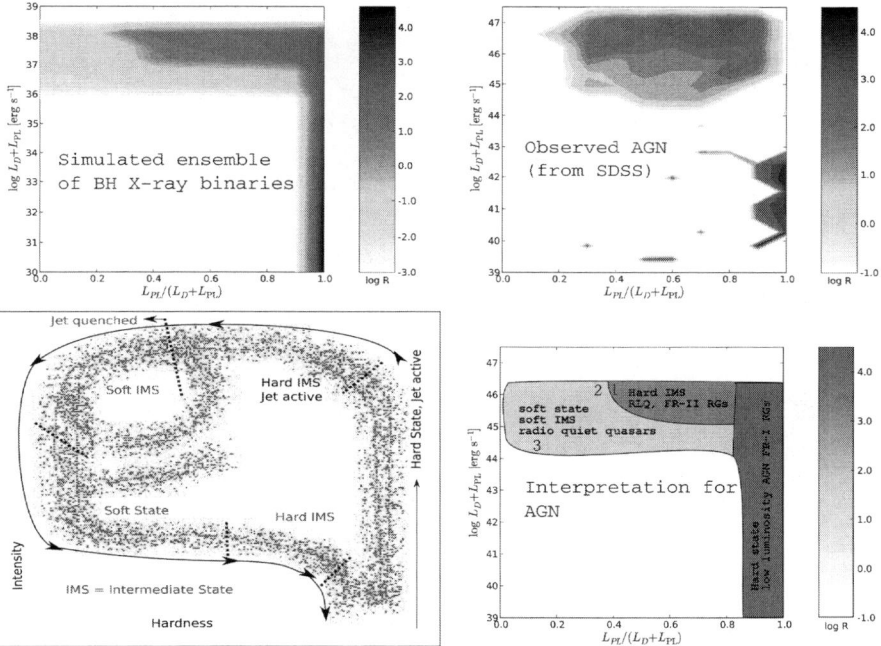

Fig. 5.12 A comparison of the disc: jet coupling in X-ray binaries with that in AGN based on the disc fraction–luminosity diagram (DFLD; see [44]). Based on the disc:jet model for black hole X-ray binaries presented in [16], an example track for which is shown in the *lower left panel*, a simulated ensemble of X-ray binaries was produced (*upper left panel*). This was compared with a sample of SDSS quasars and low-luminosity AGN (*upper right panel*) and a striking similarity was revealed. The lower right panel offers an explanation for the similarities between the different classes of object

in such loops could possibly be dominated by disc instabilities which may not apply
to AGN (although in the Fender, Belloni and Gallo interpretation, major radio flares
would require right → left transitions in order to produce the internal shocks). What
it does indicate is that when an AGN finds itself in a particular accretion "state",
whether disc or corona dominated or some mix of the two, the jet it produces will
be comparable to that which a XRB would make in the same state. [36] further
discuss possible similarities and differences between feeding cycles in AGN and
X-ray binaries.

It is worth noting that although the DFLD was chosen to both provide easy and
understandable comparison with the HID and to provide a clear method of physical
comparison with accretion states in AGN, it does show some differences with the
HID even for X-ray binaries.

First, the work of [9] who have produced the first real DFLDs for an X-ray binary
(GX 339-4) shows that the hysteretical zone of the HID becomes rather less square
in shape when transferred to the DFLD. This seems to be because as the disc cools
on the soft branch, the disc fraction starts dropping (not as obvious as it sounds – it
depends upon how fast the hard component is also dropping). Second, the work of
[53, 54] has demonstrated that by the time sources are in quiescence, the radiative
luminosity of the disc is one to two orders of magnitude greater than the X-ray
luminosity – i.e., at the lowest accretion rates discs are once more dominant in
terms of the radiation. Of course at these low accretion rates we still estimate that
the kinetic power of the jet dominates over both of these terms. Figure 5.13 presents

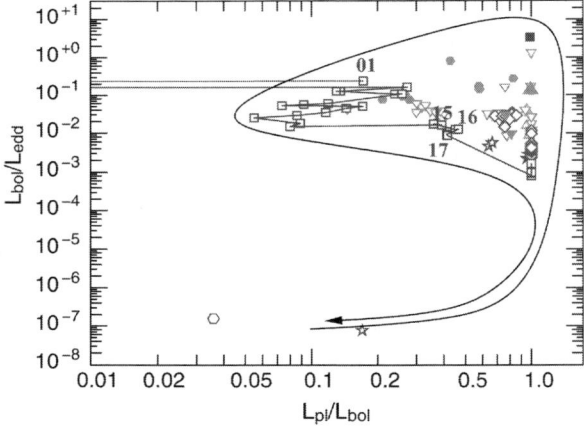

Fig. 5.13 The disc fraction–luminosity diagram (DFLD) for black hole X-ray binary systems
where the disc component is well measured in the X-ray band (Swift or SAX) or optical bands. The
connected *brown* points correspond to the Swift observations of XTE J1817-330. The *black line*
with arrowhead indicates a possible path of a transient black hole from quiescence to outburst and
back again. It is well established, but not widely appreciated, that as well as in soft X-ray states, in
quiescence systems are also completely dominated in their radiative output by the accretion disc
(see [54]). From [6], all X-ray data have been independently reanalyzed, while optical fluxes are
from the literature

the DFLD for a set of black hole binaries for which the disc component strength has been (relatively) well measured either in the soft X-ray, ultraviolet, or optical bands (from [6]). The (designed) similarity with the HID at high luminosities is apparent, but interestingly the path of a source loops back to the disc-dominated state in quiescence. It is not clear that we would expect to see this lower disc-dominated branch in AGN since it rather reflects the conditions of the outer accretion flow.

Further similarities in patterns of behavior have been noted in the past. In Fig. 5.14 patterns of X-ray temporal and spectral behavior are related to directly imaged relativistic ejection events in the AGN 3C 120 ([64]; see also Chap. 7).

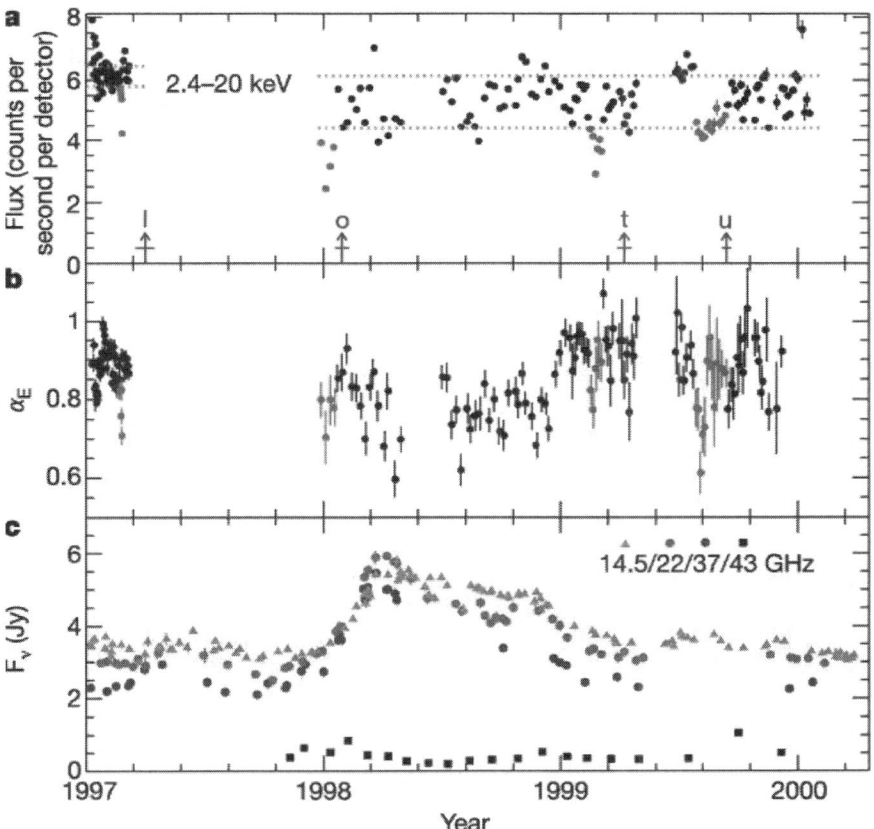

Fig. 5.14 Possible disc–jet coupling in the AGN 3C 120, from [64]. *Top panel:* X-ray light curve in photon counts per second per detector; for many data points, the error bars are smaller than the symbols. Times of ejections of superluminal radio knots, as resolved with the VLBA, are indicated by upward arrows, with the horizontal crossbar giving the uncertainty. *Middle panel* X-ray hardness; lower value is harder. *Lower panel* Radio light curves at 14.5 GHz (*triangles*); 22 GHz (*lighter circles*) and 37 GHz (*darker circles*); and 43 GHz (core only, *squares*)

5.3.4 Differences

Despite the clear similarities, there are clear differences, both expected and observed, between X-ray binaries and AGN. Some of these result from the fact that these systems are more than just the central black hole. In particular the *environment* may play a role in

1. supplying matter with varying distributions of angular momentum, possibly even changing direction of rotation, depending on the merger history of the host galaxy and the path taken by the gas in reaching the region of the black hole (e.g., [11] and references therein). This clearly does not occur in Roche-lobe overflow binary systems, in which mass and angular momentum transfer takes place at a more or less constant rate;
2. obscuration and inclination play major roles in the appearance of AGN (e.g., [89]; however, these authors do not really consider the relation of appearance to accretion rate/state). There may be some parallels with this in the case of some X-ray binaries, but in the majority it is not the case;
3. jets from AGN in dense environments will probably dissipate their energy on smaller physical scales as they inflate the medium around them. There may be some parallel here between, e.g., GHz-peaked AGN (e.g., [79]) and the interaction of powerful jets and strong stellar winds in the systems like Cygnus X-3 and SS 433.

In addition to environmental effects, there is (at least) one key region in the geometry of AGN which appears simply not to be present in X-ray binary systems: the broad line region (BLR). The broad emission lines in AGN have been shown by reverberation mapping techniques to arise within typically a few days of the central black hole ([81] and references therein). The precise velocity field of the BLR is however uncertain. Reference [10] presents a clear case for a BLR that originates in a relatively thin conical outflow which has its origin in the accretion disc, but whose major component of motion is away from the disc. However, [80] argue that the variable components to BLR emission lines show the expected virial relation between velocity V and radius r, $V \propto r^{-1/2}$, implying that the motions of the line-emitting regions are dominated by the gravity of the central black hole. This facilitates the use of reverberation mapping techniques to estimate central black hole masses ([81] and references therein, but see [50] for a critique of the method). Other works, e.g. [92], also conclude that the motion in the BLR is primarily circular/orbital. Reference [75–77] have studied in detail the appearance of emission lines from combined accretion disc plus wind systems, with application to both binary systems (cataclysmic variables) and AGN and have shown that single-peaked lines can arise from rotating flows. Why then is there no clear evidence for an outflowing BLR in black hole X-ray binaries, when there are so many other apparently scale-free similarities in black hole accretion and jet formation?

First, [82] have already demonstrated on the basis of detailed calculations a strong line-driven wind is unlikely in the case of accretion discs around X-ray binaries. The problem is essentially that the central X-ray source is too strong a source of ionizing radiation, which strongly inhibits line-driving. In AGN, in contrast, the cooler central disc is a UV source which is suitable for driving such winds. What if there are significant winds driven off by other means such as large-scale magnetic fields (e.g., [4, 73]) or thermal expansion [6]? In this case we can ignore the line-driving requirement. Nevertheless, the strongly ionizing X-ray source still prevents line emission from the inner, fast moving, regions of this wind. For example, the BLR for a 10^8 $M_\odot$ black hole in an AGN, accreting at close to the Eddington limit is typically estimated to be at a distance of several tens of light days (10^{16} cm $\leq R_{BLR} \leq 10^{17}$ cm) or 300–3000 gravitational radii [81]. The line-emitting region of the disc is fairly close to this [96]. On the other hand, X-ray to optical delays in BHXBs accreting at comparable Eddington ratios (e.g., [35]) indicate lags of order seconds, or $>10^4$ gravitational radii. The difference in the radii of the line-emitting regions of the accretion discs is due, once again, to the much hotter environments around stellar-mass black holes in BHXBs ($T \propto M^{-1/4}$ at the same Eddington ratio and for a radiatively efficient flow). Therefore it seems likely, albeit in this case without detailed calculations, that even if a strong wind could be launched without line-driving, if it were launched from close to the black hole then the gas would be too highly ionized to produce BLR-like lines.

5.3.5 What About Spin?

There has been much speculation over the past few decades about the possible role of black hole spin in jet formation (e.g., [94, 48]) and any relation it might have to the apparent radio loud:radio quiet dichotomy in AGN (listed as one of the ten major questions for the field by [89]). There have, more recently, been strong claims for clear measurements of black hole spin in AGN (e.g., [93]) and X-ray binaries (e.g., [72]). It is this author's view that to date there is no clear evidence either way for the role of spin-powering of jets in black hole X-ray binaries (contrary to the view put forward in, e.g., [95]). Some points to consider, however, are the following:

- In black hole binaries both "radio loud" and "radio quiet" states are observed from the same source, repeatedly, and are related to the accretion state, not spin changes.
- In AGN there may be a larger range of black hole spins relating to the merger history, whereas all black holes in binary systems have (presumably) formed from the collapse of a massive star and will have some intermediate value of spin.

5.4 Using, Testing, and Exploring

Now that we have established the first good quantitative scalings, as well as a new set of qualitative similarities, between black holes on all mass scales, what do we use them for? The goal expressed by many for several years was to use the insights we gain from the disc-jet coupling in stellar-mass black holes to understand better the process of accretion and feedback in AGN. The importance of this connection has been greatly strengthened in recent years by the discovery that feedback from black holes is likely to be directly linked to both the formation of galaxy bulges [27, 18], and to the heating of the inner regions of cooling flows (e.g., [3, 91]).

An obvious and important place to start is therefore with the kinetic feedback of AGN, or kinetic luminosity function. Reference [69] and, more directly, [45] have estimated the kinetic luminosity function of AGN based on both power and spectral state scalings from black hole binaries. Both groups conclude that kinetic feedback into the IGM is actually likely to be dominated by supermassive black holes accreting at relatively low Eddington ratios. Future studies need a better regulated sample with mass measurements and more precise measurements of core coronal, disc, and jet components.

Another example relates to the argument originally noted by [84] (see also, e.g., [22] and references therein) that the local mass density of black holes is consistent with the growth of black holes via *radiatively efficient* accretion (current limits place a factor of a few on this consistency). However, as is clearly the case for binary black holes, radiatively efficient accretion above $\sim 1\%$ Eddington can occur with powerful jets (hard and hard intermediate states) or without a jet (soft state). Taking extremes, if most of the X-ray background results from accretion in hard or hard intermediate states, then $\sim 10^{67}\,\mathrm{erg\,Gpc^{-3}}$ may have been injected into the ambient medium in the form of kinetic energy over the lifetime of black hole growth. If most of the X-ray background results from accretion in jet-free soft X-ray states then the figure will be much smaller, probably by a factor of a 100 or more.

5.4.1 Somewhere in the Middle: Using Radio Emission to Look for Intermediate Mass Black Holes

The publication of the fundamental planes of black hole activity by [67] and [12] coincided with a period of greatly renewed interest in "ultraluminous" X-ray sources (ULXs) and the suggestion that some of these objects could be "intermediate mass" black holes ($10^2 M_\odot \le M \le 10^4 M_\odot$). These ULXs were not located at the dynamical centers of their host galaxies and could sometimes produce X-ray luminosities in excess of $10^{41}\,\mathrm{erg\,s^{-1}}$, or the Eddington limit for a $\ge 70 M_\odot$ black hole, if the X-ray emission was isotropic. Alternatives based on anisotropy of the emission, whether intrinsic or associated with relativistic aberration, have been put forward (e.g., [38, 41]).

The fundamental planes immediately implied two things which were potentially good for this field:

1. If you could measure the X-ray and radio luminosities of a ULX, you could infer its mass.
2. Radio observations were the best way to find intermediate mass black holes accreting at low rates from the ambient medium.

Point [1] has had mixed success. Several ULXs do appear to have radio counterparts, and simply plugging these radio luminosities into the fundamental plane implies large black hole masses, $>> 100 M_\odot$. However, in most cases this radio emission is in fact resolved and looks like a large-scale nebula (e.g., [78, 85]), and no very compact and variable radio counterparts have yet been found.

Point [2] has been explored in some detail by [59] and [61] (see also [60]). Put simply, black holes accreting at low rates should be easier to find in the radio band. This situation is enhanced the greater the mass of the black hole, as the radiative efficiency seems to fall off with Eddington ratioed accretion rate. Recent confirmation of this approach may have come in the form of the detection of a radio source at the center of the globular cluster G1 by [88], which is consistent with a black hole of mass $\geq 100 M_\odot$.

5.4.2 Comparison with Neutron Stars and White Dwarfs

A control sample exists which allows us to test whether properties unique to black holes, such as the presence of an event horizon, are in anyway essential for any of the observed phenomena. Neutron stars are collapsed stellar remnants with masses $1 M_\odot \leq M_{NS} \leq 2 M_\odot$ and sizes only a factor of 2–3 larger than their Schwarzschild radii (see Fig. 5.2). Squeeze them to make them half as large as they are and they would collapse to form black holes. The gravitational potential energy released per unit mass of matter accreted onto a neutron star is therefore very similar to that of a black hole. In addition, we find them in X-ray binary systems with very similar patterns of accretion, including outbursts and extended periods of quiescence, to the black hole binaries. A full review of the properties of jets from neutron star binaries is beyond the scope of this chapter, but the key points from the comparison are as follows:

1. Neutron stars seem to produce both steady and transient jets, just like black holes, with a similar relation to hard states and outbursts (although poorly sampled to date).
2. The radio to X-ray ratio is in general lower for neutron star binaries.
3. Neutron star jets are less "quenched" in soft states than black hole jets.
4. Neutron star jets may be just as, or even more, relativistic (in terms of bulk velocity) than black hole binary jets.
5. The radio: X-ray correlation, although much less well measured, appears to be steeper than that for black holes by a factor of 2 or more, i.e., $L_{\text{radio, NS}} \propto L_{\text{X, NS}}^{\geq 1.4}$ compared to $L_{\text{radio, BH}} \propto L_{\text{X, BH}}^{\sim 0.7}$.
6. Two neutron star jet systems (Sco X-1, Cir X-1; possibly also the 'odd' system SS 433) appear to show unseen but highly relativistic flows which energize slower-moving bulk flows further out.

Reference [70] present the most comprehensive review to date of the properties of jets from neutron star binaries (see also [43] and Chap. 4 in this volume and references therein); Fig 5.16 presents a comparison of BH and NS binaries in the radio vs. X-ray plane.

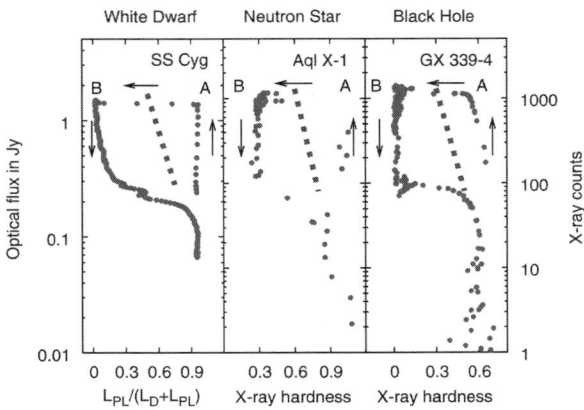

Fig. 5.15 HID for a black hole, a neutron star, and the disc-accreting CV SS Cyg. The *arrows* indicate the temporal evolution of an outburst. The *dotted lines* indicate the jet line observed in black hole and neutron star XRBs: On its right side, one generally observes a compact jet; the crossing of this line usually coincides with a radio flare. For SS Cyg, we show a disc fraction–luminosity diagram. We plotted optical flux against the power-law fraction measuring the prominence of the "power-law component" in the hard X-ray emission in relation to the boundary layer/accretion disk luminosity. This power-law fraction has similar properties to the X-ray hardness used for XRBs. Radio flaring has been observed during the high-luminosity state transition in SS Cyg just like in black hole systems

Several of the points listed above strongly imply that the basics of relativistic jet formation do not require any unique property of black holes, such as an event horizon. However, as discussed earlier (and illustrated in Fig. 5.2), in terms of gravitational potential, neutron stars are almost black holes, and so perhaps the similarities are not so surprising. However, there is another major class of accreting binary systems, which contain accreting objects which are very different: cataclysmic variables (CVs). In these systems a white dwarf, with a M/R ratio about a thousand times smaller than for a black hole or neutron star, accretes matter from a companion (white dwarf accretion is less efficient than nuclear fusion).

However, [47] have recently shown that there may even be direct similarities between accretion state changes at high luminosity in disc-accreting CVs and black holes and neutron stars. Figure 5.15 presents HIDs or equivalent for the black hole binary GX 339-4, the neutron star binary Aql X-1, and the disc-accreting CV SS Cyg. The similarity is striking, but what is key is that observations with the VLA have revealed radio flaring at around the point of the high-luminosity state transition

Fig. 5.16 A comparison of black hole and neutron star X-ray binaries in terms of radio vs X-ray emission. The top panel shows the overall picture, in which it is clear that black hole sources (e.g. GX 339-4) are more radio-loud than neutron stars at a given X-ray luminosity, although otherwise the overall pattern looks similar. The lower panel zooms in to examples of both BH and NS systems in 'softer' states - in both cases the radio emission seems to be 'suppressed' by this effect seems to be larger for the BH. From Migliari and Fender [68]

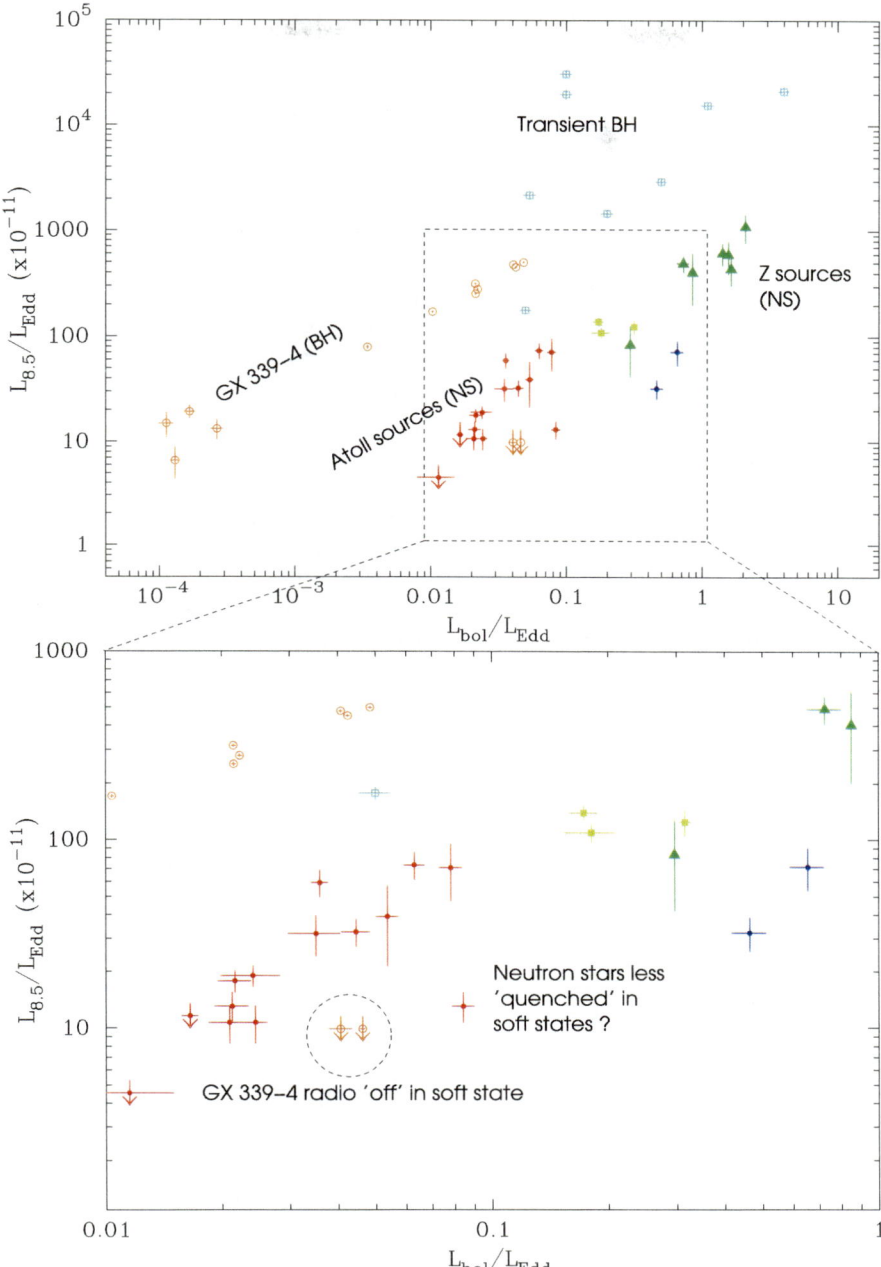

Fig. 5.16 (continued)

in SS Cyg, just as seen in GX 339-4 and other black hole binaries. This strongly suggests the transient formation of a jet during this phase, further suggesting that patterns of disc: jet coupling relate solely to the behavior of the disc and not the central accretor (which may, however, affect, e.g., jet velocity).

To conclude, there are clear similarities, both quantitative and qualitative, between accretion (luminosity, state) and jets (formation, steady, or transient) in black holes of all masses. These links provide us with tools to understand feedback from AGN and its role in the formation of galaxies and evolution of clusters, and with insight into the process of relativistic jet formation. However, we should be cautious about drawing wide-ranging conclusions about the effect of black hole properties, e.g., spin, on jet formation, when we find comparable patterns of behavior in related (neutron star) and also very different (white dwarf) classes of object.

5.4.3 Further Reading

This review has necessarily been both brief and biased, and the literature associated with this area of research is vast and ever-expanding. In particular, I have approached the area of disc-jet coupling from the realm of X-ray binaries, looking for scalings and comparisons with AGN, and have not even scratched the surface of the wide phenomenology associated with accreting supermassive black holes.

For a broader introduction to AGN, I would recommend [49] and references therein. For a more AGN-centric view of the disc–jet coupling around accreting black holes, see papers by Marscher, e.g., [65]. For an up-to-date reference list of all things related to black holes, see [21].

Acknowledgments I owe much of my understanding of this field of research to collaborations and conversations with a large number of people, most notably those who are coauthors with me on a wide range of papers. They know who they are, and if *you* do not, use ADS. In addition, I would like to thank the participants at the ISSI workshop in January 2008 for helping to further clarify ideas, and Christian Knigge, Mike Goad, and Daniel Proga for discussions about broad line regions and line-driven winds.

References

1. M.C. Begelman, C.F. McKee, G.A. Shields: ApJ, **271**, 70 (1983)
2. T. Belloni, I. Parolin, M. Del Santo et al.: MNRAS, **367**, 1113 (2006)
3. P.N. Best, G. Kauffmann, T.M. Heckman et al.: MNRAS, **362**, 25 (2005)
4. R.D. Blandford, D.G. Payne: MNRAS, **199**, 883 (1982)
5. C. Brocksopp, R.P. Fender, M. McCollough et al.: MNRAS, **331**, 765 (2002)
6. C. Cabanac, R.P. Fender, R.H.J. Dunn, E.G. Koerding: MNRAS, **396**, 1415 (2009)
7. S. Corbel, R.P. Fender, A.K. Tzioumis et al.: A&A, **359**, 251 (2000)
8. S. Corbel, M. Nowak, R.P. Fender et al.: A&A, **400**, 1007 (2003)
9. R.J.H. Dunn, R.P. Fender, E.G. Körding et al.: MNRAS, **387**, 545 (2008)
10. M. Elvis: ApJ, **545**, 63 (2000)
11. E. Emsellem, M. Cappellari, D. Krajnović et al.: MNRAS, **379**, 401 (2007)

12. H. Falcke, E. Körding, S. Markoff: A&A, **414**, 895 (2004)
13. R.P. Fender, S. Corbel, T. Tzioumis et al.: ApJ, **519**, L165 (1999)
14. R.P. Fender: MNRAS, **322**, 31 (2001)
15. R.P. Fender, E. Gallo, P.G. Jonker: MNRAS, **343**, L99 (2003)
16. R.P. Fender, T.M. Belloni, E. Gallo: MNRAS, **355**, 1105 (2004)
17. R.P. Fender, J. Homan, T.M. Belloni: MNRAS, **396**, 1370 (2009)
18. L. Ferrarese, D. Merritt: ApJ, **539**, L9 (2000)
19. J. Frank, A. King, D.J. Raine: Accretion Power in Astrophysics, third edition, Cambridge University Press, Cambridge (2002)
20. E. Gallo: AIPC, **924**, 715 (2007)
21. E. Gallo, D. Marolf: Am. J. Phys., Resource Letter (arXiv:0806.2316) (2008)
22. E. Gallo, R.P. Fender, G.G. Pooley: MNRAS, **344**, 60 (2003)
23. E. Gallo, S. Corbel, R.P. Fender et al.: MNRAS, **347**, L52 (2004)
24. E. Gallo, R.P. Fender, C. Kaiser et al.: Nature, **436**, 819 (2005)
25. E. Gallo, R.P. Fender, J.C.A. Miller-Jones et al.: MNRAS, **370**, 1351 (2006)
26. M. Garcia, J.E. McClintock, R. Narayan et al.: ApJ, **553**, L47 (2001)
27. K. Gebhardt, R. Bender, G. Bower et al.: ApJ, **539**, L13 (2000)
28. D.C. Hannikainen, R.W. Hunstead, D. Campbell-Wilson et al.: A&A, **337**, 460 (1998)
29. S. Heinz, H.-J. Grimm: ApJ, **633**, 384 (2005)
30. S. Heinz, A. Merloni, J. Schwab: ApJ, **658**, L9 (2007)
31. R.M. Hjellming, X. Han: Radio properties of X-ray binaries. In: *X*-ray Binaries, Cambridge University Press, Cambridge (1995)
32. R.M. Hjellming, M.P. Rupen: Nature, **375**, 464 (1995)
33. J. Homan, T. Belloni: Ap&SS, **300**, 107 (2005)
34. R.I. Hynes, C.A. Haswell, W. Cui et al.: MNRAS, **345**, 292 (2003)
35. R.I. Hynes, E.L. Robinson, K.J. Pearson et al.: ApJ, **651**, 401 (2006)
36. C.R. Kaiser, P.N. Best: MNRAS, **381**, 1548 (2007)
37. G. Kanbach, C. Straubmeier, H. Spruit et al.: Nature, **414**, 180 (2001)
38. A.R. King, M.B. Davies, M.J. Ward, G. Fabbiano, M. Elvis: ApJ, **552**, L109 (2001)
39. M. Klein-Wolt, M. van der Klis: ApJ, **675**, 1407 (2008)
40. M. Klein-Wolt, R.P. Fender, G.G. Pooley et al.: MNRAS, **331**, 745 (2002)
41. E. Koerding, H. Falcke, S. Markoff: A&A, **382**, L13 (2002)
42. E. Koerding, H. Falcke, S. Corbel: A&A, **456**, 439 (2006)
43. E.G. Koerding, R.P. Fender, S. Migliari: MNRAS, **369**, 1451 (2006)
44. E. Koerding, S. Jester, R.P. Fender: MNRAS, **372**, 1366 (2006)
45. E. Koerding, A. Jester, R. Fender: MNRAS, **383**, 277 (2008)
46. E.G. Koerding, S. Migliari, R. Fender et al.: MNRAS, **380**, 301 (2007)
47. E. Koerding, M. Rupen, C. Knigge et al.: Science, **320**, 1318 (2008)
48. S. Koide, K. Shibata, T. Kudoh et al.: Science, **295**, 1688 (2002)
49. J.H. Krolik: *Active Galactic Nuclei: From the Central Black Hole to the Galactic Environment*, Princeton Series in Astrophysics, Princeton University Press, Princeton (1999)
50. J.H. Krolik: ApJ, **551**, 72 (2001)
51. A. Lawrence, M.G. Watson, K.A. Pounds et al.: Nature, **325**, 694 (1987)
52. J.E. McClintock, R.A. Remillard: Black hole binaries. In: W. Lewin, M. van der Klis (eds.) Compact Stellar X-ray Sources, pp. 157–213. Cambridge Astrophysics Series No. 39, Cambridge, (2006)
53. J.E. McClintock, K. Horne, R.A. Remillard: ApJ, **442**, 358 (1995)
54. J.E. McClintock, R. Narayan, M.R. Garcia et al.: ApJ, **593**, 435 (2003)
55. I. McHardy, B. Czerny: Nature, **325**, 696 (1987)
56. I.M. McHardy, I.E. Papadakis, P. Uttley et al.: MNRAS, **348**, 783 (2004)
57. I. McHardy, E. Körding, C. Knigge et al.: Nature, **444**, 730 (2006)
58. I. McHardy: MemSAIt, **59**, 239 (1988)
59. T.J. Maccarone: MNRAS, **360**, L30 (2005)

60. T.J. Maccarone, M. Servillat: MNRAS, **389**, 379 (2009)
61. T.J. Maccarone, R.P. Fender, A.K. Tzioumis: MNRAS, **356**, L17 (2005)
62. M. Machida, K.E. Nakamura, R. Matsumoto: PASJ, **58**, 193 (2006)
63. J. Malzac, A. Merloni, A.C. Fabian: MNRAS, **351**, 253 (2004)
64. A.P. Marscher, S. Jorstad, J.-L. Gomez et al.: Nature, **417**, 625 (2002)
65. A. Marscher: Astron. Nachr. **327**, 217 (2006)
66. A. Markowitz, R. Edelson, S. Vaughan et al.: ApJ, **593**, 96 (2003)
67. A. Merloni, S. Heinz, T. di Matteo: MNRAS, **345**, 1057 (2003)
68. A. Merloni, E. Körding, S. Heinz et al.: New Astronomy, **11**, 567 (2006)
69. A. Merloni, S. Heinz: MNRAS, **381**, 589 (2007)
70. S. Migliari, R. Fender: MNRAS, **366**, 79 (2006)
71. S. Migliari, R. Fender, M. van der Klis: MNRAS, **363**, 112 (2005)
72. J.M. Miller, A.C. Fabian, R. Wijnands et al.: ApJ, **570**, L69 (2002)
73. J.M. Miller, J. Raymond, A. Fabian et al.: Nature, **441**, 953 (2006)
74. I.F. Mirabel, L.F. Rodriguez: Nature, **371**, 46 (1994)
75. N. Murray, J. Chiang: Nature, **382**, 789 (1996)
76. N. Murray, J. Chiang: ApJ, **474**, 91 (1997)
77. N. Murray, J. Chiang: ApJ, **494**, 125 (1998)
78. R. Mushotsky: Progress of Theoretical Physics Supplement, No. 155, pp. 27–44 (2004)
79. C.P. O'Dea: PASP, **110**, 493 (1988)
80. B.M. Peterson, A. Wandel: ApJ, **521**, L95 (1999)
81. B.M. Peterson, L. Ferrarese, K.M. Gilbert et al.: ApJ, **613**, 682 (2004)
82. D. Proga, T.R. Kallman: ApJ, **465**, 455 (2002)
83. N.I. Shakura, R.A. Sunyaev: A&A, **24**, 337 (1973)
84. A. Soltan: MNRAS, **200**, 115 (1982)
85. R. Soria, R.P. Fender, D.C. Hannikainen et al.: MNRAS, **368**, 1527 (2006)
86. H. Tanabaum, H. Gursky, E. Kellogg et al.: ApJ, **177**, L5 (1972)
87. G. 't Hooft: Int. J. Mod. Phys. D15 1587-1602 (arXiv:gr-qc/0606026) (2006)
88. J.S. Ulvestad, J.E. Greene, L.C. Ho: ApJ, **661**, L151 (2007)
89. C.M. Urry, P. Padovani: PASP, **107**, 803 (1995)
90. P. Uttley, I.M. McHardy, I.E. Papadakis: MNRAS, **332**, 231 (2002)
91. J.C. Vernaleo, C.S. Reynolds: ApJ, **645**, 83 (2006)
92. I. Wanders, M.R. Goad, K.T. Korista et al.: ApJ, **453**, L87 (1995)
93. J. Wilms, C.S. Reynolds, M.C. Begelman et al.: MNRAS, **328**, L27 (2001)
94. A.S. Wilson, E.J.M. Colbert: ApJ, **438**, 62 (1995)
95. S.N. Zhang, W. Cui, W. Chen: ApJ, **482**, L155 (1997)
96. X.-G. Zhang, D. Dultzin-Hacyan, T.-G. Wang: MNRAS, **374**, 691 (2007)

Chapter 6
From Multiwavelength to Mass Scaling: Accretion and Ejection in Microquasars and AGN

S. Markoff

Abstract A solid theoretical understanding of how inflowing, accreting plasma around black holes and other compact objects gives rise to outflowing winds and jets is still lacking, despite decades of observations. The fact that similar processes and morphologies are observed in both X-ray binaries and active galactic nuclei has led to suggestions that the underlying physics could scale with black hole mass, which could provide a new handle on the problem. In the last decade, simultaneous broadband campaigns of the fast-varying X-ray binaries particularly in their microquasar state have driven the development of, and in some cases altered, our ideas about the inflow/outflow connection in accreting black holes. Specifically, the discovery of correlations between the radio, infrared, and X-ray bands has revealed a remarkable connectivity between the various emission regions and argued for a more holistic approach to tackling questions about accretion. This chapter reviews the recent major observational and theoretical advances that focus specifically on the relation between the two "sides" of the accretion process in black holes, with an emphasis on how new tools can be derived for comparisons across the mass scale.

6.1 Introduction

The process of disentangling inflow from outflow in accreting black holes is something of an exercise in semantics as we approach the event horizon. What we have traditionally thought of as "inflow" solutions, such as the various flavors of radiatively inefficient accretion flows (RIAFs; e.g., [81]), are now thought to become unbound, resulting in windy "outflows" from their surfaces, without even taking jet formation into consideration (e.g., [8]). Thus the problem is far from simple, and the jets which are somehow eventually launched could tap some combination of bound flow, unbound flow, and black hole spin energy for plasma and power. Until we understand which class of solution dominates, correctly interpreting the

S. Markoff (✉)

Astronomical Institute "Anton Pannekoek", University of Amsterdam, Science Park 904, 1098 XH Amsterdam, The Netherlands, s.b.markoff@uva.nl

Markoff, S.: *From Multiwavelength to Mass Scaling: Accretion and Ejection in Microquasars and AGN*. Lect. Notes Phys. **794**, 143–172 (2010)

DOI 10.1007/978-3-540-76937-8_6

emission from the innermost regions of the accretion flow(s) will remain a challenge. Yet this region is precisely where the most extreme physics is lurking, and thus what interests many scientists in the field of astrophysical accretion rather acutely.

From the outermost reaches of the jets to the inner boundary of the accretion flow, the ensuing radiation spans the entire broadband spectrum, including the lowest frequency radio waves through TeV γ-rays. Such a range, of almost 20 decades in frequency, can provide an enormous lever arm for theoretical modeling if the accreting system is viewed holistically. As I will discuss further, the lever arm supplied by multiwavelength data is likely our best hope for actually breaking the theoretical degeneracy currently hindering our progress. However, because we are not yet at the point of achieving fully self-consistent a priori physical solutions (though that is certainly on the horizon, see section in Chap. 9), we must first make a set of assumptions relating the two "sides" of accretion before we can build models to test those assumptions against the data. Eventually, such approaches can provide the groundwork and boundary conditions for more exact numerical methods.

This chapter will in some ways take over where the last left off, with an emphasis on the recent development and evolution of our theoretical ideas specifically in the context of the increasing prevalence of multiwavelength campaigns. I will focus on documenting the process by which our ideas have undergone shifts, based on this new approach to observations, and the far-reaching consequences that black hole binaries (BHBs) may now have for deconstructing galaxy evolution on cosmological timescales, assuming mass scaling is really all it is cut out to be. My obvious bias is toward understanding jets in these systems, but I try to touch on all scenarios and give references for further reading whenever possible.

6.2 Changing Paradigms

Up until surprisingly recently, the standard view of BHBs did not include jets, even though one of the most quoted papers in the field of accretion shows a nice figure of an outflow! – I am referring to Fig. 9 in [92], which shows an X-ray emitting "cone". For example, in the now canonical textbook *X-ray Binaries* released well after radio jets were a commonly known feature of BHBs [56], only one article discusses them at all, and then only as a separate feature not integrated into the basic picture. Historically speaking, this oversight is not so surprising. When X-ray binaries (XRBs) were first discovered via sounding rockets in the late 1960s, accretion was the obvious culprit and the idea of viscously dissipating flows quickly took over as the predominant paradigm. The discovery of variable, hard X-ray emission necessitated the invention of a region called the corona (see, e.g., [25, 93]), where hot electrons reside and upscatter thermal disk photons into a hard power-law tail. The concept was eventually developed into a fairly self-consistent picture by [49]. With many varieties now spanning the parameter space of configurations including reflection (e.g., [95, 99, 100, 59, 83, 18]), this idea of a region of hot plasma in stasis above or within the cooler accretion flow, radiating relatively isotropically, has been the canonical viewpoint for decades and remains so for many researchers today.

The problem, as I see it anyway, is that these models do not take into account the role of strong magnetic fields in the inner regions of the accretion flow. At the time of their development their omission was understandable, because the study of magnetic phenomena in astrophysics was significantly less developed than it is today. However, in the intervening years we have strong evidence that magnetic fields can be generated in accretion flows via the magneto-rotational instability (MRI), as well as brought in from the environment via the accreting gas (see more about magneto-hydrodynamics, or MHD, in Chap. 9). The inclusion of magnetic fields will not only change many of the underlying assumptions in the classical thermal Comptonization picture, but is also necessary in order to begin to address the relationship between the inflow and the obviously magnetic phenomena of jets.

Interestingly, the necessity of strong magnetic fields in various BHBs was realized decades ago by [27], who proposed that the fast optical flaring observed in the BHB GX 339-4 was the result of cyclotron emission, as mentioned also in Chap. 9. Building on this approach, more recent works have considered these so-called magnetic coronae [23, 102, 75, 84], all of which involve similar distributions of hot electrons as the static corona models, but which in addition also include magnetic effects. It is hard to imagine a world in which such a plasma has no relationship to the hot, magnetized plasma required at larger scales to power the nonthermal synchrotron emission observed in BHBs in the hard and intermediate states. One approach suggested by [6] and later by [61] is that radiation pressure drives the magnetic flares away from the disk at mildly relativistic speeds, a scenario that begins to sound rather like the base of a jet. However, this idea was motivated in part by the need to reduce the problematically high reflection fraction resulting from static disk–corona models [48], because the mild beaming reduces the fraction and energy of disk photons available for Compton scattering in the jet frame. The authors also did not consider a relationship between this beamed plasma to the outflows explicitly. If one considers the presence of large-scale ordered magnetic fields at the inner edge of the accretion flow, as will likely result if any strong field is brought in or generated, then once the accreting plasma is ionized, it will more or less follow the field line configurations and the idea of a static corona is fairly unlikely.

By now it is generally recognized that there must be some kind of relationship between the corona and the jet outflows, as evidenced by the tight radio/infrared(IR)/X-ray correlations discussed in more detail in other chapters of this volume. The exact nature of this connection is where the uncertainty creeps in. Is it simply a matter of reservoirs of magnetic energy being tapped by both the corona and jet respectively, as suggested by, e.g., [62], or does a magnetized corona actually directly feed the jets as my collaborators and I have explored [70, 71]? The difficulty in determining just how absolute the inner disk/coronal outflow/jet connection is (i.e., are they both feeding from the same trough or does the jet feed off of the corona alone) can be attributed to the problems we currently have in disentangling the associated spectral contributions and timing signatures of the various components.

I think all of us writing in this volume believe that the quality of the data currently available already has the potential to address these gaps in our understanding, if only we could figure out the right questions to ask! The combination of timing (see Chaps. 3 and 8) and spectral (see other chapters) features is providing many clues, and one of the primary goals of this chapter is to summarize the recent history of how these observations have defined and, in some recent cases, altered our ideas about the accretion/outflow connection in accreting BHs. As I discuss the process of discovery itself, I emphasize that while collecting phenomenology is useful, the major steps forward did require having some theoretical frameworks already in place. We should keep this fact in mind as we consider our future lines of attack.

6.3 The Driving Observations and Some Interpretation

As described in other chapters of this volume, there are two types of jets observed in BHBs: the compact, steady, and self-absorbed flows associated with the low-hard state (LHS) and at least the brighter quiescent states and the optically thin, more discrete ejecta associated with the transition from the hard to soft intermediate states (HIMS to SIMS), near what we call the "jet line" (see Chaps. 3 and 5). The steadiness of the LHS jets, as well as their compactness, makes them much easier to model since we are essentially observing their outer layers, on a size scale selected by the stratification of frequency-dependent synchrotron self-absorption (see Chap. 7). Therefore, in this state we can treat variability, at least to first order, as due to changes in the input power to a steady flow where the same essential geometry persists at all times. Such an assumption, if merited, simplifies the physics significantly. Second, we can justify ignoring radial stratification in the jets, because while it is likely an indisputable fact of jet evolution (Chaps. 7 and 9), to first order we are only able to observe the outermost layers that photons can escape, though for timing analysis stratification can no longer be neglected. For SED modeling, however, these two factors explain why the LHS is the favorite target of theorists so far, and indeed despite our ignorance about many important factors of jet internal physics, we are still able to make headway and some sensible predictions.

In this section, I will summarize briefly how jets have been incorporated into the standard BHB picture, to a great extent based on earlier work modeling active galactic nuclei (AGN). I will then complement the discussion of scaling in Chap. 5 to include some basic jet physics and how it has been used to interpret the radio/IR/X-ray correlations and has led directly to the discovery of the fundamental plane of black hole accretion ([76, 31], and see Chap. 5). Finally, I will describe where I think new work needs to be focused, and along the way touch on other salient points regarding the inflow/outflow connection, with a distinct outflow-biased perspective.

6.3.1 Multiwavelength Correlations

The galactic source GX 339-4 has been one of the most important testbeds for the development of our ideas about accretion/ejection in BHBs. After the first detection of correlated optical/X-ray flaring [78], the observational community became more interested in looking for simultaneous variability at multiple wavelengths. The results were extremely fruitful, starting with a campaign by [50] using the Molonglo Observatory Synthesis Telescope (MOST) to monitor GX 339-4 in the radio over the course of several years. The authors reported initial evidence for a correlation between the radio flux and the X-ray fluxes from both the *RXTE* and *BATSE* orbital observatories. The log-linear correlation was later confirmed to appear in the LHS, with the exact power-law index constrained by, e.g., [19, 21]. Further developments are discussed in Chaps. 4 and 5, but most recently it has been found that the radio/X-ray correlation is also echoed by correlations between the X-rays and the near-infrared (NIR) [89] and also shows deviations from a single power law. Regardless of the exact slope of the relation, the correlations between the radio, NIR, and X-rays belie an intimate connection between the inner accretion flow and the jets and has been the driver behind much of the recent development in our understanding of accretion physics.

In retrospect, some form of correlation should not have been very surprising; matter falls in, replete with angular momentum, and is somehow subsequently expelled or some portion of its momentum serves to spin up the black hole which then provides energy for jet-powering Blandford–Znajek [10] processes. In either scenario, the two processes of inflow and outflow are clearly related. What is more surprising is that, even within the observed deviations, there is a relatively stringent relationship that holds in as many sources for which we can obtain good simultaneous data. As soon as this bond was hinted at by work in [42], the radio/X-ray correlation in particular presented itself as a chance to explore the physics driving accretion/ejection in BHBs as a class and as a new way to probe the conditions responsible for jet launching.

As discussed elsewhere in this volume, we used to think that the slope of the radio/X-ray correlation was fixed at $L_R \propto L_X^m$, where $m \sim 0.7$, though now we see evidence for scatter (even for the same source) around that value, in a range of more like 0.6–0.8. Even so, such a small range adhered to by all sources with simultaneous radio and X-ray (or IR) data is suggestive of something universal, and it turned out to be an important baseline in extending our understanding of jets to incorporate potential scaling of the underlying physics with mass. A predictable scaling with mass for the physics of accretion around black holes would automatically elevate BHBs from tiny, ISM-churning contributors to the X-ray background (in the eyes of the AGN community) to tiny, but very interesting replicas of AGN with smaller dimensions and faster timescales. Given the importance of AGN cycles for galaxy growth and cluster evolution, it is no wonder that the hint of such usefulness has recently raised the profile of BHB studies with researchers working on these problems.

In the following sections I will review how we established the first concrete mapping of a BHB accretion state to at least one AGN class and how this sequence can serve as a model for any potential further mappings for the more complex, powerful, and more transient accretion states.

6.3.2 Accretion/Ejection Correlations, Simple Models, and Scalings

6.3.2.1 The Compact Jet Paradigm

The compact jets observed in the LHS appear quite similar to those seen in the weakly accreting class of low-luminosity AGN (LLAGN; e.g., [52]) or in the inner cores of extended AGN jets. Their radio spectra are flat to inverted ($F_\nu \propto \nu^\alpha$, where $\alpha \sim 0.0$–0.3), a spectral characteristic that together with high brightness temperatures and polarization are the "hallmark" signatures of compact jet synchrotron emission. It is actually quite difficult to get such a spectrum from other scenarios, because self-absorbed synchrotron emission from a given distribution of leptons moving relativistically in a magnetic field results in a peaked spectrum, with a $\nu^{5/2}$ spectrum in the optically thick portion. Similarly, beyond the optically thick-to-thin break one usually detects a declining power law that reflects the underlying lepton distribution in the optically thin regime (see, e.g., [91]). Thus a flat spectrum requires the radiating medium to be not only optically thick but stratified; that is, the self-absorption frequency must vary with the spatial scale of its origin in the system. Ever since the classical work by [9] (and see also Chap. 7) we have understood that the flat/inverted spectrum is a natural result of a steady jet where plasma properties are conserved, and energy partition between the magnetic field the and hot and cold plasma is fixed, in a smoothly expanding flow. These initial conditions provide the right "conspiracy" where each region of the jet contributes roughly the same spectral shape, with peak flux occurring lower in frequency the further out in the jet it originates (see Fig. 6.1). As a direct result, the photosphere changes in location on the jet as a function of observing frequency. Such an effect has been empirically tested and is known as core shift. The extent of inversion in the radio slope depends on the exact scaling of density and magnetic field with distance from the launching point, the radiating lepton particle distribution, the cooling and reacceleration functions, and the jet dynamics.

Generally, imaging the jets responsible for detected flat/inverted radio spectra is challenging because of the optical depth. At a given frequency, one cannot observe the entire jet but just the photosphere, which will look rather more like a Gaussian ellipse implying elongation beyond what is expected from, e.g., an accretion flow. For instruments with very good sensitivity, a deep look at a flat spectrum source will often turn up a resolved jet. For instance, a VLBI survey of over 100 AGN showing flat core features resulted in imaged jets in all brighter sources [55]. I therefore think it is a safe assumption that any source showing a flat/inverted radio/NIR spectrum with a high brightness temperature is very likely due to a compact jet.

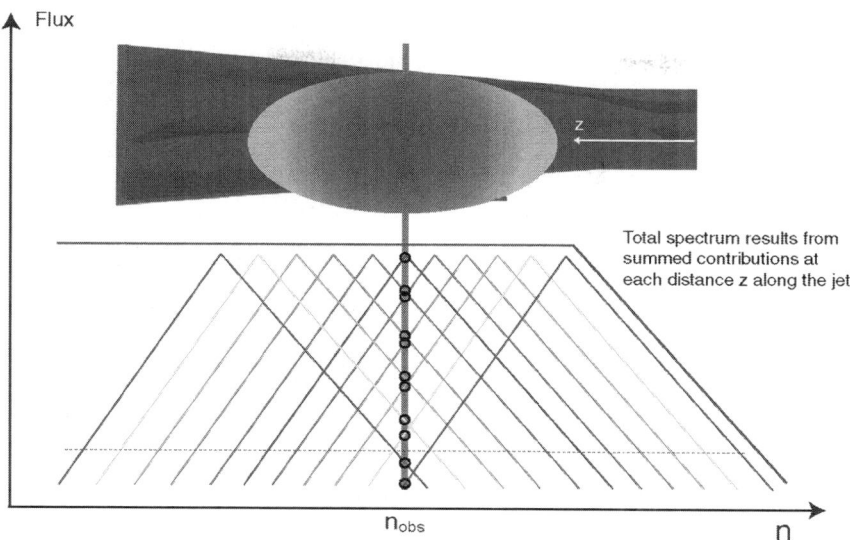

Fig. 6.1 Schematic illustrating the stratified spectrum and frequency-dependent photosphere of an idealized, self-absorbed synchrotron jet. Each segment of the jet contributes approximately the same peaked, self-absorbed spectrum that combine to give the "hallmark" flat total spectrum. A telescope observing at ν_{obs} will see the largest contribution from the segment whose individual spectrum peaks at that frequency, and increasingly smaller contributions from neighboring segments (fluxes of each segment at ν_{obs} indicated by *black circles*), generally producing an elongated Gaussian ellipse-like photosphere. The visible jet at a particular frequency is thus much smaller in scale than the actual jet in the case of high optical depth. Figure clearer in color (electronic version)

In LLAGN the radio spectra tend to peak, sometimes with an additional excess, or "bump," in the submillimeter (submm) band. If one wanted to naively begin considering any sort of mass scaling, one could imagine what a jet would look like if the same relative (in Eddington accretion units $\dot{m}_{Edd}$) power were fed into the same relatively sized region (expressed in $r_g = 2GM/c^2$), but for a $10M_\odot$ black hole instead. The most general prediction is that BHBs would exhibit a flat/inverted spectrum up to a break frequency somewhere well above the submm regime, reflecting the more compact scales and thus higher particle and magnetic energy densities for the same fraction of Eddington power. It turns out that this prediction is indeed empirically confirmed, but it took some time before the necessary observations were in place.

6.3.2.2 Flux Correlations and Mass Scaling

It is easy to take for granted that multiwavelength campaigns have always been the norm for BHB studies, as is the case for AGN blazars, but it was really only at the time that the radio/X-ray correlation was being determined that observers were in general awakening to the realization that simultaneous broadband spectra of BHBs were not only interesting but vital to capture the quick transitions observed in these sources. In the year 2000, not even a decade before the writing of this volume, a new BHB was discovered that resided at high enough galactic latitude to be visible to the *EUVE* instrument, XTE J1118+480 [85]. An extensive campaign was performed, really the first of its kind, involving quasi-simultaneous radio through submm [35] and IR through X-ray [53] observations. It seemed like an opportune chance to test whether (and how well) a jet model developed originally for a LLAGN (in this case, Sgr A*; [32, 67]) could work to describe a BHB in the LHS, believed to be accreting at a comparably low accretion rate. The original idea was to model only the radio through IR with a compact jet, since observed variations suggested that the flat/inverted spectrum continued at least up until this frequency (e.g., [33, 34]). After reworking the original model to include all size and power scales in mass-scaling units of r_g and L_{Edd} only, we replaced the supermassive black hole with a stellar remnant of around $10M_\odot$ and easily found a solution that went through the radio through IR data, indeed even extending into the optical and possibly the UV depending on the compactness of the jet nozzle region ([66], and see Fig. 6.2).

A potential high-energy contribution from jet synchrotron radiation was an interesting result, and because of simple scaling arguments that I will summarize below, not altogether surprising. A further result that we did not originally anticipate was that the broad continuum features of almost the entire spectrum could be fit with jet synchrotron emission, after only one change in assumptions between this application and the model for Sgr A*. That change was to assume that a significant fraction of the particles entering the jet in a quasi-thermal distribution, as assumed for Sgr

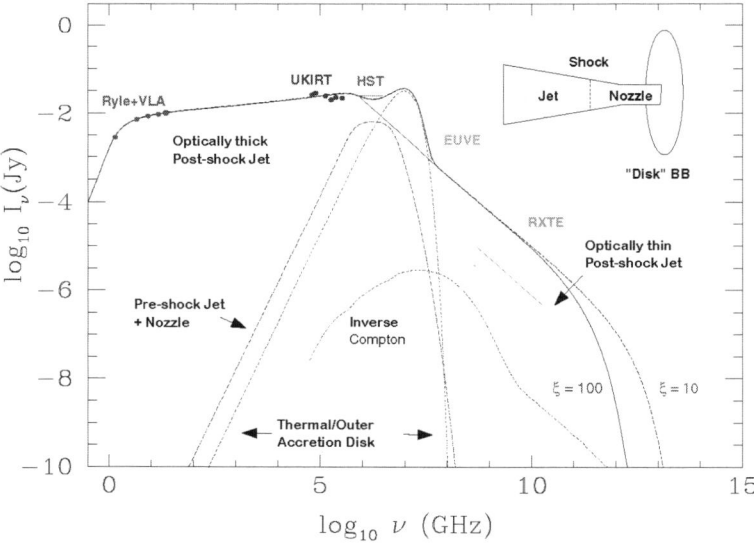

Fig. 6.2 Application of a mass-scaling jet model to the first quasi-simultaneous multiwavelength data set of a BHB including UV data, for the Galactic source XTE J1118+480. The smaller scales compared to AGN results in a prediction of the jet synchrotron spectral break in the OIR band and a significant contribution of jet synchrotron to the X-rays (see [66] for details)

A*, are subsequently accelerated into a power law starting at a location around 100–1000 r_g from the jet base. In Sgr A*, the source is accreting at $\leq 10^{-9} \dot{M}_{Edd}$ and particle acceleration in the jets must be extremely weak or absent during quiescence (flaring may be another story; [67]), as indicated by the stringent limits on any power-law component in the optically thin IR regime (e.g., [74]). One of the main results of [66] was to demonstrate that the combination of empirical fact (jets accelerate particles into power laws with energy index 2–2.4) with conservative assumptions about the acceleration rate (see, e.g., [54]), taken together with calculated cooling rates from synchrotron and Compton processes, implies that it is very difficult to suppress some level of synchrotron emission extending into the X-ray band for BHBs, even for "low-luminosity" systems. For the case of XTE J1118+480 we showed an extreme scenario, although not in conflict with the actual data set, which displays no indications of disk reflection or spectral cutoff. There are good reasons, however, to think that in general there is also a Compton scattered emission component (see later).

6.3.3 Interpreting the Correlations

Because the synchrotron jet model presents an obvious connection between emission over decades in frequency, it seemed worth exploring its relevance for explaining the radio/X-ray correlation. We decided to try fitting the same data that had established the correlation, using a variation of this mass-scaling jet model, and interestingly found that we could explain the correlation by only changing the power (as a fraction of accretion energy) input into the jets [68]. The correlation suddenly presented itself as a tool with which we could probe scalings of various emission processes with power as well as mass.

It turns out that the reason this class of model is so successful at explaining the correlations follows from some fairly straightforward radiative physics within the framework of self-absorbed jet models, as well as some accretion flows. Given how complicated the physics internal to jets seems likely to be, with turbulent fields, stratified layers of different velocities, and distributed acceleration processes, it is somehow rather amazing that the spectra we observe fall into such a small range of slopes, remarkably close to the predictions of the idealized case of [9]. It is reminiscent of the situation with thermal accretion disks: we know that these should be turbulent, dissipative, and complex inflows, yet they appear to follow the simple multicolor black body paradigm rather well. Ironically, the phenomenon of high opacity allows us to understand global properties of compact jets, despite our lack of knowledge about their detailed internal properties.

To understand how a synchrotron jet model reproduces the $m = \sim 0.7$ index of the radio/X-ray correlations, one can think first of an idealized jet spectrum which has flux $F_\nu \propto \nu^{\alpha_R}$ up to the synchrotron self-absorption break frequency, ν_{SSA}, where the most compact radiating region of the jet (containing a power law of leptons with distribution $N_e \propto E_e^{-p}$) becomes optically thin and thus afterward has a $\nu^{-\alpha_X}$ spectrum, where $\alpha_X = (p - 1)/2$ is the synchrotron spectrum.

The total spectrum is thus a broken power law (see Fig. 6.1) and if one knows how the optically thick flux and the break frequency scale with jet power, Q_j, it then comes down to algebra to derive the predicted correlation index $m \equiv d \log L_R / d \log L_X$. In the original derivation of the dependence of radio luminosity on power for an idealized, compact, conical jet, [9] found L_R (i.e., $F_{R,\nu}$) $\propto Q_j^{17/12}$, and the same scaling was subsequently derived in later models [29, 68] because of similar underlying assumptions about the dependence of internal pressures with distance along the jet.

A more generic treatment of the scaling can be found in [51], who explore a broader range of assumptions for the various scaling accretion models. For instance, whether a jet is assumed to launch specifically from a classical ADAF [81] or any other scenario where the input power is assumed to be simply proportional to the disk luminosity, similar scalings are predicted. The magnetic energy density U_B and lepton energy density U_e have a fixed partition ratio (the simplest case being the assumption of equipartition) proportional to $\dot{m}/M_{BH}$, where $\dot{m} \equiv \dot{M}/M_{Edd}$ and is thus dependent on black hole mass M_{BH}. By using these scalings, with all distances in addition scaled in terms of $r_g(M_{BH})$, in the expressions for self-absorbed synchrotron flux from, e.g., [91], one can directly calculate the total flux at a given frequency from the entire jet:

$$F_\nu = \int_{r_g}^{\infty} dr\, R(r) S_\nu(r), \tag{6.1}$$

where $R(r)$ is the jet diameter at r and S_ν is the synchrotron source function. For either case of an ADAF or $Q_j \propto M\dot{m}$, following Eq. (12b) in [51], one finds the dependence on $\dot{m}$:

$$\frac{\partial \ln F_{\nu,R}}{\partial \ln \dot{m}} = \frac{2p + (p+6)\alpha_R + 13}{2(p+4)}. \tag{6.2}$$

For the "canonical" accelerated particle index $p = 2$ one finds $F_\nu \propto \dot{m}^{17/12 + 2\alpha_R/3}$ which for a perfectly flat self-absorbed spectrum reproduces exactly the original Blandford and Königl result.

To derive the correlation index m, we need to derive the predicted ratio of radio to X-ray fluxes. For the case of a model such as that presented in Fig. 6.2, this index is just the ratio of the (log) fluxes above and below ν_{SSA}, because the X-rays are due to optically thin synchrotron alone. However, for an arbitrary X-ray process we need to insert its dependence on $\dot{m}$ as in (6.2):

$$m = \frac{\partial \ln F_{\nu,R}}{\partial \ln F_{\nu,X}} = \frac{\partial \ln F_{\nu,R}}{\partial \ln \dot{m}} \frac{\partial \ln \dot{m}}{\partial \ln F_{\nu,X}}, \tag{6.3}$$

which for $p = 2$ and an ADAF or $Q_j \propto M\dot{m}$ case gives $m = (17 + 8\alpha_R)/21$ and reproduces the observed $m \sim 0.7$ for a flattish radio spectrum.

Clearly the exact value of the correlation will vary slightly from source to source, because we expect variations in the local acceleration particle index p and in the geometry and compactness effecting α_R and α_X. The power of this simplistic analysis is that even accounting for such scatter, certain accretion processes can be ruled out entirely already, based solely on the data. For instance, if the X-rays originate in a perfectly efficient corona where $F_\nu, X \propto \dot{m}$ then $m \sim 1.4$, which has never been observed. Not surprisingly, our conclusion has to be that the radiative efficiency of the emission process responsible for the weak LHS X-ray flux is low. The point here is that regardless of how well (or not, as the case may be) we understand the exact inner workings of the jets or accretion flows, we are probably not wildly off in terms of their dependencies and global scalings, which are relatively model independent as long as certain basic assumptions about conservation and power budgets hold. Perhaps more importantly, these results demonstrate that understanding the accretion/ejection process in black holes, of all scales, irrefutably requires a multiwavelength perspective. With single-band or non-simultaneous studies, we would never have discovered the correlation, nor would we have been able to derive such exacting constraints on accretion flow efficiency very close to the black hole. It is interesting that in order to understand the regions closest to the black hole, we seem to require information about emission from regions which lie *well beyond the gravitational reach of the black hole itself.* Such a scenario would likely have not been envisioned even a decade ago, and the studies building on the discovery of the multiband correlations represent the significant evolution in our outlook over a relatively short period of time. On the other hand, the premise behind AGN feedback's potential role in galaxy evolution necessitates such a link between the inner gravitational radii and the largest system scales, obviously in the form of the jets. It would naturally be appealing if physical trends such as this correlation could be extended to the larger physical scales, in order to place better constraints on accretion processes in galactic nuclei.

6.3.4 Mass Scaling and the Fundamental Plane

Because most physical quantities in accretion models can in general be expressed via mass-scaling variables, it is possible to make predictions for the effect of mass on, e.g., the radio/X-ray correlation. It turns out that all black holes are predicted to follow a similar radio/X-ray correlation – at least as long as they are in the equivalent of the LHS – but the normalization of the relation will be inversely proportional to the mass. The exact dependence can be predicted theoretically using scalings as above and then tested against a sample of relevant (in this case, weakly accreting) AGN such as LLAGN, and likely FR Is and BL Lacs. This process was first carried out by two independent groups [76, 31], and the relationship between radiative power, black hole mass, and accretion rate was dubbed "the fundamental plane of black hole accretion." More details about the FP are provided in Chap. 5; however, there is quite a bit of scatter in the correlation for AGN even after "mass correction" for comparison with BHBs, and it is difficult to separate out the possible effects of spin in AGN samples, which obviously does not play a role in correlations measured from individual sources. Despite the complications, the fundamental plane has been a rousing success for theory and has opened up a new avenue for studying accretion because it allows us to compare behavior at very different size- and timescales. We can essentially take the best of both worlds, studying for instance jet formation and evolution directly in the fast-evolving BHBs, while actually imaging regions close to the event horizon in nearby AGN.

There are, however, important caveats to using data from AGN samples, not least of which are the lack of simultaneity and the lack of comparable spatial resolution. Even though the timescales are much longer than in BHBs, there is enough variability particularly for AGN at the smaller end of the mass range, that non-simultaneous data can skew estimates of the indices for the fundamental plane. For instance, LLAGN routinely display at least 20% variability on a timescale of months [28, 79], sometimes up to 50% as seen in the UV cores of several AGN [63]. In [69] we found that repeating an earlier analysis of the mass-dependent coefficient in the fundamental plane, using simultaneous data instead of averages derived from the literature, resulted in a 50% change in its value! Combined with the different size scales leading to different predicted lags between the various bands of emission, it is quite difficult to be sure that we are always comparing apples to apples. In a BHB we are likely averaging over waves of variability in the radio/IR emission of the jets, while in a LLAGN we can detect this motion explicitly over timescales of weeks to months. Eventually with better data and statistics, we should be able to factor any lags into the analysis, but it is something to keep aware of as a factor in divergent values for correlation slopes, as well as other trends.

For the brightest AGN with the largest scale jets, there is a more significant problem. Many of the radio measurements are still taken with instruments like the VLA, which does not provide sufficient spatial resolution for sources residing at very high redshifts. When we study a BHB, the jets are small enough (AU to pc) that achieving causality between the core and outer jets is never much of a concern; we can be assured that all jet activity is related to a recent outburst or flaring. However, the radio flux we measure from a distant AGN may include emission from plasma far out in the jets that is associated with activity in the nucleus thousands or more years ago, even from entirely different outburst cycles of accretion activity and reactivated by, e.g., internal shocks. Accidentally integrating over several outbursts in bright radio-loud AGN could impact our attempts at associating them with BHB states. For instance, one of the major outstanding questions is why radio-loud jets are present in some high accretion rate active nuclei, when this does not seem to be the case in BHBs except for a small period of transitional activity between the HIMS and SIMS. There are two possible explanations, assuming one believes the states are shared in all black holes: either these are short-lived states (by AGN standards, still steady for us) that we are catching in the act or we are indeed associating prior radio activity with a currently "HSS-type" nucleus that is no longer powering the jets. Therefore, it is important to push for VLBI imaging whenever possible for distant luminous radio galaxies used in comparisons with BHBs.

As a final note, I think it is important to recognize the process by which the fundamental plane was discovered. Interesting trends using new observational techniques were discovered, but a theoretical framework for interpreting and extending the results was necessary to fully exploit the data. At the moment we are at something of an impasse in extending our mappings of particularly the transitional BHB states to AGN classes, because our theory has yet to catch up with the data. The LHS/LLAGN (+FRI/BL Lac?) mapping was in many senses the easiest one, because we could assume a semi-steady state and high optical depth. In order to proceed further, I believe that we will first need theoretical models that can explain the transitional states, and that means understanding significantly more than we do about internal jet physics and launching processes. So while it is important to pursue better quality samples of AGN with good multiwavelength coverage, ultimately there is still hefty progress to be made on the accretion/ejection modeling.

6.4 Modeling and Hysteresis

6.4.1 A Brief Overview of Recent Models

So far I have not said much about modeling, other than to try to document how relatively simple models have been rather successful at helping us understand overall trends in the data. Obviously there are a myriad of models at the moment for various aspects of accretion theory and only so many pages in this chapter, so I will focus on the ones which seek to explicitly address, in some quantitative way, the accretion/ejection connection. Because the radio emission is generally agreed to originate in the jets, this bias naturally selects out that subset of models which attempt to link the X-ray producing processes directly to the outflows in order to reproduce observables such as the radio/X-ray correlations. At the time of writing, all such models are necessarily somewhat contrived, reflecting our rather limited basic understanding of the exact physical mechanisms linking the inflow and outflow. In order to bridge this gap, any model must make various strategic assumptions for the physics of jet launching, in the hopes that their validity can be gauged by comparison with the data.

The most "fundamental" approach is to attempt to model the plasma physics of the accretion process, often requiring the use of complex (M)HD codes, although significant progress can sometimes be made analytically. Both techniques also involve approximations to the physics, for simulations because otherwise they are too computationally expensive and for more analytical calculations because otherwise they are not solvable without a simulation. In general these approaches focus usually more on the plasma dynamics rather than radiative physics and are thus not optimized for direct comparison with spectral data. Because this chapter focuses more on spectral calculations, I will only touch on some examples of these two approaches, which are discussed in detail in Chaps. 9 and 10.

Most models so far generally focus on some form of shared power budget as a way to link the jets and the accretion inflow, emphasizing uncovering the process that binds them and accounting for overall trends rather than explicit spectral/timing models. However, some classes of models start by assuming a specific energetic relationship and then derive exact spectral predictions. Having complementary approaches is always valuable, and so in the following I will briefly summarize the various modeling strategies currently underway, roughly grouped by conceptual framework.

6.4.1.1 Jet Launching Scenarios

After the discovery of jet quenching in BHBs during outbursts, as well as indications of a possible link between observed accretion states and recession of the thin accretion disk [26], it seemed logical that the geometry of the accretion disk should be linked somehow to jet production. Two papers earlier this decade explored, using rather simple and elegant arguments, how the disk geometry together with magnetic field configuration could affect the resultant jet power, and have strongly influenced the thinking of many subsequent groups. In [72], Meier considered the case in which jet power is mostly dependent on the poloidal magnetic field strength, and to some extent also the field angular velocity in the inner regions, and explored the results for thin and thick disk geometries. He argued that only geometrically thick flows such as RIAFs (or magnetic coronae, see below) could produce poloidal fields powerful enough to lead to the efficient radio-producing jets observed in, e.g., hard states. A similar conclusion was derived by [57], in the context of exploring accretion disk dissipation and the variability of GRS 1915+105, and to this date most models rely on the interplay of accretion disk geometries to aid in explaining the launching scenario.

Several groups have in the meantime been working toward more detailed semi-analytical scenarios, in order to try to quantify more specifically not only the jet powers but also to understand the various BHB states and their relation to accretion disk physics. One of these, [101] based on an earlier work by [96], suggest that an "accretion-ejection instability" in magnetized accretion disks can solve the problem of how disks transfer angular momentum vertically into jets rather than only outward within the plane of the disk. If the disk is assumed to be threaded by a poloidal magnetic field, then a spiral density wave instability can generate a Rossby vortex at the corotation radius. The motion of the field footpoints due to the magnetic stress transports accretion energy and angular momentum from the inner disk to the corona via Alfvén waves. If this corona feeds the jets, this mechanism can provide a link between the inflow and the outflow. It has also been proposed as a source of the QPOs observed in BHBs [98, 97]. While this is a very interesting scenario, it has yet to be verified to occur in 3D (i.e., simulations are not yet showing such instabilities). Similarly it is difficult to compare to data as most of the conclusions are not yet predictive for generic sources. However, it is important to note that such instabilities likely must play a role in some of the explosive phenomena we observe, and the complex physics involved makes it challenging to calculate exact predictions.

Another line of approach is a magnetic corona, as mentioned in Sect. 6.2. In the scenario explored by [75], a standard thin disk [92] at low accretion rates experiencing magneto-rotational instability (MRI) is "sandwiched" by a patchy magnetic corona, into which it dissipates a significant amount of its gravitational energy via buoyant flux tubes. The rate at which energy is fed into the corona via dissipation increases with higher viscosity, and the speed at which magnetic structures are buoyantly transported upward. Both of these increase at lower accretion rates, and the dissipated energy will either heat the corona or launch outflows. Coronae with larger radii and scale heights will result in more powerful outflows, and both will contribute to the total spectrum via synchrotron emission. The authors present some sample spectra that already indicate the expected correlation between X-ray flux and hardness with optically thick jet emission (Fig. 6.3). This baseline model was expanded upon in [62], where the consequences of this "commmon reservoir" of magnetic energy feeding both the corona and the jets were applied to explain the puzzling optical/UV/X-ray variability in XTE J1118+480. In their proposed sce-

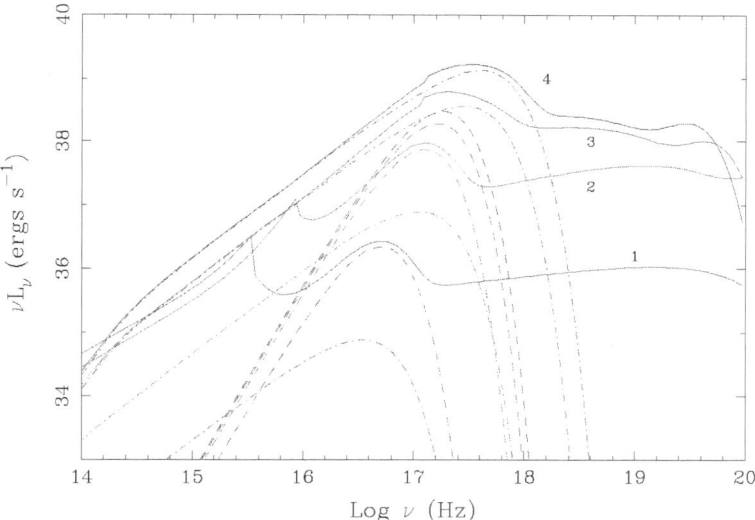

Fig. 6.3 Predicted broadband SEDs for a thermally driven jet combined with a magnetic corona model of a $10M_\odot$ black hole [75]. *Solid lines* show the total spectra, *dashed lines* show the contribution of thermal reprocessed radiation from the accretion disk, and *dot-dashed* lines show the multicolor blackbody disk emission. The accretion rate is increasing from 1 to 4 and colors (electronic only) indicate component/spectrum groupings. The X-rays originate from the inverse Compton upscattering of thermal disk and coronal synchrotron photons in the magnetic corona. The jet luminosity has been estimated assuming a flat spectrum up to the coronal self-absorption frequency and a radiative efficiency of 10%. The spiky peak at 10^{15}–10^{16} Hz in spectra 1 and 2 is the self-absorbed synchrotron emission from the magnetic corona itself, dominating the seed photons for Comptonization in spectrum 1 and still significant for spectrum 2. These SEDs do not include coronal bulk motion as suggested by [6], which would significantly reduce the reprocessed emission (*dashed lines*) and thus extend the domination of synchrotron photons as Comptonization seed photons up to higher accretion rates

nario, the corona produces the X-ray via standard Comptonization of lower energy seed photons from the geometrically thin disk, while the jets produce the optical via synchrotron radiation. In this way, the lag between the optical and the X-ray is explained by the displacement of plasma outward from the corona into the jets, while energetically they share a common input of power from the magnetic processes generated in the thin disk.

Finally there is the series of works building on [37, 38, 36], culminating most recently in [39], which describes a complicated phenomenology to tackle the various states observed in BHBs. Within a standard accretion disk, they posit that a magnetized region forms where the jet is launched, called the "jet-emitting disk," or JED. The inner part of the jet is a pure pair outflow that is accelerated to high bulk Lorentz factors, surrounded and confined by a slower, matter-dominated jet, itself self-confining. This scenario bears some similarity to the "magnetically dominated accretion flow" (MDAF) solution described in [73] and explored for the first time numerically in 2D using a fully general relativistic MHD code including radiative cooling (COSMOS++; [3]) by [40]. It is interesting to note that if radiative cooling is self-consistently included, magnetic pressure dominates the gas pressure near the inner regions of the inflow, and synchrotron cooling becomes very important for the dynamics. The advantage of the simulations is that they can study jet formation directly from the given initial conditions, but linking them to reality can be challenging. Often it can be too computationally expensive to run the simulations long enough to see a steady-state form. My colleagues and I are currently developing several new methods, both semi-analytical and using simulations, to try to address this gap.

6.4.1.2 Jet-Disk Symbiosis

In the meantime, there is always a need to understand the boundary conditions and the environment in order to guide the necessary assumptions in the above classes of approach. From this more phenomenological point of view, aimed at fitting observations, some of the earliest quantitative work on the self-coined "jet–disk symbiosis" can be found in [29, 30]. By assuming that the jet power is linearly proportional to the accretion rate in the inner disk and assuming a freely expanding jet with maximal efficiency, they were able to extend the earlier work on very idealized jets by [9] by solving for the velocity profile and including the resultant cooling terms. The lack of knowledge about the exact launching geometry and mechanism is channelled into several free parameters determining the jet initial conditions, but the exact nature of the jet/disk link cannot be derived from this model because the disk power mainly serves as an "accretion power reservoir" for the jets. Although mostly applied to AGN, this work provided the groundwork theory for the scalings used to determine the fundamental plane described above.

The model explored for BHBs in [66], and the first attempt at explaining the radio/X-ray correlation [68] was based on these same general assumptions but included more detailed internal physics and radiative processes. One of the most important conclusions of this work, as mentioned in Sect. 6.3.2, is that for reasonable energy inputs the compactness of BHBs compared to AGN even at low luminosities leads to a prediction of significant synchrotron X-ray emission from the jets in BHBs. At the same time, the presence of reflection features observed in BHB hard states (e.g. [110]) argues that a significant portion of the hard X-rays cannot be too strongly beamed or originate too distant from the disk. The weakness of the reflection in the LHS, however, does imply that the covering fraction cannot be too high. Reference [70] explored this issue and determined that to reproduce the observed reflection fractions in the context of a jet-dominated model, the base of the jets would need to be compact enough to produce high-energy photons via Comptonization of either internal synchrotron photons or external disk photons. This idea was explored in more detail in [71], where we showed that this base region can in fact mimic the observational features attributed to an accretion disk–corona, however with different assumptions than those required for thermal disk Comptonization models [e.g., 49]. Specifically, because of the relativistic flow in a direction away from the disk, the photon field for upscattering is quickly dominated by synchrotron photons, and thus the thermal disk spectrum is hard to constrain via its weak spectral signature alone. On the other hand, the strength of this model is how it demonstrates that a magnetized corona or RIAF-like flow at small radii can be consistent with directly feeding the jets, while also providing a very good description of the radio through X-ray data (see also [43, 77, 60]). The soft X-rays contain a significant synchrotron radiation component, becoming increasingly inverse Compton dominated toward the hard X-rays. Thus one prediction of this model is that the exact slope of the radio/X-ray correlation will vary depending on which component dominates the X-ray band under consideration. While synchrotron predicts a slope of 0.7, synchrotron self-Compton SSC will produce a slope closer to 0.5, thus combinations can give intermediate values, and this may contribute to some of the variations in slope recently reported (Corbel et al. 2008, in prep.). The limitation of this model is that, by assuming the disk/jet linkage a priori, a detailed understanding of their relationship cannot be tested beyond a consistency check. Similarly, the fact that the model assumes a steady state precludes it from being tested against timing data such as frequency-dependent lags. Another steady-state approach to fitting broadband SEDs can be found in [105] and [104], who have developed new models based on earlier work [106, 107], that pair a simple jet model with a more detailed treatment of a RIAF. In this case, however, the Comptonizing region belongs predominantly to the accretion inflow. The strength of both of these model classes is their ability to test basic assumptions about the link between inflow and outflow components against the high-quality broadband data now available (Fig. 6.4).

6.4.1.3 Other Modeling Approaches

Several other groups have honed in on the modeling of more specific behaviors. For instance, [16] have an almost inverted approach to those described above, in that they begin by assuming the presence of a pair jet at the inner accretion flow and explore the energetics and mechanisms involved in accelerating and confining it via radiation pressure from the accretion flow. The authors focus thus in detail only on

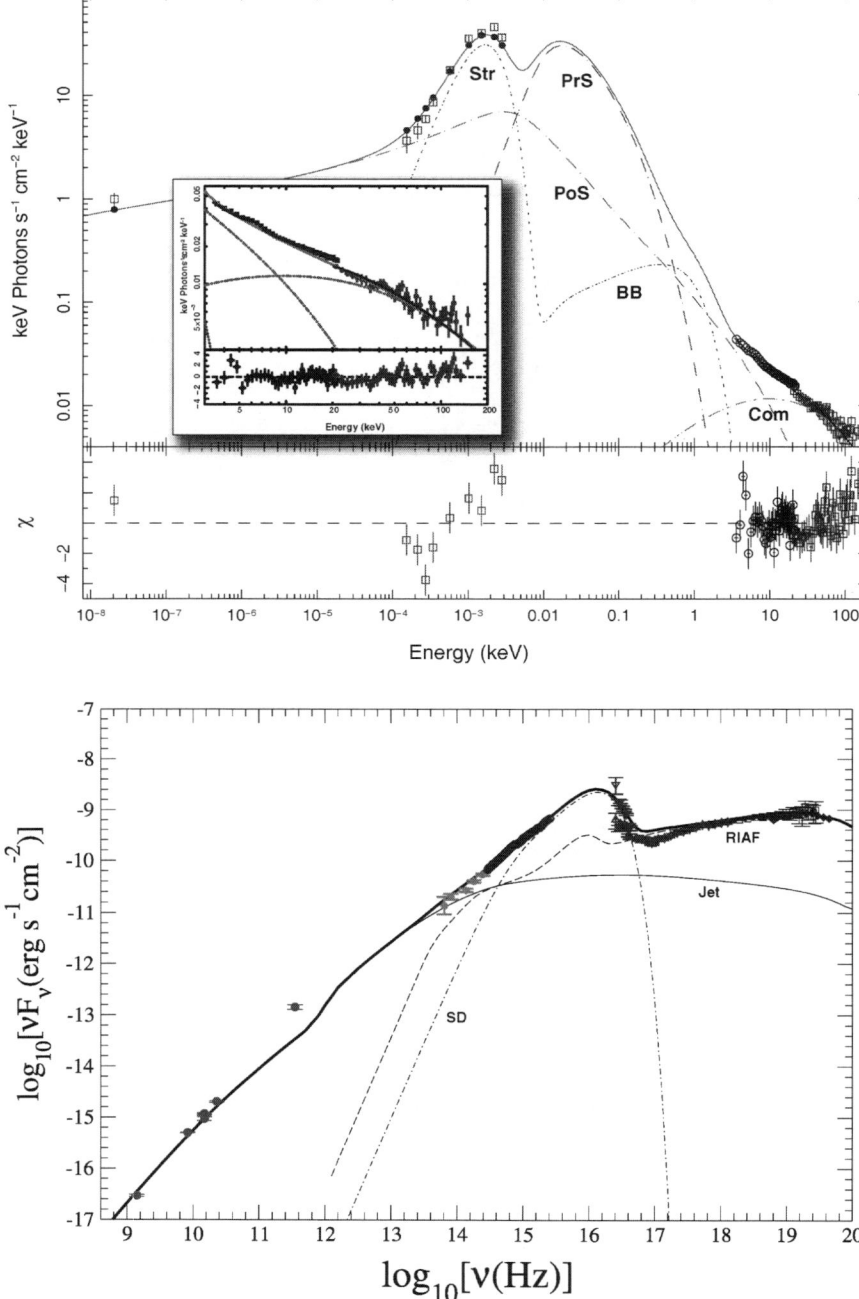

Fig. 6.4 (continued)

the accretion flow rather than the jets, in the context of a two-component advective flow (TCAF; [15, 14]), resulting in the formation of what they term a centrifugal boundary layer (CENBOL), that acts as the "corona" for producing the hard X-rays via Comptonization. This class of models has been invoked to explain the observed pivoting in the X-rays that appears to be correlated with the radio flux in Cyg X-3 and GRS 1915+105 [17]. The physics and launching of the jets is not addressed explicitly.

One of the more radical proposals comes from [86], who suggested that any jet-producing object (driven by magnetic fields) could be interpreted instead as a "magnetospheric eternally collapsing object," or MECO, as long as the magnetic moment is sufficiently high. For sources we interpret as BHBs, they predict a field strength around 10^8 G, quite a bit higher than derived from standard equipartition arguments in plasma, but not inconsistent with the assumed energy budget. The idea is a new application of magnetic propeller theory, where the inner disk is coupled to the interior magnetosphere, and in [87] the same authors actually derive the scaling laws for the radio/X-ray correlation and dependence on mass for their favored scenario. They show that for certain assumptions, they can indeed reproduce the relations determined empirically as discussed above and in Chaps. 4 and 5. Detailed spectral fits have, however, thus far not been presented, nor is it clear that this model can survive frequency-dependent size constraints such as those now possible with Sgr A* [12, 94, 13, 24].

Another model that focuses more on explaining the timing features of BHBs than detailed spectra is that of [47]. Taking a simplistic formulation for the jet, they obtain results that are consistent with both the soft-hard X-ray lags and energy-dependent trends in the autocorrelation function [58], while also reproducing trends in the evolving X-ray spectra. In a follow-up paper, [46] consider the effects of a power-law distribution of electrons in the jets and demonstrate that this model can also account for the multiwavelength features in BHB spectra. However, it is not clear if this particular choice of configuration can successfully reproduce the radio/X-ray correlations.

Fig. 6.4 *Top:* Outflow-dominated model [71] fit to the simultaneous broadband data from a hard state observation of GRO J1655-40 taken on 24 Sep 2005, with the VLA, Spitzer, SMARTs, and RXTE. The *dark* (*green* in electronic version) *dashed line* (labeled PrS in print version) is the pre-shock synchrotron component, the lighter (*green*) *dash-dotted line* (labeled PoS in print version) is the post-shock synchrotron component, the (*orange*) *dash-dotted line* (labeled Com in print version) represents the SSC plus the disk external Compton component, the (*magenta*) *dotted line* (with peaks labeled BB and Str in print version) is the multi-temperature disk black body plus a black body representing the companion star in the binary system. The *solid* (*red*) *line/points* indicate the total model after including absorption and convolution with reflection and the RXTE response matrices, while the individual components and (*gray*) total spectrum do not take these into account. See [77] for more details. *Bottom:* A joined accretion disk–jet model applied to the broadband data from XTE J1118+480 (color in electronic version). The *dashed* (labeled RIAF) and *dot-dashed* (labeled SD) *lines* show the emission from the hot (RIAF) and cool accretion flows, respectively. The *thin solid line* shows the emission from the jet, and the *thick solid line* indicates the total spectrum. For details see [105]

And most recently, another range of jet models have been developed with the aim especially to tackle the issue of whether or not microquasars can produce γ-rays, as some recent papers have claimed [e.g., 82, 1]. While high-energy protons are considered explicitly as sources of high-energy secondary particles, via their collisions, in [88] (as compared to the Falcke and Biermann, Markoff et al. papers, where protons are included only as kinetic energy bearers), the focus was not on the broadband spectrum or the link to the accretion flow. However, recent work by [11], and several subsequent papers based on this model, explores several potential scenarios for high-energy photon production, where the jet power is assumed proportional to the inflowing accretion rate. In general they consider cases with a sub-equipartition magnetic field that may influence their findings in favor of external photons as sources of Comptonized flux, but it is interesting nonetheless, especially with the recent launch of the Fermi observatory. With new broadband campaigns being planned that include Fermi as well as the ground-based TeV instruments H.E.S.S. and MAGIC (and eventually the CTA), along with particle detectors like Auger, ANTARES, and IceCube, it is extremely important to understand whether BHBs can be significant sources of cosmic rays and neutrinos.

6.4.2 Hysteresis

As discussed elsewhere in this volume, the hysteresis observed during the typical BHB outburst cycle, as evidenced especially by the "q" shape in the hardness-intensity diagram, signals that at least some other factor besides the accretion rate is playing a major role in the source evolution. One strong candidate would be magnetic field configurations in the accretion disk, but this question is by no means settled. Interestingly, hysteresis has also recently been discovered in the radio/X-ray and IR/X-ray correlations found in the hard state during outbursts; some examples of this are shown in Fig. 6.5. It is now clear that the system does not return during decay to the same hard state configuration that it had during the rise stage.

Figure 6.5 shows two examples of hysteresis, in the radio/X-ray and IR/X-ray correlations for two BHBs. The difference in IR flux between the hard and soft state could be interpreted as due to the quenching of the compact jet synchrotron emission during the hard-to-soft transition. Within the hard states themselves, however, the IR flux is clearly higher during the decay compared to the rise in outburst, a trend that is also seen in the radio/X-ray correlation hysteresis. In general, the slope of the correlation appears to steepen below around 10^{-4} $L_{\rm Edd}$, at least during the decay phase, but these results are new and such hysteresis has not been measured for many sources at this point. It is already clear that the variation in correlation slope as the system "jumps" tracks of slope $\sim$0.6–0.7, as well as the difference in normalization for the two tracks, will have implications for the fundamental plane if AGN experience something similar. Perhaps some of the scatter observed in that correlation could be due to a similar effect. But perhaps more importantly, this hysteresis must provide information about the second physical parameter driving the outburst in

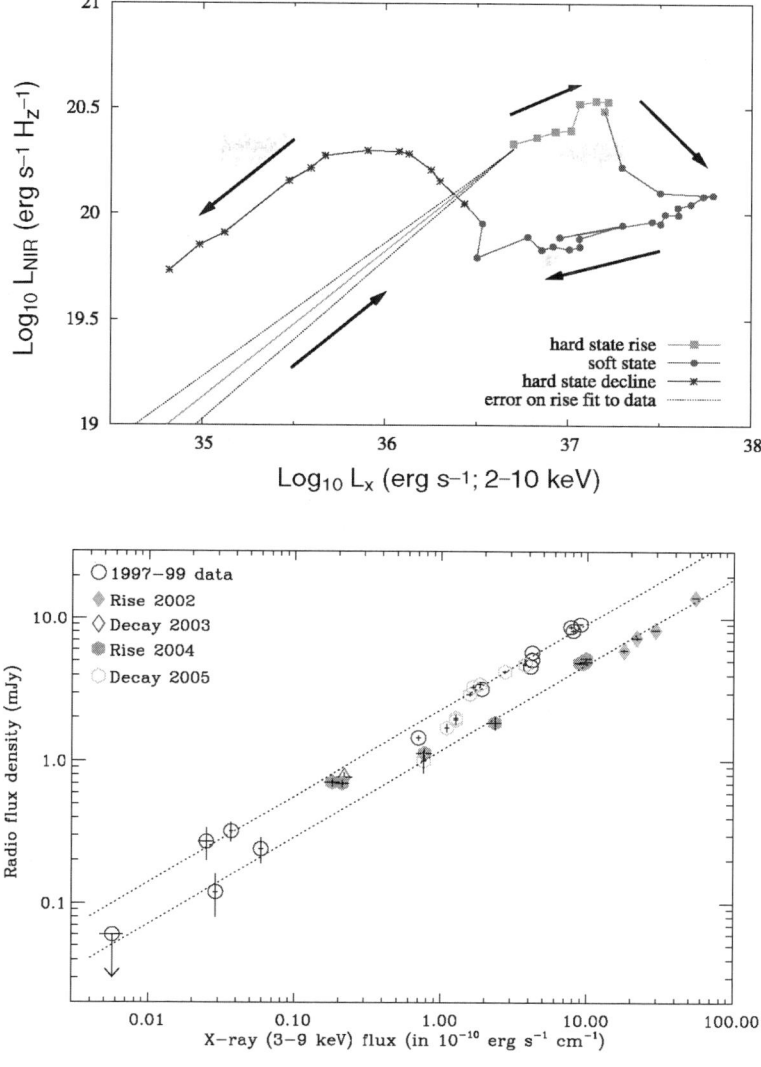

Fig. 6.5 *Top:* The hysteresis in the infrared vs. X-ray quasi-simultaneous luminosities of the BHB XTE J1550-564 during its outburst of 2000. In at least bright hard states, the (N)IR emission is dominated by the jet. The jet is quenched in the soft state so the infrared drops and is then dominated by light from the accretion disk. At a given X-ray luminosity, the infrared emission appears to be weaker in the rise of the outburst compared with the decline. The power law indicating the slope of the rise correlation is derived from fits to the X-ray and infrared exponential rise light curves. From [90]. *Bottom:* Hysteresis in the radio vs. X-ray simultaneous fluxes of the BHB GX 339-4. The rise phase always seems to follow a lower luminosity track for the same X-ray flux compared to the decay phase (from Corbel, Coriat et al., in prep.)

addition to the accretion rate. Furthermore, there is preliminary evidence that for some source outbursts, the hysteresis is not seen at all in the IR/X-ray correlation, although present in the radio (Corbel et al., in prep.).

As discussed in Sect. 6.3.3, changes in the correlation slope could be rooted in a change in the efficiency of the radiative process. One possibility could be that the X-ray band contains emission from two different components that are swapping dominance during the decay. In their analysis of ~200 pointed RXTE observations of Cyg X-1, [103] found the best fits with a broken power law where the indices of the two components were strongly correlated. In my colleagues' and my interpretation, such a correlation would be an indication of correlated synchrotron and inverse Compton radiation, where synchrotron photons dominate the Compton photon pool. A flattening in the slope of the radio/X-ray correlation at low luminosities (i.e., from 0.7 to 0.5) could for instance indicate that SSC cooling was the preferred channel rather than synchrotron cooling. But this explanation does not account for the hysteresis in correlation normalization itself. Most likely, the change in normalization relates to physical differences between the inner disk/corona and jets during the rise and decay, such as different magnetic field configurations perhaps resulting from changes in the inner disk structure. At the time of writing we do not yet have an explanation for this phenomenology, but interestingly this trend may be consistent with some other phenomena observed at the lowest luminosities, as discussed briefly below.

6.4.3 Is Quiescence Distinct from the Hard State? Sgr A* vs. A 0620 - 00

As a final point, I would like to mention some additional aspects of jet physics where we might glean some insight by studying two sources at the very bottom of the accretion range, Sgr A* and A 0620-00. By virtue of the FP, we can also compare these black holes to each other in order to derive a better picture of the buildup of accretion activity at the lowest levels.

The change in the correlation toward quiescent luminosities, as well as reports of X-ray spectral steepening [22, 20], has led to some speculation that quiescence is not simply a smooth continuation of the hard state down to lower accretion rates. On the one hand, it seems that steady radio jets continue to exist down to at least X-ray luminosities of $\sim 10^{-8.5}$ L_{Edd} [41] and that the correlation is maintained to quiescent-level accretion rates, at least for the one source for which there are good data. The persistence of the same correlation down into quiescence would argue against scenarios such as [104] mentioned above, whose coupled RIAF-jet model predicts a turnover to a form $L_R \propto L_X^{1.23}$ somewhere below 10^{-5}–10^{-6} L_{Edd}. On the other hand, in some sources we do see the radio begin to drop faster compared to the X-ray emission at higher accretion rates than classically considered quiescent, though not exactly with the same 1.23 index. However, the data are currently not sufficient to determine if there is also a mass dependence of this turnover, which could explain some of these differences from source to source.

The rise phase out of quiescence can also be very steep in the other direction: in several outbursts, the X-rays are seen to rise very rapidly while almost no change is observed in the radio band. Such an extreme X-ray rise is especially interesting in the context of similar behavior in Sgr A*. Sgr A* is the only accreting black hole, for which we have good radio and X-ray data, that does not fit on the FP. Instead, it sits in quiescence several orders of magnitude below the mass-scaled FP correlation, in the direction of weak X-ray flux. However, during the approximately daily X-ray flares [e.g., 4, 5], Sgr A* approaches the X-ray flux at which the FP radio/X-ray correlation would be predicted to take hold, though it has yet to come within an order of magnitude of actually achieving the emissivity required to test this idea (see, e.g., [64] and Fig. 6.6). Even so, the sharp X-ray rise seems suggestively analogous to that displayed by the 2004 rising data for GX 339-4, shown in the bottom panel of Fig. 6.5. If Sgr A*'s behavior is indeed a reflection of the same physics occurring in BHBs, then we could be witnessing its flickering attempts at outburst activity from out of quiescence, but that it is never sufficiently fueled to reach the sustainable levels of the FP. Although we do not yet have an understanding of the mass dependence of any steepening of the FP correlation, it seems likely that Sgr A* is sitting far below (in X-ray flux) the observations of the rise data from a BHB, when scaled to a comparable mass. Unfortunately equivalent luminosities ($L_X \leq 10^{-10} L_{\mathrm{Edd}}$) would be difficult to observe for BHBs, and in observing long enough to obtain a significant measure of such an X-ray flux we would invariably be integrating well over any similar fast flaring timescales, when scaled for the smaller dynamical times. Potentially, the only way to test this hypothesis is to wait (and hope we observe) a flare in Sgr A* at least an order of magnitude brighter than any seen so far, assumedly quite a rare event if one assumes a power-law distribution in flare magnitude. Perhaps we will get lucky and one of the S-stars will travel a bit too close during periastron, suffering tidal disruption and triggering a major accretion event (e.g., [2])!

6.4.3.1 Jet Launching Physics: Clues from the Lowest Luminosities

Aside from this speculation, Sgr A* is very interesting for understanding jet formation at the lowest accretion rates because it is so close that we actually have very high-quality data, despite its weak (in Eddington units) emission relative to any other black hole we know. Its radio through IR emission is clearly due to synchrotron radiation, given the polarization and high brightness temperature (see [74], and references therein), and the presence of the signature flat/inverted radio spectrum likely indicates a weak, collimated jet outflow. Such a jet has not yet been imaged, however, leading to suggestions that the radio emission originates in an uncollimated disk wind, despite many difficulties with this interpretation. The opaque and stratified geometry required to reproduce a flat/inverted spectrum can in principle be fulfilled by a RIAF with synchrotron-emitting thermal particles, but the radial profile of such a flow would normally result in a steep falloff in the radio flux with decreasing frequency (see, for example, [80]). The only way to avoid this problem is to posit the existence of an additional nonthermal power-law tail

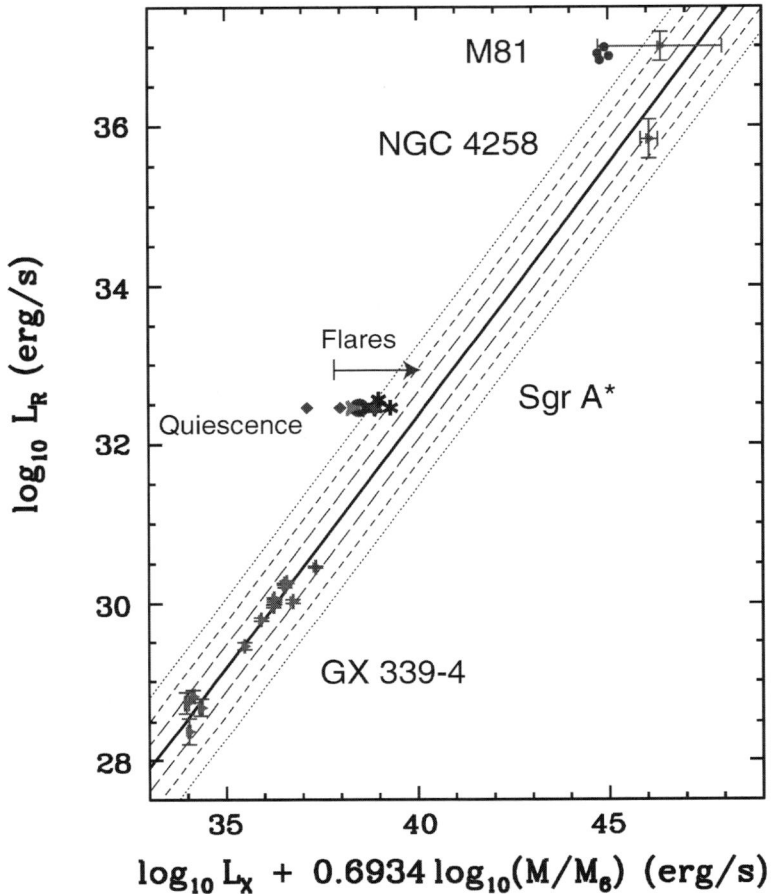

Fig. 6.6 The fundamental plane (FP) radio/X-ray luminosity correlation for the three best measured sources bracketing (in scaled flux) the Galactic center supermassive black hole, Sgr A*: the BHB GX 339-4, and the LLAGN NGC 4358 and M 81. Both quiescent and flaring states of Sgr A* are indicated. The two LLAGN data points and their respective error bars represent the average and rms variation of all prior non-simultaneous observations. For M 81, the four dark (*red* in electronic version) points indicate results from simultaneous radio/X-ray observations [69]. The solid line indicates the best fit correlation using linear regression from Monte Carlo simulations of the data, with contours in average scatter $< \sigma >$ from the correlation represented as increasingly finer dashed lines (figure modified from [64])

in the electron distribution, that is somehow engendered in the thermal accretion flow [108]. To be fair, the amount of electrons required in the tail is very small, and thus minor magnetic events in the flow could potentially account for such limited particle acceleration. However, this idea still seems rather ad hoc, given that it has only been necessary to invoke for the one source where any jets would be difficult to resolve through scattering effects in the Galactic plane. Jets naturally provide a solution to the problem, as they are known particle accelerators, and in radio spectral shape and polarization Sgr A* resembles another LLAGN in a similar grand design spiral galaxy, M 81*, where weak jets have been resolved [7, 69].

The physics necessary to explain the lack of extended jets in Sgr A* also fits in with our understanding of particle acceleration near the black hole. It is interesting to note that Sgr A* also provides the most stringent limits on its internal particle distributions than any other black hole we can currently study. While optical depth effects allow us to derive important scaling relations despite our lack of precise knowledge of such particle distributions in most sources, it hides details of the internal physics that would allow us to constrain particle acceleration or derive the total jet internal power due to lack of information about the minimum energy in the radiating lepton distribution. In Sgr A* we can deduce directly from measurements that the particles responsible for the highest energy emission are predominantly thermal in nature, because any significant accelerated power-law tail would reveal itself prominently in the IR band. In contrast, the observed IR spectrum reflects an extremely steep drop more consistent with weak to absent acceleration [44, 45]. Thanks to advances in adaptive optics, along with identifying Sgr A* for the first time in the IR, these observations also led to the discovery of sporadic flaring events with marginal evidence for hardening in a power-law tail in the underlying particle distribution. However, statistics are still a challenge, and the two main observational groups do not yet agree on the exact spectral index during quiescence and flares.

I propose that the flaring in both IR and X-ray is an indication that while plasma jets are still formed at low luminosities, the internal structures necessary to accelerate particles efficiently within them are not yet able to maintain themselves. The resulting lack of a stable high-energy particle population compared to accreting black holes at higher accretion rates leads directly to a prediction of very compact jet photospheres. Going back to Fig. 6.1, the measured size of the photosphere at a given radio-observing frequency is dependent on the broadband extent of the optically thin emission, in accreting black holes represented by power-law emission. Therefore, if the jets in Sgr A* contain predominantly thermal distributions of particles, the lack of a strong power-law tail would predict a more compact photosphere and result in a less elongated Gaussian, difficult to distinguish from predictions of a RIAF. Even small elongation would be easily hidden behind the scattering screen of intervening Galactic Plane material (see discussion in [65]). One interesting test for the future will be to look for elongation in the photosphere measured with VLBI during bright X-ray flares, ideally at high frequencies such as reported in [24].

As for the flares themselves, they are providing hints that nonthermal processes also contribute significantly to the X-ray emission in both weak LLAGN and hard-state BHBs. Because of the lack of a thermal accretion disk as a source of thermal seed photons in Sgr A*, the most likely explanation for the simultaneous nonthermal IR/X-ray flares is the synchrotron self-Compton process [67, 109], although a direct synchrotron contribution may be present for very short episodes. It is notable that all "flavors" of theory currently agree on this point, regardless of whether they favor inflow or outflow accretion solutions. The submm/IR "bump" and nonthermal X-ray flaring emission must originate in a region of hot, magnetized plasma very close to the supermassive black hole, because of the short variability timescales (e.g., [4]). If the same geometry is present in BHBs at low accretion rates, as seems likely given the success of various scaling arguments, then a synchrotron/SSC contribution to the X-rays is even more likely because of the more compact relative scales.

By looking at actual similarly weakly accreting BHBs when possible, we can attempt to test this theory. As detailed in Sect. 6.4.1, the same physical model for brighter weakly accreting states seems to well describe data across the mass scale. To really probe quiescence is much more challenging in BHBs. For instance, A 0620-00 is only several orders of magnitude brighter (in Eddington units) than Sgr A* in the X-ray band, but for a Galactic BHB this power translates into a very low count rate. In [43] we reported on preliminary results attempting to fit the rather poor-quality X-ray spectrum, where we did find that an SSC-dominated model was favored, similar to the case in Sgr A*. We also found evidence for weaker particle acceleration in comparison with other BHBs at higher accretion rates, supporting the interpretation that the processes responsible for particle acceleration in the jets break down below some critical $\dot{m}_c$ and which may bear some relevance for the behavior rising out of and decaying back into quiescence. If acceleration is indeed associated with internal shocks, perhaps below $\dot{m}_c$ the flow does not have the initial conditions required for the necessary multiple-velocity ejecta. Or it may be that the jet plasma simply can no longer be accelerated to bulk velocities beyond the fast magnetosonic point where shock-like disruptions could form. In any event, it is an intriguing possibility that somewhere between A 0620-00 and Sgr A*, or between $L_X = 10^{-10}$ and $10^{-7} L_{Edd}$, the particle acceleration process in the jets begins to break down. With new all-sky monitors such as LOFAR soon discovering many new transients, we can hope to find an even more massive, closer BHB in quiescence to test some of these premises, at the same time that we are currently investigating some of the more theoretical aspects (Polko et al., in prep.).

6.5 Conclusions

As is often the case with theoretical studies, there is no definitive conclusion to this chapter, other than to say that the story continues. The last decade has been extremely productive in terms of furthering our understanding of the systematic behavior of BHBs, and in particular simultaneous multiwavelength campaigns have for the first time allowed us to connect accretion behavior over the extreme ranges of the black hole mass scale. While the fundamental plane of black hole accretion provides a powerful tool for comparing the fast-varying trends in BHBs to the protracted duty cycles of AGN, it represents just the tip of the iceberg in terms of understanding the subtleties of potential mass and power scalings. First of all, we now know that the underlying correlation for BHBs does not have exactly the same slope for all sources, though its spread is still rather narrow, and we see different normalization "tracks" during rise and outburst that has yet to be factored into the AGN results. But perhaps more importantly, when comparing BHB cycles based on observing individual sources to populations of AGN selected from samples, there are enormous elephants in the room such as how to account for the role of spin and the kinds of differences in accretion phases expected in AGN of various galaxy morphologies that should not be relevant for BHBs. In order to assess the extent to which mass scaling translates into an exact AGN equivalent for every BHB state, we need to move away from the comfortable lower right-hand side of the hardness–intensity diagram, away from steady-state jets and toward a better understanding of the more extreme transitional behavior that occurs between the hard and soft accretion states. Several groups are now currently tackling this progression, and armed with the experience and knowledge described here and elsewhere in this volume, I anticipate that we will be able to address these open issues within the next decade. At the same time, it is important to note that the impending loss of the RXTE ASM with no immediate replacement will be an undeniable detriment to our ability to obtain the high-quality multiwavelength monitoring data that enabled these discoveries in the first place. On the other hand, the all-sky monitoring to find new classes of sources and transient behavior, now with γ-rays using Fermi as well as ground-based radio facilities like LOFAR and other SKA pathfinders, will undoubtably flesh out the picture we have now of black hole accretion in transition. But aside from the high-quality multiwavelength data, I would like to end by returning once again to emphasize the theoretical framework that was necessary before the observed trends could be "translated" into a more global picture. We need a new foothold on the theory of low-to-high accretion transition pathway, and once we have it this foothold will hopefully help open the door to our understanding not only of jet quenching in BHBs but also cycles of AGN activity producing super-cavities in clusters and potentially feedback. I think that the small scales still have quite a bit left to teach us about the largest and that we have a long but interesting way to go before we can consider our paradigms complete.

References

1. J. Albert, E. Aliu, H. Anderhub, et al.: ApJ, **665**, L51–L54 (Aug. 2007)
2. T. Alexander, M. Livio: ApJ, **560**, L143–L146 (Oct. 2001)
3. P. Anninos, P.C. Fragile, J.D. Salmonson: ApJ, **635**, 723–740 (Dec. 2005)
4. F.K. Baganoff, M.W. Bautz, W.N. Brandt, G. Chartas, E.D. Feigelson, G.P. Garmire, Y. Maeda, M. Morris, G.R. Ricker, L.K. Townsley, F. Walter: Nature, **413**, 45–48 (Sept. 2001)
5. F.K. Baganoff, Y. Maeda, M. Morris, M.W. Bautz, W.N. Brandt, W. Cui, J.P. Doty, E.D. Feigelson, G.P. Garmire, S.H. Pravdo, G.R. Ricker, L.K. Townsley: ApJ, **591**, 891–915 (July 2003)
6. A.M. Beloborodov: ApJ, **510**, L123–L126 (Jan. 1999)
7. M.F. Bietenholz, N. Bartel, M. P. Rupen: ApJ, **615**, 173–180 (Nov. 2004)
8. R.D. Blandford, M.C. Begelman: MNRAS, **303**, L1–L5 (Feb. 1999)
9. R.D. Blandford, A. Königl: ApJ, **232**, 34–48 (Aug. 1979)
10. R.D. Blandford, R.L. Znajek: MNRAS, **179**, 433–456 (May 1977)
11. V. Bosch-Ramon, J.M. Paredes, M. Ribó, J.M. Miller, P. Reig, J. Martí: ApJ, **628**, 388–394 (July 2005)

12. G.C. Bower, H. Falcke, R.M. Herrnstein, J. Zhao, W.M. Goss, D.C. Backer: Science, **304**, 704–708 (Apr. 2004)
13. G.C. Bower, W.M. Goss, H. Falcke, D.C. Backer, Y. Lithwick: ApJ, **648**, L127–L130 (Sept. 2006)
14. S. Chakrabarti, L.G. Titarchuk: ApJ, **455**, 623–639 (Dec. 1995)
15. S.K. Chakrabarti: MNRAS, **283**, 325–335 (Nov. 1996)
16. I. Chattopadhyay, S. Das, S.K. Chakrabarti: MNRAS, **348**, 846–856 (Mar. 2004)
17. M. Choudhury, A.R. Rao, S.V. Vadawale, A.K. Jain: ApJ, **593**, 452–462 (Aug. 2003)
18. P.S. Coppi: The physics of hybrid thermal/non-thermal plasmas. In: ASP Conf. Ser. 161: High Energy Processes in Accreting Black Holes, p. 375 (1999)
19. S. Corbel, R.P. Fender, A.K. Tzioumis, M. Nowak, V. McIntyre, P. Durouchoux, R. Sood: A&A, **359**, 251–268 (July 2000)
20. S. Corbel, E. Koerding, P. Kaaret: MNRAS, **389**, 1697–1702 (Oct. 2008)
21. S. Corbel, M. Nowak, R.P. Fender, A.K. Tzioumis, S. Markoff: A&A, **400**, 1007 (2003)
22. S. Corbel, J.A. Tomsick, P. Kaaret: ApJ, **636**, 971–978 (Jan. 2006)
23. T. di Matteo, A. Celotti, A.C. Fabian: MNRAS, **291**, 805–810 (Nov. 1997)
24. S.S. Doeleman, J. Weintroub, A.E.E. Rogers, R. Plambeck, R. Freund, R.P.J. Tilanus, P. Friberg, L.M. Ziurys, J.M. Moran, B. Corey, K.H. Young, D.L. Smythe, M. Titus, D.P. Marrone, R.J. Cappallo, D.C.-J. Bock, G.C. Bower, R. Chamberlin, G.R. Davis, T.P. Krichbaum, J. Lamb, H. Maness, A.E. Niell, A. Roy, P. Strittmatter, D. Werthimer, A.R. Whitney, D. Woody: Nature, **455**, 78–80 (Sept. 2008)
25. D.M. Eardley, A.P. Lightman, S.L. Shapiro: ApJ, **199**, L153–L155 (Aug. 1975)
26. A.A. Esin, J.E. McClintock, R. Narayan: ApJ, **489**, 865 (Nov. 1997)
27. A.C. Fabian, P.W. Guilbert, C. Motch, M. Ricketts, S.A. Ilovaisky, C. Chevalier: A&A, **111**, L9+ (July 1982)
28. H. Falcke, T. Beckert, S. Markoff, E. Körding, R. Fender, G.C. Bower: The power of jets: new clues from radio circular polarization and x-rays. In: R. Sunyaev, M. Gilfanov, E. Churazov (eds.) Lighthouses of the Universe, Springer Verlag, ESO Astrophysics Symposia, 428 (2002)
29. H. Falcke, P.L. Biermann: A&A, **293**, 665–682 (Jan. 1995)
30. H. Falcke, P.L. Biermann: A&A, **342**, 49–56 (Feb. 1999)
31. H. Falcke, E. Körding, S. Markoff: A&A, **414**, 895 (2004)
32. H. Falcke, S. Markoff: A&A, **362**, 113 (Oct. 2000)
33. R. Fender: Jets from X-Ray Binaries, pp. 381–419. Compact stellar X-ray sources (Apr. 2006)
34. R.P. Fender: MNRAS, **322**, 31–42 (Mar. 2001)
35. R.P. Fender, R.M. Hjellming, R.P.J. Tilanus, G.G. Pooley, J.R. Deane, R.N. Ogley, R.E. Spencer: MNRAS, **322**, L23–L27 (Apr. 2001)
36. J. Ferreira: A&A, **319**, 340–359 (Mar. 1997)
37. J. Ferreira, G. Pelletier: A&A, **276**, 625–636 (Sept. 1993)
38. J. Ferreira, G. Pelletier: A&A, **295**, 807–832 (Mar. 1995)
39. J. Ferreira, P.-O. Petrucci, G. Henri, L. Saugé, G. Pelletier: A&A, **447**, 813–825 (Mar. 2006)
40. P.C. Fragile, D.L. Meier: ApJ, **693**, 771–783 (Mar. 2009)
41. E. Gallo, R.P. Fender, J.C.A. Miller-Jones, A. Merloni, P.G. Jonker, S. Heinz, T.J. Maccarone, M. van der Klis: MNRAS, **370**, 1351–1360 (Aug. 2006)
42. E. Gallo, R.P. Fender, G.G. Pooley: MNRAS, **344**, 60–72 (Sept. 2003)
43. E. Gallo, S. Migliari, S. Markoff, J.A. Tomsick, C.D. Bailyn, S. Berta, R. Fender, J.C.A. Miller-Jones: ApJ, **670**, 600–609 (Nov. 2007)
44. R. Genzel, R. Schödel, T. Ott, A. Eckart, T. Alexander, F. Lacombe, D. Rouan, B. Aschenbach: Nature, **425**, 934–937 (Oct. 2003)
45. A.M. Ghez, S.A. Wright, K. Matthews, D. Thompson, D. Le Mignant, A. Tanner, S.D. Hornstein, M. Morris, E.E. Becklin, B.T. Soifer: ApJ, **601**, L159–L162 (Feb. 2004)
46. D. Giannios: A&A, **437**, 1007–1015 (July 2005)
47. D. Giannios, N.D. Kylafis, D. Psaltis: A&A, **425**, 163–169 (Oct. 2004)

48. M. Gierlinski, A.A. Zdziarski, C. Done, W.N. Johnson, K. Ebisawa, Y. Ueda, F. Haardt, B.F. Phlips: MNRAS, **288**, 958–964 (July 1997)
49. F. Haardt, L. Maraschi: ApJ, **380**, L51–L54 (Oct. 1991)
50. D.C. Hannikainen, R.W. Hunstead, D. Campbell-Wilson, R.K. Sood: A&A, **337**, 460–464 (Sept. 1998)
51. S. Heinz, R.A. Sunyaev: MNRAS, **343**, L59–L64 (Aug. 2003)
52. L.C. Ho: ApJ, **516**, 672–682 (May 1999)
53. R.I. Hynes, C.W. Mauche, C.A. Haswell, C.R. Shrader, W. Cui, S. Chaty: ApJ, **539**, L37 (Aug. 2000)
54. J.R. Jokipii: ApJ, **313**, 842 (Feb. 1987)
55. K.I. Kellermann, R.C. Vermeulen, J.A. Zensus, M.H. Cohen: AJ, **115**, 1295–1318 (Apr. 1998)
56. W.H.G. Lewin, J. van Paradijs, E.P.J. van den Heuvel (eds.): X-ray binaries (1995)
57. M. Livio, J.E. Pringle, A.R. King: ApJ, **593**, 184–188 (Aug. 2003)
58. T.J. Maccarone, P.S. Coppi, J. Poutanen: ApJ, **537**, L107–L110 (July 2000)
59. P. Magdziarz, A.A. Zdziarski: MNRAS, **273**, 837–848 (Apr. 1995)
60. D. Maitra, S. Markoff, C. Brocksopp, M. Noble, M. Nowak, J. Wilms: MNRAS, **398**, 1638 (2009)
61. J. Malzac, A.M. Beloborodov, J. Poutanen: MNRAS, **326**, 417–427 (Sept. 2001)
62. J. Malzac, A. Merloni, A.C. Fabian: MNRAS, **351**, 253–264 (June 2004)
63. D. Maoz, N.M. Nagar, H. Falcke, A.S. Wilson: ApJ, **625**, 699–715 (June 2005)
64. S. Markoff: ApJ, **618**, L103 (2005)
65. S. Markoff, G.C. Bower, H. Falcke: MNRAS, **379**, 1519–1532 (Aug. 2007)
66. S. Markoff, H. Falcke, R. Fender: A&A, **372**, L25–L28 (2001)
67. S. Markoff, H. Falcke, F. Yuan, P.L. Biermann: A&A, **379**, L13–L16 (2001)
68. S. Markoff, M. Nowak, S. Corbel, R. Fender, H. Falcke: A&A, **397**, 645 (2003)
69. S. Markoff, M. Nowak, A. Young, H.L. Marshall, C.R. Canizares, A. Peck, M. Krips, G. Petitpas, R. Schödel, G.C. Bower, P. Chandra, A. Ray, M. Muno, S. Gallagher, S. Hornstein, C.C. Cheung: ApJ, **681**, 905–924 (July 2008)
70. S. Markoff, M.A. Nowak: ApJ, **609**, 972–976 (July 2004)
71. S. Markoff, M.A. Nowak, J. Wilms: ApJ, **635**, 1203–1216 (Dec. 2005)
72. D.L. Meier: ApJ, **548**, L9–L12 (Feb. 2001)
73. D.L. Meier: Ap&SS, **300**, 55–65 (Nov. 2005)
74. F. Melia, H. Falcke: ARA&A, **39**, 309–352 (2001)
75. A. Merloni, A.C. Fabian: MNRAS, **332**, 165–175 (May 2002)
76. A. Merloni, S. Heinz, T. di Matteo: MNRAS, **345**, 1057–1076 (Nov. 2003)
77. S. Migliari, J.A. Tomsick, S. Markoff, E. Kalemci, C.D. Bailyn, M. Buxton, S. Corbel, R.P. Fender, P. Kaaret: ApJ, **670**, 610–623 (Nov. 2007)
78. C. Motch, S.A. Ilovaisky, C. Chevalier: A&A, **109**, L1–L4 (May 1982)
79. N.M. Nagar, H. Falcke, A.S. Wilson, J.S. Ulvestad: A&A, **392**, 53–82 (Sept. 2002)
80. R. Narayan, R. Mahadevan, J.E. Grindlay, R.G. Popham, C. Gammie: ApJ, **492**, 554–568 (Jan. 1998)
81. R. Narayan, I. Yi: ApJ, **428**, L13–L16 (June 1994)
82. J.M. Paredes, J. Martí, M. Ribó, M. Massi: Science, **288**, 2340–2342 (June 2000)
83. J. Poutanen, R. Svensson: ApJ, **470**, 249–268 (Oct. 1996)
84. J. Poutanen, I. Vurm: ApJ, **690**, L97 (2009)
85. R. Remillard, E. Morgan, D. Smith, E. Smith: Xte j1118+480. IAU Circ, 7389 (Mar. 2000)
86. S.L. Robertson, D.J. Leiter: ApJ, **565**, 447–454 (Jan. 2002)
87. S.L. Robertson, D.J. Leiter: MNRAS, **350**, 1391–1396 (June 2004)
88. G.E. Romero, H.R. Christiansen, M. Orellana: ApJ, **632**, 1093–1098 (Oct. 2005)
89. D.M. Russell, R.P. Fender, R.I. Hynes, C. Brocksopp, J. Homan, P.G. Jonker, M.M. Buxton: MNRAS, **371**, 1334–1350 (Sept. 2006)

90. D.M. Russell, T.J. Maccarone, E.G. Körding, J. Homan: MNRAS, **379**, 1401–1408 (Aug. 2007)
91. G.B. Rybicki, A.P. Lightman: Radiative Processes in Astrophysics. Wiley-Interscience, New York, p. 393 (1979)
92. N.I. Shakura, R.A. Sunyaev: A&A, **24**, 337–355 (1973)
93. S.L. Shapiro, A.P. Lightman, D.M. Eardley: ApJ, **204**, 187–199 (Feb. 1976)
94. Z.-Q. Shen, K.Y. Lo, M.-C. Liang, P.T.P. Ho, J.-H. Zhao: Nature, **438**, 62–64 (Nov. 2005)
95. R.A. Sunyaev, L.G. Titarchuk: A&A, **86**, 121–138 (June 1980)
96. M. Tagger, R. Pellat: A&A, **349**, 1003–1016 (Sept. 1999)
97. M. Tagger, P. Varnière: ApJ, **652**, 1457–1465 (Dec. 2006)
98. M. Tagger, P. Varnière, J. Rodriguez, R. Pellat: ApJ, **607**, 410–419 (May 2004)
99. L. Titarchuk: ApJ, **434**, 570–586 (Oct. 1994)
100. L. Titarchuk, Y. Lyubarskij: ApJ, **450**, 876–882 (Sept. 1995)
101. P. Varnière, M. Tagger: A&A, **394**, 329–338 (Oct. 2002)
102. G. Wardziński, A.A. Zdziarski: MNRAS, **314**, 183–198 (May 2000)
103. J. Wilms, M.A. Nowak, K. Pottschmidt, G.G. Pooley, S. Fritz: A&A, **447**, 245–261 (Feb. 2006)
104. F. Yuan, W. Cui: ApJ, **629**, 408–413 (Aug. 2005)
105. F. Yuan, W. Cui, R. Narayan: ApJ, **620**, 905–914 (Feb. 2005)
106. F. Yuan, S. Markoff, H. Falcke: A&A, **383**, 854–863 (Mar. 2002)
107. F. Yuan, S. Markoff, H. Falcke, P.L. Biermann: A&A, **391**, 139 (2002)
108. F. Yuan, E. Quataert, R. Narayan: ApJ, **598**, 301–312 (Nov. 2003)
109. F. Yuan, E. Quataert, R. Narayan: ApJ, **606**, 894–899 (May 2004)
110. A.A. Zdziarski, P. Lubiński, D.A. Smith: MNRAS, **303**, L11–L15 (1999)

Chapter 7
Jets in Active Galactic Nuclei

A.P. Marscher

Abstract The jets of active galactic nuclei can carry a large fraction of the accreted power of the black hole system into interstellar and even extragalactic space. They radiate profusely from radio to X-ray and γ-ray frequencies. In the most extreme cases, the outward flow speeds correspond to high Lorentz factors that can reach 40 or more. This chapter displays images at various wavebands as well as light curves and continuum spectra that illustrate the variability with location, time, and frequency of the emission from compact, parsec-, and subparsec-scale jets. It presents a physical framework for investigating many aspects of the structure and dynamical processes from such data.

7.1 Introduction

Since accretion onto stellar-mass black holes is often accompanied by a pair of jets emanating from the rotational poles, we have the right to expect an even more spectacular brand of jets in active galactic nuclei (AGN). Indeed, AGN jets were the first to be observed. These systems, with their supermassive black holes of millions to billions of solar masses, can produce ultraluminous jets that bore their way through the interstellar medium and into intergalactic space, ending in huge, billowy lobes punctuated by hotspots. Even the less luminous versions disturb the gas in their host galaxies. During the epoch of galaxy formation, this may have controlled the infall of gas so that the central black hole and galactic bulge grew in step with each other.

The wide variety of AGN is reflected in the diversity of their jets. These range from relatively slow, weak, and poorly collimated flows in Seyfert galaxies to strong jets with relativistic speeds in Fanaroff–Riley (FR) I radio galaxies and BL Lacertae (BL Lac) objects to the most luminous, highly focused, and relativistic beams in FR II radio galaxies and radio-loud quasars. The reason for this dichotomy is

A.P. Marscher (✉)

Institute for Astrophysical Research, Boston University, Boston, MA 02215, USA,
marscher@bu.edu

Marscher, A.P.: *Jets in Active Galactic Nuclei.* Lect. Notes Phys. **794**, 173–201 (2010)
DOI 10.1007/978-3-540-76937-8_7

unclear, although it probably relates to the galactic environment, with Seyfert galaxies usually of spiral morphology, FR I sources hosted by giant elliptical galaxies in rich clusters, and FR II objects in elliptical galaxies lying in somewhat less dense groupings. This might influence the rate of accretion of gas or spin of the black hole, one (or both) of which could be the factor that determines how fast and well focused an outflow is propelled in the two polar directions.

The most extreme jets with highly relativistic flow velocities are in fact the best-studied variety. This is because relativistic beaming of the radiation amplifies the brightness of these jets so that they can be prominent even at relatively low luminosities if one of the jets points within several degrees of the line of sight. Emission from the jet often dominates the spectral energy distribution of the source in this case, so that observations across the electromagnetic spectrum serve to define the properties of the jet. For this reason, and because one often learns the most by studying extreme cosmic phenomena, this chapter will emphasize relativistic jets pointing nearly at us. AGN with such jets are termed "blazars."

Because of space limitations, this chapter discusses only a fraction of the phenomenology and physics of AGN jets. The interested reader should consult other sources of information, especially [6, 9, 36, 49, 25].

7.2 Observed Properties of Jets

The jets of blazars are indeed extreme, with kinetic powers that can exceed the luminosities of entire galaxies and with flow speeds sometimes greater than $0.999c$. This section contains a brief overview of the characteristics of jets compiled from a wealth of imaging, spectral, and multiwaveband monitoring and polarization data. Subsequent sections will examine these properties in more detail from a physical standpoint.

7.2.1 Images at Different Wavebands

Figure 7.1 shows an example of a jet on both parsec and kiloparsec scales, while Fig. 7.2 compares the X-ray and radio emission from a jet on kiloparsec scales. The jet, which broadens with distance from the nucleus, is defined by knots of emission. There is an overall likeness on the different scales, which indicates that jet flows are roughly self-similar over several orders of magnitude in distance from the black hole. This is easiest to understand if the jets are mainly fluid phenomena so that they are governed by the laws of gas dynamics and magnetohydrodynamics.

Images of jets contain a bright, nearly unresolved "core" at the upstream end. This roughly stationary feature can represent a physical structure, such as a standing shock wave [23, 60, 16], or simply the most compact bright emission that is visible at the frequency of observation. In the latter case, the core can be at a different position at different wavelengths. The remainder of the jet tends to be knotty, although

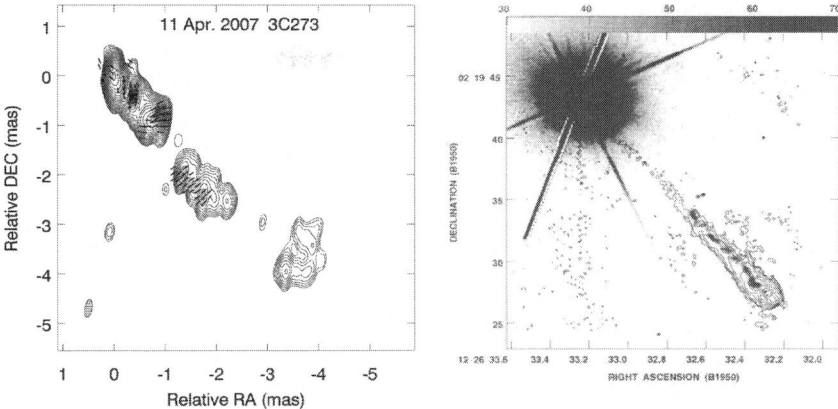

Fig. 7.1 Example of an AGN jet, in the quasar 3C 273. *Left*: on parsec scales at radio frequency 43 GHz; *right*: on kiloparsec scales at radio frequency 1.7 GHz (contours) and optical wavelengths centered on 606 nm (grayscale). The core (brightest feature, at the northeast end) of the parsec-scale jet is located very close to the center of the overexposed nucleus in the upper-left corner of the right-hand panel. The scale of the image on the right is 4000 times larger than that on the left. Contours of the radio images in this and subsequent figures are in increments of a factor of 2. At the redshift of 0.158, $1'' = 2.7\,\text{kpc}$. *Left*: author's image, derived from data obtained with the Very Long Baseline Array (VLBA); *right*: from [4], from data obtained with the NASA's Chandra X-ray Observatory and the Very Large Array (VLA). The VLBA and VLA are instruments of the National Radio Astronomy Observatory, a facility of the National Science Foundation operated under cooperative agreement by Associated Universities, Inc.

regions of smoothly varying intensity are sometimes present. On parsec scales imaged by very long baseline interferometry (VLBI), we find that many of the knots move, while other emission features appear stationary. Motions probably occur on kiloparsec scales as well, but monitoring over very long times is needed to detect such motions.

7.2.2 Motions Inside Jets and Relativistic Beaming of Radiation

Strong evidence that some jets have relativistic flow velocities comes from sequences of VLBI images that show apparent superluminal motions of bright knots away from the core, as illustrated in Fig. 7.3. The apparent motion is faster than the actual speed βc because the emitting plasma approaches us at nearly the speed of light c. Therefore, it moves almost as fast as the radio waves it emits, so that the arrival times at Earth are compressed relative to those in the rest frame of the plasma. The apparent velocity in the plane of the sky is given by

$$v_{\text{app}} = \frac{\beta c \sin\,\theta}{1 - \beta \cos\,\theta}, \qquad (7.1)$$

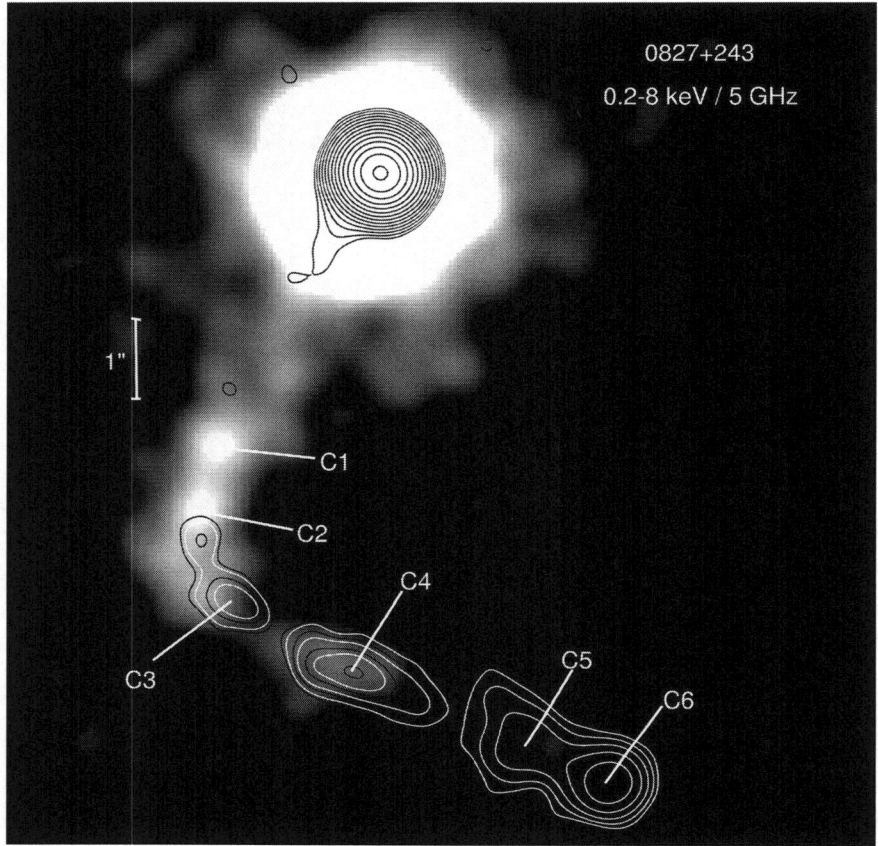

Fig. 7.2 Image of the jet of the quasar 0827+243 (OJ 248) at radio frequency 5 GHz (contours) and at X-ray photon energies (*grayscale*). The nucleus is the bright region near the *top*. At the redshift (0.939) of this quasar, $1'' = 7.2$ kpc on the sky, but $\sim$100 kpc when de-projected. From [42]

where θ is the angle between the velocity vector and the line of sight. The apparent velocity is a maximum at $\theta = \sin^{-1}(1/\Gamma)$, where $\Gamma = (1 - \beta^2)^{-1/2}$ is the Lorentz factor of the centroid of the "blob" of emitting plasma.

While the time scales involved make superluminal motion easiest to detect on parsec scales, a superluminal feature has been seen at least 120 parsecs from the nucleus in the relatively nearby radio galaxy M 87 [20].

The relativistic motion produces a strong Doppler effect that increases the observed frequency and decreases the time scales of variability by a factor

$$\delta = [\Gamma(1 - \beta \cos \theta)]^{-1}, \tag{7.2}$$

which is termed the "Doppler factor." It is a maximum of $(1 + \beta)\Gamma$ when $\theta = 0$ and equals Γ when $\theta = \sin^{-1}(1/\Gamma)$. Because of relativistic aberration, most of

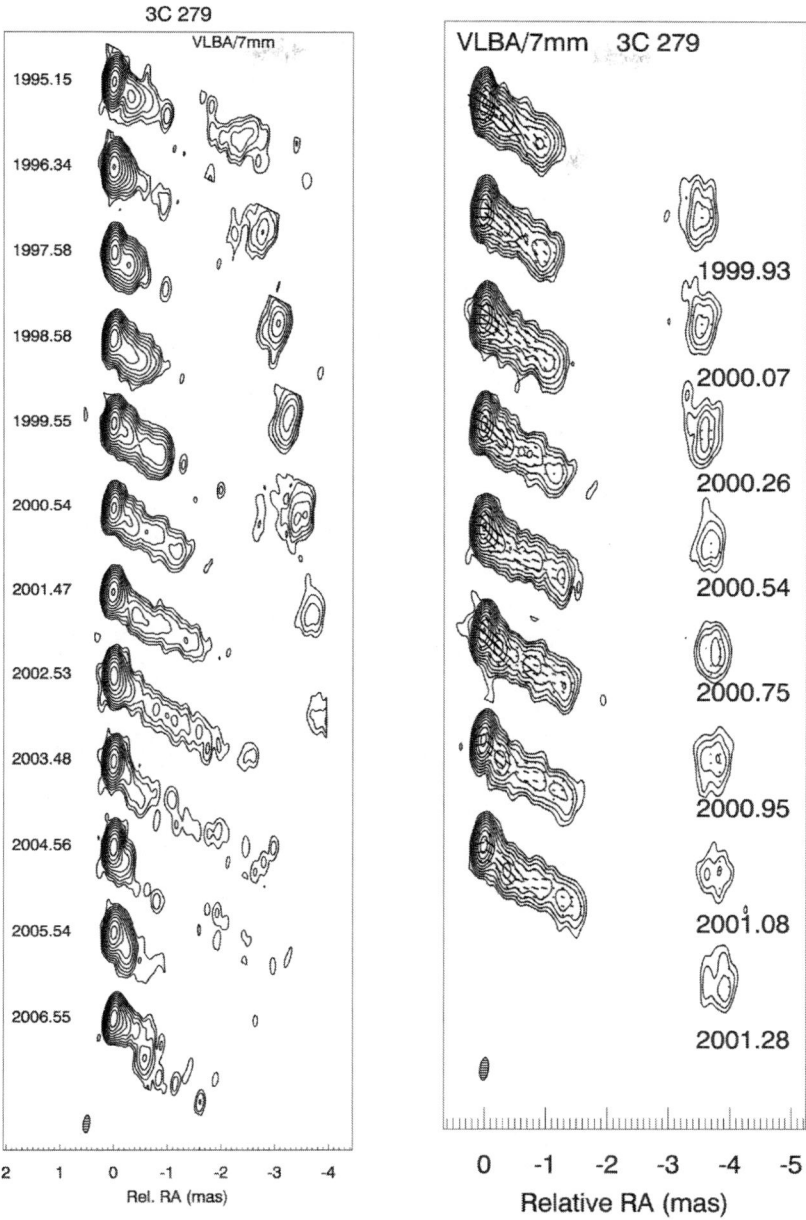

Fig. 7.3 *Left*: Twelve-year sequence of annual VLBI images of the quasar 3C 279 at 43 GHz. At the redshift of the quasar, 1 mas = 6.3 pc. The feature at the narrow end of the jet, closest to the supermassive black hole, is referred to as the "core" and is assumed to be stationary. The motion of the outlying knot has an apparent superluminal speed of ∼10*c*. Note the change in direction of the innermost jet with time. *Right*: Sequence of roughly bimonthly images of 3C 279 showing the systematic growth of the length of the inner jet at a speed of 13*c*. The linear polarization at 43 GHz is indicated by the sticks inside the contours. Images from author and S. Jorstad

the emission is beamed into a cone with an opening half-angle $\sim \Gamma^{-1}$, so that the flux density F_ν of the plasma is boosted by a factor of δ^2. Since the frequency ν is also increased, if $F_\nu \propto \nu^{-\alpha}$, the flux density is a factor of $\delta^{2+\alpha}$ higher than it would be if the velocity were much less than the speed of light. This applies if the emission comes from a feature in a steady state – a standing shock wave, for example. However, if the emitting structure moves, as does a superluminal knot, there is another factor of δ from the time compression ($\Delta T \propto \nu^{-1} \propto \delta^{-1}$). We then have

$$F_\nu \propto \delta^{2+\alpha} \nu^{-\alpha} \quad \text{(steady state)}, \tag{7.3}$$

$$F_\nu \propto \delta^{3+\alpha} \nu^{-\alpha} \quad \text{(moving structure)}. \tag{7.4}$$

Jets have basically conical shapes, with opening half-angles ϕ ranging from $\sim 5°$ to considerably less than $1°$. A study of jets in 15 AGN indicates that $\phi \propto \Gamma^{-1}$ [43]. This implies that acceleration of jets to the highest velocities is related to their collimation toward a nearly cylindrical geometry.

7.2.3 Polarization

Jets are usually linearly polarized at a level that can be as high as 50% or greater for the more extended features, but more typically a few percent in the core. On the other hand, the circular polarization, with few exceptions, is generally less than 1%. This is the signature of incoherent synchrotron radiation, with relativistic electrons over a wide range of pitch angles spiraling in a magnetic field. The field can be nearly uniform over the resolution beam of the interferometer in the case of linear polarization of tens of percent. Lower levels of polarization indicate a chaotic magnetic field, which can be described roughly in terms of N cells of equal size, each with uniform but randomly oriented field. The degree of polarization is then

$$p = p_{max} N^{-1/2} \pm p_{max}(2N)^{-1/2}, \tag{7.5}$$

where $p_{max} = (\alpha + 1)/(\alpha + 5/3)$ is usually in the range of 0.7–0.75 [15].

The electric vector position angle (EVPA, or χ) of linear polarization is perpendicular to the projected direction of the magnetic field if the emission is from optically thin synchrotron radiation. However, relativistic motion aberrates the angles, an effect that causes the observed value of χ to align more closely with the jet direction than is the case in the plasma rest frame. Because of this, polarization electric vectors observed to be roughly perpendicular to the jet axis correspond to magnetic fields that lie essentially parallel to the flow direction, but EVPAs that are modestly misaligned with the jet actually imply that the rest-frame magnetic field is quite oblique to the jet axis.

Faraday rotation alters the EVPA as well. If the synchrotron emission from the jet passes through a Faraday "screen" – e.g., a sheath of thermal plasma surrounding the jet – then the EVPA rotates by an amount

$$\Delta\chi = 7.27 \times 10^4 [\nu_{\mathrm{GHz}}(1+z)]^{-2} \int n_{\mathrm{e}} B_{\parallel} ds_{\mathrm{pc}} \text{ radians,} \tag{7.6}$$

where z is the redshift of the AGN, n_{e} is the electron density of the screen in cm^{-3}, $B_{\parallel}$ is the component of the screen's magnetic field along the line of sight in gauss, and s_{pc} is the path length through the screen in parsecs. Any electron–proton plasma inside the jet also causes rotation, although the magnitude of $\Delta\chi$ is reduced by the inverse of the average electron Lorentz factor $\langle\gamma\rangle$, where $\gamma \equiv E/(m_{\mathrm{e}}c^2)$. Furthermore, when the rotation is distributed through the emission region, $\Delta\chi$ is different for the different path lengths to the emission sites along the line of sight. Because of this, the net effect is partial depolarization of the radiation through cancellation of mutually orthogonal components of the rotated polarization vectors from the different sites.

The linear polarization of any part of the source that is optically thick to synchrotron self-absorption is reduced by a factor ~7 from the maximum value – to $\sim10\%$ in the uniform field case. Further reductions caused by a lower degree of order of the magnetic field render self-absorbed synchrotron emission nearly unpolarized in most cases. The EVPA is transverse to that of optically thin emission and is therefore parallel to the direction of the mean magnetic field.

Incoherent synchrotron radiation is circularly polarized at a level $p_{\mathrm{c}} \sim \gamma^{-1}(\nu)$, where $\gamma(\nu)$ is the Lorentz factor of the electrons whose critical frequency is the frequency of observation ν: $\gamma(\nu) = 6.0 \times 10^{-4}[\nu(1+z)/(B\delta)]^{1/2}$, where B is the magnetic field strength. This is generally much less than 1%. The circular polarization observed in the compact jets, on the other hand, can be as high as 1–3% [83]. The likely cause is Faraday conversion of linear to circular polarization by electrons (and positrons, if present) with $\gamma \sim 1$–10 [40].

7.2.4 Spectral Energy Distribution

A striking property of jets is their ability to radiate profusely from radio to γ-ray frequencies. As Figs. 7.1 and 7.2 illustrate, jets can be prominent in optical and X-ray images even at distances of hundreds of kiloparsecs from the nucleus. As seen in the quasar 3C 279 (Figs. 7.3 and 7.4 and [32]), the compact, parsec-scale jets of prominent blazars imaged by VLBI are the sites of bright radio through γ-ray emission that varies – often quite dramatically – on time scales as short as hours or even minutes. In some BL Lac objects, the fluctuating emission extends up to TeV γ-ray energies. The variations are often correlated across wavebands, sometimes with time delays.

Figure 7.5 displays the full electromagnetic continuum spectrum of 3C 279, with the main physical components of the emission identified. Synchrotron emission from the extended jet dominates at the lowest radio frequencies, while synchrotron radiation from the compact jet supplies most of the flux from GHz to optical frequencies. Thermal black body radiation from the accretion disk is prominent at ultraviolet frequencies, while inverse Compton scattering by relativistic electrons

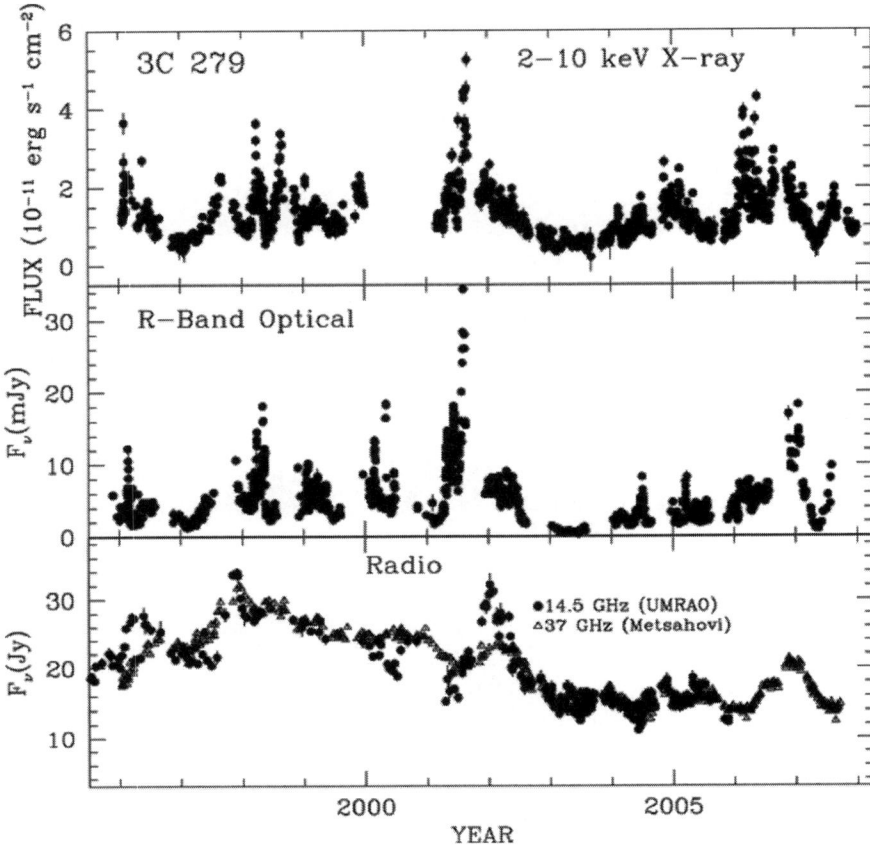

Fig. 7.4 Variation of flux vs. time of the quasar 3C 279 at X-ray, optical, and radio frequencies. Adapted from [19]

in the compact jet produces most of the X-rays at >1 keV and γ rays at <100 GeV energies in this quasar.

The plot in the right panel of Fig. 7.5 emphasizes the spectral bands containing most of the *apparent* luminosity. Translation to *actual* luminosity, however, requires that we remove the effects of relativistic beaming by dividing by $\delta^{2+\alpha}$ if the flux is steady and by $\delta^{3+\alpha}$ if the spectrum corresponds to a temporary outburst. Since the Doppler factor of 3C 279 is $\delta = 24 \pm 6$ [43], this requires division of the apparent luminosity of the jet by a factor exceeding 1000. In fact, the accretion disk is the most intrinsically luminous component in most quasars. This does not, however, imply that the *kinetic power* of jets is a negligible fraction of the overall luminosity, since jets are rather radiatively inefficient.

We can calculate the optically thin synchrotron flux density in the power-law portion of the spectrum of the steady-state jet by adopting a conical geom-

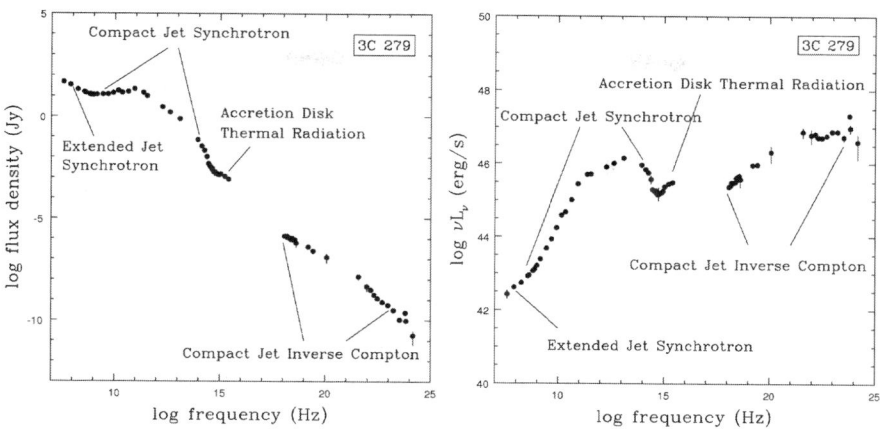

Fig. 7.5 *Left*: Continuum spectrum of the quasar 3C 279. *Right*: Same plot but with spectral luminosity along the vertical axis. The main emission components are indicated. From data compiled by the NASA Extragalactic Database plus data of the author and collaborators

etry with a magnetic field whose strength decreases with cross-sectional radius R, $B = B_0(R/R_0)^{-b}$, and with a power-law electron density per unit energy, $N(E) = N_0(R/R_0)^{-a}E^{-(1+2\alpha)}$. Here R_0 is the radius at the location where the emission turns on, presumably the point farthest upstream where electrons are accelerated to energies high enough to radiate at the rest frequency $v(1+z)/\delta$. Because of the gradients, we need to integrate the emission coefficient [68] over volume, divide by the square of the luminosity distance D_l, and apply appropriate relativistic corrections:

$$F_v \approx (D_l^2)^{-2}c_1(\alpha)N_0B_0^{1+\alpha}\delta^{2+\alpha}(1+z)^{1-\alpha}v^{-\alpha}\pi\phi^2\int_{Z_0}^{\infty}Z^{2-a-(\alpha+1)b}dZ, \quad (7.7)$$

where, in CGS units, $c_1(\alpha) = 1.01\times10^{-18}$, 3.54×10^{-14}, 1.44×10^{-9}, 6.30×10^{-5}, and 1.5×10^5 for $\alpha = 0.25$, 0.5, 0.75, 1.0, and 1.5, respectively, incorporates numerical and physical constants (see [42, 59] for a tabulation), z is the redshift of the host galaxy, ϕ is the opening half-angle of the jet in radians, Z is distance in pc from the black hole along the jet axis, and $Z_0 = R_0/\sin\phi$. The upper limit to the integral is actually $Z_{\max}(v)$, the point where there are no longer substantial numbers of electrons that can radiate at rest frequency $v(1+z)/\delta$. However, if $a > 2$ and $b > 1$, we can use infinity in practice as long as $Z_{\max}(v) \gg Z_0$. Equation (7.7) is valid under the approximation that the Doppler factor is essentially the same across the jet, which requires that Γ does not depend on R and that either the viewing angle $\theta \gg \phi$ or $\phi \ll \Gamma^{-1}$. A more accurate calculation requires numerical integration over polar and azimuthal angles inside the jet.

Note the flatness of the radio spectrum in the left panel of Fig. 7.5, a property caused by opacity plus gradients in magnetic field strength and density of relativistic electrons. We can derive the shape of the spectrum by considering the ideal case of a jet pointing directly at us with an opening angle $\phi \ll \Gamma^{-1}$ and a Doppler factor that does not vary significantly with R. We further approximate that emission with optical depth $\tau > 1$ is completely opaque, while that at lower optical depths is completely transparent. This means that all the emission comes from radial distances $Z > Z_m$, where Z_m corresponds to the position in the jet where the optical depth along the line of sight is unity. In order to derive this quantity, we again need to integrate

$$\tau \sim 1 \approx c_2(\alpha) N_0 B_0^{0.5(2\alpha+3)} \left[\frac{\delta}{\nu(1+z)}\right]^{0.5(2\alpha+5)} \int_{Z_m}^{\infty} Z^{-a-0.5(2\alpha+3)b} dZ, \qquad (7.8)$$

where $c_2(\alpha)$ is 2.29×10^{12}, 1.17×10^{17}, 6.42×10^{21}, and 3.5×10^{26}, for $\alpha = 0.25$, 0.5, 0.75, and 1.0, respectively. We can solve for Z_m after evaluating the integral

$$Z_m \approx \left(c_2(\alpha)[a + 0.5(2\alpha + 3)b - 1]^{-1} N_0 B_0^{0.5(2\alpha+3)} \times \right.$$
$$\left.[\frac{\delta}{\nu(1+z)}]^{0.5(2\alpha+5)}\right)^{1/[a+0.5(2\alpha+3)b-1]}. \qquad (7.9)$$

We can now calculate the approximate flux at frequency $\nu < \nu_m$, where ν_m is the frequency at which the source is transparent down to Z_0, by replacing Z_0 by Z_m in the lower limit of the integral in (7.7). (We can derive an expression for ν_m by replacing Z_m with Z_0 in (7.9) and solving for ν.) Since the main result we want is the dependence on frequency, we express the answer as a proportionality:

$$F_\nu(\nu < \nu_m) \propto \nu^\zeta, \qquad (7.10)$$

where

$$\zeta = \frac{4\alpha(b-1) + 5(a+b-3)}{2(a-1) + (2\alpha+3)}. \qquad (7.11)$$

A completely flat spectrum, with $\zeta = 0$, is possible if $a = 2$ and $b = 1$, as proposed by Königl [47]. While this value for b corresponds to the expected dependence of a transverse magnetic field on R, the value $a = 2$ would only hold if adiabatic expansion cooling did not occur. If, on the other hand, expansion cooling is the main energy loss mechanism for the electrons, we expect $a = (4/3)\alpha + 2$ in the case of a jet with constant flow Lorentz factor Γ. Since actual jets usually contain a number of knots in addition to the core, the flat spectrum must be caused by a superposition of the individual spectra of the various features, each peaking at different frequencies [21]. The fact that these tend to produce composite spectra that are, to a rough approximation, flat implies that Königl's ansatz is globally correct, even if it might not apply locally. For example, standing and/or moving shock waves in the jet can convert flow energy into plasma energy, whereas adiabatic expansion cooling does the reverse. As long as only an insignificant fraction of the total flow energy is lost to radiation, the overall spectrum will follow the $a = 2$, $b = 1$ case. In fact, careful analysis of the jet in 3C 120 over several orders of magnitude in length scale indicates that it maintains the values $a = 2$ and $b = 1$ [82].

7.3 Physical Processes in AGN Jets

Considerable progress has been made in the interpretation of observations of jets by assuming that the main physical processes governing jets are those of gas dynamics and magnetohydrodynamics (MHD). Plasma physics in the relativistic regime must also play a role in the heating and motions of the particles that compose jets. Unfortunately, our understanding of the relevant plasma phenomena is hindered by insufficient observations coupled with the likely complexity involved when the mean particle energy is relativistic. In addition, electrodynamics might govern the dynamics close to the black hole, and electrical currents can play an important role there and farther downstream. Here we will discuss some of the physical processes that have been successfully applied to explain a wide range of observations of jets.

7.3.1 Launching of Jets

The basic requirement for producing a supersonic, highly collimated flow is a confinement mechanism coupled with a strong negative outward pressure gradient. In principle, this is possible with gas dynamics if the pressure of the interstellar medium in the nucleus drops by many orders of magnitude from the central regions to the point where the jet reaches its asymptotic Lorentz factor [11]. However, the most extreme jets are accelerated to $\Gamma > 10$ and collimated to $\phi < 1°$ within ~ 1 pc from the central engine [43]. At this location, the pressure in the jet is typically of order 0.1 dyne cm^{-2} (e.g., [64]). Gas dynamical acceleration and collimation would then require an external pressure $\sim 10^4$ dyne cm^{-2} within ~ 0.01 pc of the black hole. If this were provided, for example, by a hot ($\sim 10^8$ K) wind emanating from the accretion disk, the X-ray luminosity would be enormous, $\sim 10^{50}$ erg s^{-1}.

For this reason, magnetic launching is considered to be the driving force behind most relativistic jets in AGN. The differential rotation of the ergosphere of a spinning black hole and/or accretion disk causes the polar component of the magnetic field lines of the accreting matter to wind up into a helix. The toroidal component of the wound-up field, B_{tor}, provides a pinching force through the hoop stress, with magnitude $F_{pinch} = B_{tor}^2/(4\pi R)$, directed toward the jet axis. The magnetic field expands with distance from the black hole, lowering the magnetic pressure. This creates a strong pressure gradient along the axis, which drives the flow. In order for the velocity to become highly relativistic, the magnetic energy density must greatly exceed the rest-mass energy density at the base of the jet. Hence, the flow is initially dominated by Poynting flux that is transferred to kinetic flux of the particles in the jet toward greater axial distances Z.

The magnetic acceleration essentially stops when the kinetic energy density of the particles reaches equipartition with the magnetic energy density. Therefore, the final flow Lorentz factor is [81]

$$\Gamma_{\max}(Z) \sim (B_i^2)/[16\pi n(Z)\langle m\rangle c^2], \tag{7.12}$$

where $n(Z)\langle m\rangle$ is the mass density of particles at position Z and B_i is the magnetic field at the base of the jet.

In the magnetic (as well as gas dynamical) acceleration model, the opening angle ϕ is inversely proportional to the final Lorentz factor, as observed [43]. The gradual acceleration implies that any radiation from the innermost jet is weakly beamed relative to that on parsec scales. Therefore, one should expect that most of the observed emission arises from regions near and downstream of the core seen on millimeter-wave VLBI images, even at high frequencies where the entire jet is transparent.

7.3.2 Gas Dynamics of Jets

Since jets are long-lived phenomena, we can assume that, to first order, they represent steady flows. Then the relativistic Bernoulli equation applies:

$$\Gamma(u + p)/n = \text{constant}, \tag{7.13}$$

where u is the energy density of the plasma, p is the pressure, and n is the number density of particles, all measured in the co-moving reference frame. If the mean energy per particle $\langle E\rangle$ greatly exceeds the mean rest-mass energy, then $u = 3p = n\langle\gamma m\rangle c^2$. In addition, the ideal gas law for a relativistic plasma can be expressed as $p \propto n^{4/3}$. We can use these expressions in (7.13) to obtain

$$\Gamma(Z) = \Gamma(Z_0)[p(Z_0)/p(Z)]^{1/4}, \tag{7.14}$$

where Z_0 is a reference point. This is the relativistic version of Bernoulli's principle, which states that the velocity of a flow is inversely related to its pressure.

We could also rearrange the ideal gas law to write $\langle\gamma\rangle \propto n^{1/3}$ and

$$\Gamma(Z) = \Gamma(Z_0)\langle\gamma(Z_0)\rangle/\langle\gamma(Z)\rangle. \tag{7.15}$$

This formulation emphasizes that the acceleration of the flow results from conversion of internal energy into bulk kinetic energy. Significant increase in the flow speed through gas dynamics is no longer possible if $\langle\gamma(Z)\rangle$ is not substantially greater than unity. Beyond this point, the jet will be ballistic, coasting at a constant velocity until it is disturbed.

The cross-sectional area of a jet will tend to expand and contract such that the pressure at the boundary matches the external pressure. The maximum opening angle ϕ, however, is the inverse of the Mach number at the point where the pressure drop occurs, $\phi_{max} = v_s/(\Gamma \beta c)$, where v_s is the local sound speed.

Since the interstellar medium is dynamic, and since the pressure in the jet is subject to time variability, a jet is likely to be subject to mismatches in pressure at the boundary. If modest in magnitude, the difference in pressure is accommodated through sound waves that communicate the pressure imbalance to the jet interior. This causes both oscillations in the cross-sectional radius and compressions and rarefactions inside the jet [12, 23, 28]. The Lorentz factor is higher in the lower pressure regions and lower where the pressure is greater. If the pressure mismatch exceeds ~50%, oblique "recollimation" shock waves form to adjust the flow more abruptly. If the jet is sufficiently circularly symmetric, the shock waves will be conical. If the symmetry is nearly perfect, the conical shock is truncated by a strong shock aligned perpendicular to the axis, called a "Mach disk." Since the role of a shock front is to decelerate a disturbed supersonic flow to subsonic velocities, the flow will cease to be highly relativistic until it reaccelerates where it encounters a pronounced rarefaction downstream of the Mach disk. Shock waves set up by pressure mismatches with the surrounding medium are often labeled "external" since they are driven by influences from outside the jet.

"Internal" shock waves can result from changes in the jet velocity or injected energy at or near the base. A major disturbance is required to generate a shock when the flow velocity is near the speed of light, since the relative velocity needs to be supersonic. If the local sound speed is the fully relativistic value, $c/\sqrt{3}$, then the Lorentz factor of the faster flow must be at least a factor ~2 higher than that of the slower flow that it overtakes.

A shock compresses the plasma so that the density rises by a factor ξ across the shock front. The magnetic field component that lies parallel to the shock front increases by the same factor, while the component transverse to this direction remains unchanged. The maximum value of ξ is 4 in a thermal plasma; it can be higher in relativistic plasma, but if the jet flow is already highly relativistic, disturbances are unlikely to be sufficiently strong relative to the quiescent flow for values of ξ greater than about 2 to be common. In the reference frame of the shock front, the undisturbed plasma flows in at a supersonic velocity and exits at a subsonic speed. Unless the sound speed is close to c – the maximum value of $c/\sqrt{3}$ requires that the plasma be highly relativistic, with $\langle \gamma \rangle \gg 1$ – this means that the shocked plasma will have a bulk Lorentz factor only slightly lower than that of the shock front.

7.3.3 Magnetohydrodynamics of AGN Jets

The gas dynamical description of a jet is valid as long as the energy density of the magnetic field is significantly less than that of the particles. In this case, the field is "frozen in" and will follow the plasma. Close to the black hole, however, the magnetic field plays a major role in the dynamics, and even on kiloparsec scales the field energy is probably roughly in equipartition with that of the particles.

Section 7.3.1 briefly summarizes the MHD forces that can accelerate and collimate a jet in the inner parsec. This depends on a strong toroidal component of the
magnetic field, which itself requires a radial electric current. Since the system has
considerable angular momentum, the streamlines that the flow follows trace out a
spiral pattern about the axis [80].

When the magnetic field is mainly toroidal, kinks that occur in the jet are subject
to a current-driven instability that causes the kinks to grow [5]. Simulations indicate
that the net result is to convert Poynting flux into internal energy of the plasma [30].

We therefore have a crude theoretical description of a jet that is collimated by a
toroidal magnetic field and accelerated by a decreasing magnetic pressure gradient,
but eventually subject to an internal instability that heats the plasma. By the time
such a jet reaches parsec scales, it has a relativistic flow velocity, contains a magnetic
field that is at least somewhat chaotic, and is loaded with relativistic electrons (and
either protons or positrons, or both). These are the necessary ingredients to produce
the jets observed on VLBI images, with bright synchrotron and inverse Compton
emission and fairly weak linear polarization.

7.3.4 Instabilities in Jets

As the parsec to kiloparsec-scale continuity of jets like the one in 3C 273 displayed
in Fig. 7.1 demonstrates, many jets are stable over extremely large distances. Jets
that are completely free, i.e., containing higher pressure than their surroundings
and propagating ballistically, are inherently stable. However, if their cross sections
are able to adjust to maintain pressure equilibrium with the external medium, they
are subject to the Kelvin–Helmholtz instability that occurs when two fluids have
different velocities adjacent to a tangential discontinuity. These instabilities have
numerous modes that are divided into two classes, body modes and surface modes.
The body modes cause departure of the cross-sectional geometry from circular symmetry as well as oscillations of the transverse radius and filamentary structure [30].
Figure 7.6 shows an example of such a jet. The surface modes generate turbulence
and velocity shear at the boundaries that can permeate the entire jet cross section as
the instability grows.

The reason why jets can remain largely intact hundreds of kiloparsecs from their
site of origin in the nucleus is that Kelvin–Helmoltz instabilities are suppressed at
higher Mach numbers and Lorentz factors and when there is a strong axial magnetic
field. These characteristics are found in the most powerful jets, those of the FR II
quasars and radio galaxies.

Since they can lead to the development of turbulence and formation of shocks,
instabilities might be the cause of much of the emission seen on parsec and kiloparsec scales [75]. In addition, much of the rich structure seen in some jets might
owe its existence to such instabilities. The turbulence in the boundary region also
promotes entrainment of material from the surrounding medium [8, 50], a process
that slows the jet down [53], eventually leading to its disruption in FR I sources.

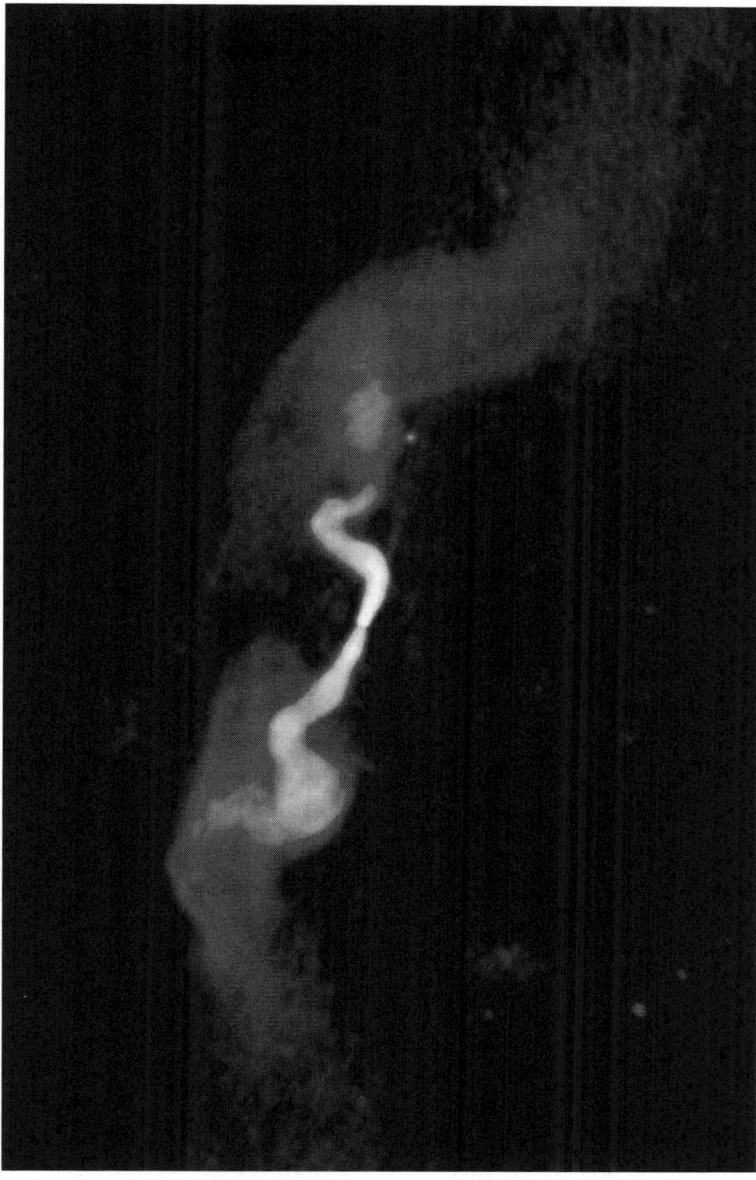

Fig. 7.6 Radio image of the FR I radio galaxy 3C 31 showing a two-side jet that becomes globally unstable. The nucleus is in the *middle* of the straight section of the jet. Image courtesy of
NRAO/AUI/NSF, data published in [51]

7.3.5 Cross-Jet Structure

Jets can possess structure transverse to the axis whether or not instabilities develop. In fact, theoretical models and simulations of jet launching generally predict such structure [77, 70, 66, 81, 33, 65]. Although there can be a smooth gradient in density and velocity [81], transverse structure is often described as an ultra-fast spine surrounded by a more slowly moving sheath. The spine is predicted by many theories to be composed of an electron–positron pair plasma, with the sheath containing a normal electron–proton mixture [77, 70, 65]. While there is considerable observational evidence for cross-jet structure [53, 71, 43], it is not yet clear whether there is a smooth gradient in properties or a two-layer structure.

7.3.6 Magnetic Field

As discussed previously, we expect the magnetic field to have a tight helical geometry close to the black hole. At the end of this acceleration and collimation zone, current-driven instabilities mix the field into a more chaotic configuration. Velocity shear can either align the magnetic field parallel to the axis or generate turbulence that disorders the field. In the former case, one can imagine closed loops of field lines that are stretched by the relative velocity between faster flow closer to the axis and slower layers closer to the boundary.

As in any spherical or conical flow in which magnetic flux is conserved, the axial component of the magnetic field decreases as R^{-2}. The component transverse to the axis should decrease as R^{-1}. There are numerous examples of extended jets in which the expected transition from longitudinal to transverse magnetic field is evident in the linear polarization [13]. However, comparison of simulated and observed polarization indicates that, while the toroidal to longitudinal magnetic field ratio does increase with distance down the jet, the gradient in the ratio is not as steep as predicted in the case of a magnetic field that is frozen into the plasma [52]. There is instead a substantial component of disordered field that has a more complex behavior.

7.3.7 Energization of Electrons

Since the radiation from AGN jets comes from electrons (and positrons, if any), the sites of strong emission are strongly related to the processes that accelerate electrons to Lorentz factors γ that can exceed 10^6 on both subparsec and kiloparsec scales in some jets. The power-law shape with slope $-\alpha$ of the optically thin synchrotron spectrum over $\sim$4 orders of magnitude (see Fig. 7.5) implies that the electron energy distribution is also a power law, with slope (on a log–log plot) $-(2\alpha + 1)$ over at least 2 orders of magnitude in γ.

A long-standing explanation for a power-law electron energy distribution is the Fermi mechanism [26]. Particles repeatedly reflect off regions of higher-than-average magnetic fields, gaining energy if the region was moving toward the particle (in the plasma's rest frame) and losing energy if it was receding. The net result is a power-law energy distribution. If a particle gains an average energy $\Delta E = \xi E$ per reflection and if the probability of staying in the acceleration zone after a reflection is η, $\alpha = \ln \xi / (2\ln \eta)$. The process can produce very high energies if η is close to unity so that a typical particle encounters many reflections before exiting the zone.

One site where the Fermi mechanism can be particularly efficient is at a shock front. In this case, $\xi = (4\Delta u)/(3c)$, where Δu is the increase in flow velocity across the shock front [7]. Plasma waves are likely to be present on both sides of the shock. These reflect the particles so that they pass back and forth across the shock multiple times. Less systematic acceleration of particles can occur in a turbulent region since, statistically, the number of approaching collisions with cells of higher-than-average magnetic field is greater than the number of receding collisions. The mean fractional energy gain per reflection is $\xi \approx 4(\Gamma_{cell} v_{cell}/c)^2$. Therefore, the process is efficient if the turbulence is relativistic, as one might expect if the jet has substantial transverse velocity gradients.

An unresolved issue with the Fermi mechanism is how it can produce a similar power-law slope in remotely separate physical regions. For example, the spectral index in the jet of 3C 120 is $\alpha \approx 0.65$ over several orders of magnitude in size scale [82]. If the bright emission regions are caused by shocks, all the shocks would need to have the same compression ratio, which would seem difficult to engineer. If turbulent processes energize the particles, the physical conditions would again need to be more similar than one might expect over such a large range of size scales.

7.4 Physical Description of Features Observed in Jets

7.4.1 The Core

The core is generally the most compact part of a jet that we can image at any given frequency. This implies that an understanding of its nature can provide insights into the physical structure of jets at and perhaps upstream of the core's position. As we have discussed above, what is seen as the core on VLBI images at radio frequencies might be the transition region between optically thick and thin emission. However, it could also be a bright, stationary emission structure a short distance downstream of that region. If the core is the surface where the optical depth is near unity, its position Z_m should move toward Z_0 and therefore toward the central engine at higher frequencies. This effect is apparent in some compact extragalactic jets [55], but not in others [67]. The implication of the latter negative result is that the jet is not just a smooth flow punctuated by superluminal knots, but also contains multiple bright stationary features. In the quasar 0420−014, observations of fluctuating polarization in the core and at optical wavelengths have been interpreted as the signature of turbulent jet plasma passing through a standing conical shock system – perhaps an example of a recollimation shock discussed in Sect. 7.3.2 – that represents the physical structure of the core [24].

In some objects the core could be a bend in the jet where it becomes more closely aligned with the line of sight. This would cause an increase in the Doppler beaming factor relative to points farther upstream. This cannot be the main reason for the appearance of cores, however, since it is less likely for a jet to curve toward than away from the line of sight.

While jets are often approximately self-similar on parsec scales, this must break down close to the central engine in the flow acceleration and collimation zone (ACZ). The point where the change must occur is indicated by the frequency above which the synchrotron spectrum steepens, signaling that the jet emission is optically thin. According to observations [39], this is in the range of tens to >1000 GHz. (It is 10^{11} Hz in Fig. 7.5.) The angular width of the jet at this point can be estimated by the value at 43 GHz, $\sim 50\,\mu$arcsec times 43 GHz/ν_m. For a quasar with redshift of order 0.5, this translates to ~ 0.3 (43 GHz/ν_m) pc, which is small (a few 10^{16} cm for $\nu_m \sim 1000$ GHz) but still hundreds of Schwarzschild radii even for a black hole mass $\sim 10^9$ $M_\odot$. The distance of the end of the ACZ from the black hole might be many times larger than this. The flux of radiation from a blazar jet inside the ACZ probably decreases toward the black hole, since the relativistic beaming is weaker owing to the progressively lower flow Lorentz factor [57]. In the jet of a radio galaxy viewed nearly side-on, however, the emission should become stronger (as long as there are energetic electrons) toward smaller distances from the black hole, since the radiation is anti-beamed. This implies that there should be a very short gap between the start of the jet and the beginning of the counterjet, except for a limited section obscured by free–free opacity in the accretion disk and dusty torus beyond the disk. Furthermore, the flow should start out fairly broad and become more collimated as it accelerates downstream. VLBA observations of the radio galaxy M87, whose relatively local distance allows a linear resolution of tens of gravitational radii, indicate that the flow is indeed quite broad near the central engine [44].

7.4.2 Quasi-stationary Features

Besides the core, there are often other bright features in blazar jets that are either stationary or move at subluminal apparent speeds [41, 46, 43]. In straight jets, these can be produced by recollimation shocks or instabilities (see Sects. 7.3.2 and 7.3.4) that compress the flow or excite turbulence. This should lead to particle acceleration and amplification of the magnetic field and therefore enhanced emission.

Bends can also cause quasi-stationary hotspots, either because the jet turns more into the line of sight [1] or from the formation of a shock. An example of the latter is the collision of the jet with an interstellar cloud [29]. In this case, an oblique shock forms to deflect the jet flow away from the cloud.

Long-lived jets propagate through their host galaxy and into the intergalactic medium. In the case of low luminosity, FR I sources (e.g., the one shown in Fig. 7.6),

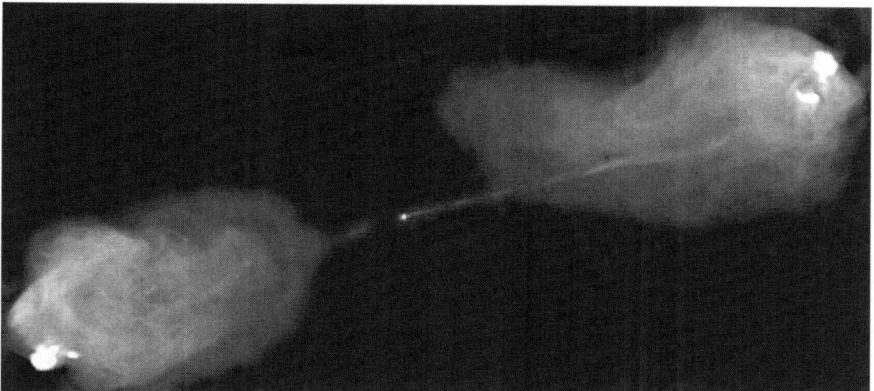

Fig. 7.7 Radio image of the FR II radio galaxy Cygnus A showing the small, bright core at the center, two-side jet, billowy lobes, and hotspots at the outer edges of the lobes. The western side approaches us, hence the lobe is farther from the core since the hotspots and lobes move outward with time and we see the eastern lobe at an earlier time owing to light-travel delay. Image courtesy of NRAO/AUI/NSF, published in [69]

instabilities at the boundary and entrainment of the external gas decelerate the jet to subsonic speeds and completely disrupt it. The more powerful FR II jets generally survive out to the point where a strong termination shock decelerates them to subsonic speeds. The shock appears as a bright, compact hotspot, while the jet plasma spreads into a giant turbulent "lobe" that emits radio synchrotron radiation (see Fig. 7.7). As inferred by the ratio of distances from the nucleus of the near and far lobes, the termination shock moves subluminally away from the galaxy as the momentum of the jet pushes away the external gas.

7.4.3 Models for Superluminal Knots in Jets

The bright knots that appear to move superluminally down compact jets (see Fig. 7.3) must represent regions of higher relativistic electron density and/or magnetic field than the ambient jet. They could be coherent structures, such as propagating shock waves caused by surges in the energy density or velocity injected into the jet at its origin near the central engine. However, they could also be "blobs" of turbulent plasma with electrons accelerated by the second-order, statistical Fermi process, i.e., random encounters with magnetic irregularities moving in various directions. Or, they might be ribbons or filaments generated by instabilities that move down the jet. It is likely that each of these possibilities occurs in some jets.

Shock waves are the most common interpretation of superluminal knots, since basic shock models have enjoyed considerable success at reproducing time variations of continuum spectra and polarization [61, 37, 38, 78, 79]. At frequencies above $\sim$10 GHz, radiative energy losses of the electrons can be important. In this

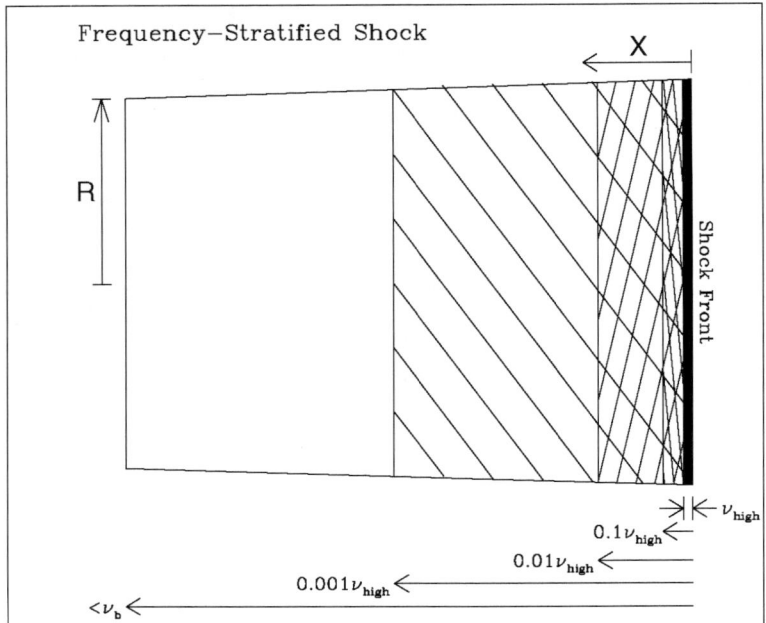

Fig. 7.8 Frequency stratification in a square-wave shock, drawn to scale. The electrons are accelerated at the shock front and drift behind the shock while losing energy to radiative losses. This example is for a forward propagating shock; the same principle is valid for a reverse or stationary shock, as well as for a shock that is oblique to the jet axis. See [61] for more details

case, if we assume that electrons are accelerated only at the shock front, they will lose their ability to produce synchrotron radiation at high frequencies as they advect away from the front and lose energy, since they cannot radiate much above their critical frequency, $\nu_{\mathrm{crit}} = 2.8 \times 10^6 B \gamma^2$ (in CGS units). This results in frequency stratification, with the highest frequency emission confined to a thin layer behind the shock front and progressively lower frequency emission occupying larger volumes. Figure 7.8 illustrates the effect. We can derive a proportionality between distance x behind the front and the frequency of emission that is limited to a shell of thickness x if we assume that the magnetic field and advection speed v_{ad} are constant out to distance x from the shock front:

$$x(\nu) \propto \nu^{-1/2}. \tag{7.16}$$

Here, we have applied the solution to the synchrotron energy loss equation, $\gamma_{\mathrm{max}} \approx (1.3 \times 10^{-9} B^2 x / v_{\mathrm{ad}})^{-1}$. The volume of the emitting region is then $\pi R^2 x \propto \nu^{-1/2}$. This steepens the optically thin spectral index from α to $\alpha + 0.5$ at frequencies $\nu \gg \nu_{\mathrm{b}}$, where ν_{b} is the frequency at which $x(\nu)$ equals the entire length of the shocked region and $-\alpha$ is the spectral slope of the emission coefficient. It also affects the time evolution of the flux density and turnover frequency as the shock expands while it propagates down the jet. In particular, at frequencies where synchrotron losses are greater than expansion cooling, the peak in the spectrum propagates toward lower frequencies while the flux density at the peak remains roughly constant [61]. This can contribute to the flatness of the overall synchrotron spectrum at radio frequencies.

The stratification results in time lags in the optically thin synchrotron radiation during the outburst caused by the formation of the shock, with higher frequency variations leading those at lower frequencies. Even beyond the point where the magnetic field is too low for synchrotron losses to compete with expansion cooling, time delays at radio frequencies continue owing to the outburst becoming transparent at progressively lower frequencies with time. Lower amplitude flares and flickering (cf. the light curves displayed in Fig. 7.4) can be explained as the consequence of shocks compressing turbulent cells that they pass [62]. In this case, the higher frequency variations in flux density and polarization occur on shorter time scales because of the smaller volumes involved.

For the sake of simplicity, shocks in jets are usually modeled as transverse square waves in attempts to reproduce the observed variability of emission and polarization. Such shocks will compress the magnetic field component lying parallel to the shock front. In most jets showing superluminal motion, the orientation of the shock front is aberrated to appear nearly transverse to the jet axis in the observer's frame. This should yield linear polarization with electric vectors aligned along the jet axis. A large sample of quasar jets [54] observed with VLBI possess a distribution of polarization directions that is inconsistent with this assumption. Instead, the magnetic field seems to be mostly turbulent, with a small uniform component that is parallel to the jet axis. Other polarization observations [2] indicate that, if most superluminal knots are shocks, the fronts must be oblique to the jet axis in most cases.

The main differences between shock and blob models are the absence of frequency stratification in the latter and that it is unnecessary for the particles and magnetic field in a blob to be directly related to those of the surrounding jet plasma. In the case of an outburst of 3C 273, an attempt to model the event in terms of an expanding blob led to the nonsensical requirement that the magnetic field would need to have decreased while the particle density increased to reproduce the time behavior of the flare's continuum spectrum [61]. On the other hand, it is common to model X-ray synchrotron and γ-ray inverse Compton flares in BL Lac objects as blobs (e.g., [45]).

7.4.4 Internal Energy and Kinetic Power

There are basically two methods for determining the energy contained in relativistic particles in extragalactic jets. The first is to assume equipartition between the magnetic and particle energy densities, which is similar to calculating the minimum energy required to produce the synchrotron emission [14]. We can express the ratio of relativistic particle to magnetic energy density as $u_p/u_{mag} \equiv A$, with $A = 1$

corresponding to equipartition. If k is the fraction of particle energy that is contained in neither electrons nor positrons, the energy density in particles is

$$u_p = (1 + k)N_0 \int_{\gamma_{min}m_ec^2}^{\gamma_{max}m_ec^2} E^{-2\alpha}dE, \qquad (7.17)$$

which we can set equal to $Au_{mag} = AB^2/(8\pi)$.

If we consider a section of volume V inside the jet, with radiation coming from some fraction f (the "filling factor") of that volume and observe it to have a flux density F_ν at a frequency ν where the emission is optically thin, we can write

$$F_\nu \approx c_1(\alpha)D_l^{-2}fVN_0B^{1+\alpha}\delta^{3+\alpha}(1+z)^{1-\alpha}\nu^{-\alpha}. \qquad (7.18)$$

Here we have assumed that the measurement is for a moving knot; if it is a section of the steady-state jet, the exponent of δ should be $2+\alpha$. We can solve this equation for N_0, substitute the result for N_0 in (7.18), and set u_p equal to $AB^2/(8\pi)$. After evaluation of the integral, some algebraic manipulation, and conversion of various parameters into more readily observable quantities, we arrive at

$$\begin{aligned} B = & \ c_4(\alpha)\delta^{-1}(1+z)^{(5+\alpha)/(3+\alpha)} \times \\ & \times \left[\tfrac{1+k}{Af}g(\alpha, \gamma_{min}, \gamma_{max})D_{Gpc}^{-1}\Theta_{as}^{-3}F_{mJy}\nu_{GHz}^{\alpha}\right]^{1/(3+\alpha)} \mu G. \end{aligned} \qquad (7.19)$$

Here D_{Gpc} is the luminosity distance in Gpc, Θ_{as} is the angular size in arcseconds (under the approximation of spherical symmetry), F_{mJy} is the flux density in milliJanskys (1 mJy = 10^{-26} erg s^{-1} cm^{-2} Hz^{-1}), ν_{GHz} is the observed frequency in GHz, and

$$g(\alpha, \gamma_{min}, \gamma_{max}) \equiv (2\alpha - 1)^{-1}[\gamma_{min}^{-(2\alpha-1)} - \gamma_{max}^{-(2\alpha-1)}], \qquad (7.20)$$

except for the case $\alpha = 0.5$ for which $g = \ln(\gamma_{max}/\gamma_{min})$. The function $c_4(\alpha)$, which is tabulated in [42], has a value of 1.9, 7.8, 25, 74, 170, and 380 for $\alpha = 0.25, 0.5,$ 0.75, 1, 1.25, and 1.5, respectively. The total energy density is $(A + 1)B^2/(8\pi)$, hence the kinetic power is $(\pi R^2c)\Gamma^2(A + 1)B^2/(8\pi)$.

For typical spectral indices in extended jets, $\alpha = 0.8 \pm 0.2$, it is mainly γ_{min} that determines the value of the function g. Unfortunately, we cannot in general determine γ_{min} accurately, since electrons at this energy radiate mainly in the MHz range or lower where the emission has many components that are difficult to disentangle. We also do not have reliable methods to estimate the parameters f, A, or k. The usual procedure is to assume $f = 1$, $A = 1$, and $k = 1$, and some fairly high value of γ_{min}, e.g., 100–1000, in order to derive a lower limit to the kinetic power. The powers thus calculated can be very high in quasar jets, exceeding 10^{46} erg s^{-1} in many cases, and as high as 10^{48} erg s^{-1} if $\gamma_{min} \sim 10$ or if the ratio of proton to electron energy density k is closer to 100, as is the case for cosmic rays in our Galaxy. The observational data are usually consistent with rough equipartition between the energy densities in electrons and in magnetic field under the assumption that $k \sim 1$.

The second method for deriving the energy content of jets is to analyze compact knots that are synchrotron self-absorbed, in which case there are two equations, one for the flux density and the other for the optical depth (which is close to unity at the frequency ν_m of maximum flux density). We can then solve separately for B and N_0 (see, e.g., [58]). The magnetic field can be expressed as

$$
\begin{aligned}
B &= 3.6 \times 10^{-5} c_b(\alpha) \Theta_{mas}^4 \nu_{m,GHz}^5 F_{m,Jy}^{-2} \delta (1+z)^{-1} \\
&= 0.011 c_b(\alpha) T_{m,11}^{-2} \nu_{m,GHz} \delta (1+z)^{-1},
\end{aligned}
\tag{7.21}
$$

where the units are indicated by the subscripts and $T_{m,11}$ is the brightness temperature at ν_m in units of 10^{11} K. The function $c_b(\alpha)$ is 0.50, 0.89, 1.0, and 1.06 for $\alpha = 0.25, 0.5, 0.75$, and 1.0, respectively. The similar equation for N_0 is

$$
\begin{aligned}
N_0 &= 0.012 c_n(\alpha) D_{Gpc}^{-1} \Theta_{mas}^{-(7+4\alpha)} \nu_{m,GHz}^{-(5+4\alpha)} F_{m,Jy}^{3+2\alpha} \delta^{-2(2+\alpha)} (1+z)^{2(3+\alpha)} \\
&= 0.012(17.7)^{-(3+2\alpha)} c_n(\alpha) D_{Gpc}^{-1} T_{m,11}^{3+2\alpha} \nu_{m,GHz} \Theta_{mas}^{-1} (1+z)^3 \delta^{-1},
\end{aligned}
\tag{7.22}
$$

where $c_n(\alpha)$ is 660, 22, 1.0, and 0.049 for $\alpha = 0.25, 0.5, 0.75$, and 1.0, respectively.

We can therefore express the ratio of the relativistic electron to magnetic energy density mainly in terms of the brightness temperature:

$$
u_{re}/u_{mag} \approx g_u(\alpha, \gamma_{min}, \gamma_{max}) T_{m,11}^{7+2\alpha} (D_{Gpc} \nu_{m,GHz} \Theta_{mas})^{-1} (1+z)^5 \delta^{-3},
\tag{7.23}
$$

where

$$
g_u(\alpha, \gamma_{min}, \gamma_{max}) \approx
\begin{array}{ll}
0.48 \gamma_{max}^{0.5} & (\alpha = 0.25) \\
0.69 \ln(\gamma_{max}/\gamma_{min}) & (\alpha = 0.5) \\
13 \gamma_{min}^{-0.5} & (\alpha = 0.75) \\
74 \gamma_{min}^{-1} & (\alpha = 1.0).
\end{array}
\tag{7.24}
$$

We can invert (7.23) to determine an "equipartition brightness temperature,"

$$
T_{eq} \approx 10^{11} [D_{Gpc} \nu_{m,GHz} \Theta_{mas} (1+z)^{-5} \delta^3 g_u^{-1}(\alpha, \gamma_{min}, \gamma_{max})]^{1/(7+2\alpha)} \text{ K.}
\tag{7.25}
$$

Features with brightness temperatures higher than this have energy densities dominated by relativistic particles.

Use of this method requires very accurate measurements of the angular size Θ, spectral turnover frequency ν_m, and flux density at the turnover F_m of individual features in jets, a difficult undertaking requiring multifrequency VLBI observations. With current VLBI, restricted to Earth-diameter baselines at high radio frequencies, only a relatively small number of well-observed knots have properties amenable to reliable estimates of the physical parameters. Application of this method to semi-compact jets (between parsec and kiloparsec scales, observed at frequencies less than 1 GHz), indicates that jets are close to equipartition on these scales [72]. The same may be true for the parsec-scale cores, although some cores and individual very compact knots can have particle energy densities greatly exceeding $B^2/(8\pi)$ [34, 48, 64].

7.4.5 Inverse Compton X-Ray and γ-Ray Emission

Derivation of the magnetic field and N_0 allows the calculation of the flux of high-energy photons from inverse Compton scattering. If the "seed" photons that the relativistic electrons scatter are produced by synchrotron radiation from the same region of the jet, we refer to the process as "synchrotron self-Compton" (SSC) emission. In this case, we can apply the same analysis as in Sect. 7.4.4 to obtain the flux density at photon energy $h\nu$ in keV, where h is Planck's constant:

$$F_\nu^{SSC} \approx c_{ssc}\, \alpha\, (h\nu)_{keV}^{-\alpha} \Theta_{mas}^{-2(3+2\alpha)} \nu_{m,GHz}^{-(5+3\alpha)} F_{m,Jy}^{2(2+\alpha)} \left(\frac{1+z}{\delta}\right)^{2(2+\alpha)} \ln\left(\frac{\nu_{max}}{\nu_m}\right)\ \mu Jy$$

$$\approx (c_{tc}\, \alpha\, T_{m,11})^{3+2\alpha}(h\nu)_{keV}^{-\alpha} F_{m,Jy}\, \nu_{m,GHz}^{1+\alpha} \left(\frac{1+z}{\delta}\right)^{2(2+\alpha)} \times$$

$$\ln\left(\frac{\nu_{max}}{\nu_m}\right)\ \mu Jy. \tag{7.26}$$

The value of c_{ssc} is 130, 43, 18, and 91 and that of c_{tc} is 0.22, 0.14, 0.11, and 0.083 for $\alpha = 0.25$, 0.5, 0.75, and 1.0, respectively. The formula applies over the frequency range $\sim\gamma_{min}^2\nu_m$ to $\sim\gamma_{max}^2\nu_{max}$ over which the SSC spectrum is a power law with slope $-\alpha$. Here, ν_{max} is the highest frequency at which the synchrotron spectrum possesses a power-law slope of $-\alpha$.

A simpler formula relates the SSC flux density at frequency ν^c to the synchrotron flux density at frequency ν^s:

$$\frac{F_\nu^c(\nu^c)}{F_\nu^s(\nu^s)} \approx c_{cs}(\alpha) R N_0 \ln\frac{\nu_2}{\nu_m}\left(\frac{\nu^s}{\nu^c}\right)^\alpha, \tag{7.27}$$

where $c_{cs}(\alpha) = 2.7 \times 10^{-22}$, 3.2×10^{-19}, 3.9×10^{-16}, 5.0×10^{-13} for $\alpha = 0.25$, 0.5, 0.75, and 1.0, respectively.

The source of seed photons that are scattered may also lie outside the jet or in some other section of the jet. If the seed photons are monochromatic with frequency ν_{seed} and energy density u_{seed}, both measured in the rest frame of the host galaxy, the "external Compton" flux density is given by

$$F_\nu^{ec} = c_{ec}(\alpha)\frac{u_{seed}}{\nu_{seed}} D_l^{-2} N_0 f V \delta^{4+2\alpha}(1+z)^{1-\alpha}\left(\frac{\nu_{seed}}{\nu_{ec}}\right)^\alpha, \tag{7.28}$$

where $c_{ec}(\alpha) \equiv (c\sigma_t/2)(mc^2)^{-2\alpha}$, with σ_t being the Thomson cross section, is tabulated in [42]. If the seed photon distribution is actually a Planck function, as is the case for the cosmic microwave background [CMB, $u_{seed} = 4.2 \times 10^{-13}(1+z)^4$], we can use the same expression if we multiply it by $c_{planck}(\alpha) = 0.46, 0.58, 0.75$, and 1.00 for $\alpha = 0.25, 0.5, 0.75$, and 1.0, respectively. Since the CMB seed photons are only of relative importance in extended jets and lobes, we express the EC/CMB flux density in units appropriate to the larger scales:

$$F_\nu^{ec}(\text{CMB}) = \tfrac{c_{cmb}(\alpha)Af}{1+k} g(\alpha, \gamma_{min}, \gamma_{max})^{-1} D_{\text{Gpc}} \Theta_{as}^3 B_{\mu G}^2 \delta^{4+2\alpha} \times$$
$$(1+z)^{-2}(h\nu_{ec})_{\text{keV}}^{-\alpha} \text{ nJy}, \qquad (7.29)$$

where $c_{cmb}(\alpha) \equiv 0.063(4.2 \times 10^6)^{-\alpha} c_{planck}(\alpha) = 6.4 \times 10^{-4}, 1.8 \times 10^{-5}, 5.1 \times 10^{-7}$, 1.5×10^{-8} for $\alpha = 0.25, 0.5, 0.75$, and 1.0, respectively.

7.4.6 Matter Content

A major, yet unsettled question is whether jet plasmas are composed of "normal matter" – electrons and protons – electrons and positrons, or a combination of the two. Theoretically, a jet (or sheath of a jet) produced by a wind blown off the accretion disk should contain mainly normal matter. However, jets that are Poynting flux dominated might be populated by electron–positron pairs created by interaction of high-energy photons with the electromagnetic fields or particles in the jet [10].

While direct detection of the 511 keV electron–positron two-photon annihilation line would be the strongest evidence for a pair plasma, this emission line is difficult to detect. One of the problems is that current detectors at this energy are relatively insensitive. Another is the relatively low cross section for annihilation, so that a strong line requires a density higher than that found in jets. In addition, the line width from such a high-velocity environment as a jet flow can cause the line to be broadened beyond recognition. If the jet material mixes with interstellar clouds, however, the positrons can become thermalized and annihilate with cold electrons, leading to a narrow line [3, 27, 31, 35, 56]. This might be occurring in the radio galaxy 3C 120, but even in this case observations have failed thus far to detect a significant spectral feature at the expected energy [64].

Interpretation of circular polarization in the core of the quasar 3C 279 suggests that the jet in this extreme blazar ($\Gamma > 20$; [43]) contains a 5:1 ratio of positrons to protons [83]. This relies on explaining the circular polarization as conversion of linear to circular polarization in an optically thick region. Tuning of the amount of Faraday rotation versus synchrotron self-absorption sets the ratio of positive particle type. However, there has been a suggestion [74] that a suitably turbulent magnetic field in a 100% normal-matter plasma could also reproduce the data.

Various other attempts to determine the matter content are less direct and have led to contradictory conclusions [18, 73, 76, 17].

7.5 AGN Jets in Context

Although only a minority of AGN possess bright, relativistic jets, jets play an important role wherever they are present. A high fraction of the mass–energy accreted onto the black hole can be transformed into highly collimated, high-speed outflows in the form of both jets and winds. This probably stems from the difficulty in squeezing magnetic fields and matter with angular momentum into small spaces, so plasma squirts out perpendicular to the plane of accretion. We do not yet understand the physical conditions that decide whether or not relativistic jets form. The main hint that nature provides is that it seems to be quite difficult in spiral galaxies, but no generally accepted solution has yet grown out of this clue.

Figure 7.9 sketches the entire AGN system as we currently understand it. The broad-line clouds are immersed in a more dilute, hotter medium, perhaps in a wind emanating from the accretion disk. The jet can be thought of as either the same

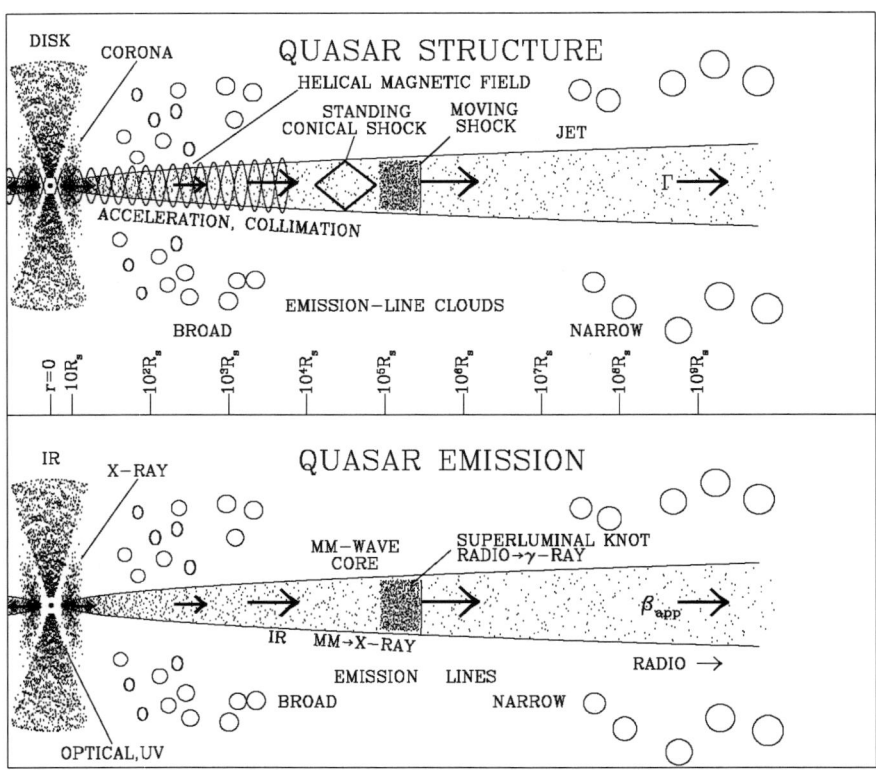

Fig. 7.9 Cartoon illustrating the various physical and emission components of a quasar or other AGN with a relativistic jet. The length scale (shown along the *bottom* of the *top panel* in terms of Schwarzschild radii of the black hole) is logarithmic in order to include phenomena over a wide range of distances from the black hole. The radiation produced in the jet is relativistically beamed, while that outside the jet is not

general outflow, with the velocity increasing toward smaller angles to the polar axes of the disk. Alternatively, the jet is a distinct flow structure that originates in the ergosphere of the black hole, surrounded by the wind but with properties quite different. For example, the jet could be populated mainly by electron–positron pairs, with normal electron–proton plasma representing a minority of the particles.

Our observational tools for probing jets continue to become more sophisticated with time. Millimeter-wave antennas in high-Earth orbit as part of a VLBI array can provide the angular resolution needed to explore regions upstream of the core [63]. Well-sampled multiwaveband flux and, where possible, polarization monitoring observations from radio to γ-ray frequencies might reveal the structure of the jet between these locations and the black hole. And the ever-surprising blazar jets are likely to provide us with singular events that unveil yet another clue in our quest to understand this exotic and enigmatic phenomenon.

References

1. A. Alberdi, J.M. Marcaide, A.P. Marscher et al.: Astrophys. J., **402**, 160 (1993)
2. M.F. Aller, H.D. Aller, P.A. Hughes: Astrophys. J., **586**, 33 (2003)
3. D.J. Axon, A. Pedlar, S.W. Unger et al.: Nature, **341**, 631 (1989)
4. J.N. Bahcall, S. Kirhakos, D.P. Schneider et al.: Astrophys. J., **452**, L91 (1995)
5. M.C. Begelman: Astrophys. J., **493**, 291 (1998)
6. M.C. Begelman, R.D. Blandford, M.J. Rees: Rev. Mod. Phys., **56**, 255 (1984)
7. A.R. Bell: Mon. Not. Royal Astron. Soc., **182**, 147 (1978)
8. G.V. Bicknell: Astrophys. J., **422**, 542 (1994)
9. R.D. Blandford, A. Königl: Astrophys. J., **232**, 34 (1979)
10. R.D. Blandford, A. Levinson: Astrophys. J., **441**, 79 (1995)
11. R.D. Blandford, M.J. Rees: Mon. Not. Royal Astron. Soc., **169**, 395 (1974)
12. S. Bogovalov, K. Tsinganos: Mon. Not. Royal Astron. Soc., **357**, 918 (2005)
13. A.H. Bridle: Astron. J., **89**, 979 (1984)
14. G.R. Burbidge: Astrophys. J., **129**, 849 (1959)
15. B.J. Burn: Mon. Not. Royal Astron. Soc., **133**, 67 (1966)
16. T.V. Cawthorne: Mon. Not. Royal Astron. Soc., **367**, 851 (2006)
17. A. Celotti: Astrophys. Space Sci., **288**, 175 (2003)
18. A. Celotti, A.C. Fabian: Mon. Not. Royal Astron. Soc., **264**, 228 (1993)
19. R. Chatterjee, A.P. Marscher, S.G. Jorstad et al.: Astrophys. J., **689**, 79 (2008)
20. C.C. Cheung, D.E. Harris, L. Stawarz: Astrophys. J., **663**, L65 (2007)
21. W.D. Cotton, J.J. Wittels, I.I. Shapiro et al.: Astrophys. J., **238**, L123 (1980)
22. R. Courant, K.O. Friedrichs: Supersonic Flow and Shock Waves, Springer-Verlag, New York, Berlin, Heidelberg, London, Paris, Tokyo, Hong Kong, Barcelona, Budapest (1948)
23. R.A. Daly, A.P. Marscher: Astrophys. J., **334**, 539 (1988)
24. F.D. D'Arcangelo, A.P. Marscher, S.G. Jorstad et al.: Astrophys. J., **659**, L107
25. D.S. De Young: The Physics of Extragalactic Radio Sources, The University of Chicago Press, Chicago (2002)
26. E. Fermi: Phys. Rev., **75**, 1169 (1949)
27. H.C. Ford, R.J. Harms, Z.I. Tsvetanov et al.: Astrophys. J., **435**, L27 (1994)
28. J.L. Gómez, J.M. Martí, A.P. Marscher et al.: Astrophys. J., **449**, L19 (1995)
29. J.L. Gómez, A.P. Marscher, A. Alberdi et al.: Science, **289**, 2317 (2000)
30. P.E. Hardee: AGN Jets: A Review of Stability and Structure. In: P.A. Hughes, J.N. Bregman (eds.) Relativistic Jets: The Common Physics of AGN, Microquasars, and Gamma-Ray

Bursts, AIP Conf. Proc., vol. 856, pp. 57–77. American Institute of Physics, Melville, New York (2006)
31. R.J. Harms, H.C. Ford, Z.I. Tsvetanov et al.: Astrophys. J., **435**, L35 (1994)
32. R.C. Hartman, D.L. Bertsch, S.D. Bloom et al.: Astrophys. J. Suppl., **123**, 79 (1999)
33. J. Hawley, J.H. Krolik: Astrophys. J., **641**, 103 (2006)
34. D.C. Homan, Y.Y. Kovalev, M.L. Lister et al.: Astrophys. J., **642**, 115 (2006)
35. C.T. Hua: Astron. J., **199**, 105 (1988)
36. P.A. Hughes (ed.): Beams and Jets in Astrophysics. Cambridge University Press, Cambridge, New York, Port Chester, Melbourne, Sydney (1991)
37. P.A. Hughes, H.D. Aller, M.F. Aller: Astrophys. J., **298**, 301 (1985)
38. P.A. Hughes, H.D. Aller, M.F. Aller: Astrophys. J., **341**, 54 (1989)
39. C. Impey, G. Neugebauer: Astron. J., **95**, 307 (1988)
40. T.W. Jones, S.L. Odell: Astrophys. J., **214**, 522 (1977)
41. S.G. Jorstad, A.P. Marscher, J.R. Mattox: Astrophys. J. Suppl., **134**, 181 (2001)
42. S.G. Jorstad, A.P. Marscher: Astrophys. J., **614**, 615 (2004)
43. S.G. Jorstad, A.P. Marscher, M.L. Lister et al.: Astron. J., **130**, 1418 (2005)
44. W. Junor, J.A. Biretta, M. Livio: Nature, **401**, 891 (1999)
45. J. Kataoka, J.R. Mattox, J. Quinn: Astrophys. J., **514**, 138 (1999)
46. K.I. Kellermann, M.L. Lister, D.C. Homan et al.: Astrophys. J., **609**, 539 (2004)
47. A. Königl: Astrophys. J., **243**, 700 (1981)
48. Y.Y. Kovalev, K.I. Kellermann, M.L. Lister et al.: Astron. J., **130**, 2473 (2005)
49. J.H. Krolik: Active Galactic Nuclei. Princeton University Press, Princeton (1999)
50. R.A. Laing: Brightness and polarization structure of decelerating relativistic jets. In: P.E. Hardee, A.H. Bridle, J.A. Zensus (eds.) Energy Transportation in Galaxies and Quasars, ASP Conf. Ser., vol. 100, pp. 241–252. Astronomical Society of the Pacific, San Francisco (1996)
51. R.A. Laing, A.H. Bridle: Mon. Not. Royal Astron. Soc., **336**, 328 (2002)
52. R.A. Laing, J.R. Canvin, A.H. Bridle: Astronomische Nachrichten, **327**, 523 (2006)
53. R.A. Laing, P. Parma, H.R. de Ruiter et al.: Mon. Not. Royal Astron. Soc., **306**, 513 (1999)
54. M.L. Lister, D.C. Homan: Astron. J., **130**, 1389
55. A.P. Lobanov: Astron. Astrophys., **330**, 79 (1998)
56. F.D. Macchetto, A. Marconi, D.J. Axon et al.: Astrophys. J., **489**, 579 (1997)
57. A.P. Marscher: Astrophys. J., **235**, 386 (1980)
58. A.P. Marscher: Astrophys. J., **264**, 296 (1983)
59. A.P. Marscher: Synchro-Compton emission from superluminal sources. In: J.A. Zensus, T.J. Pearson (eds.) Superluminal Radio Sources, pp. 280–300. Cambridge University Press, Cambridge (1987)
60. A.P. Marscher: Proc. Nat. Acad. Sci., **92**, 11439 (1995)
61. A.P. Marscher, W.K. Gear: Astrophys. J., **298**, 114 (1985)
62. A.P. Marscher, W.K. Gear, J.P. Travis: Variability of nonthermal continuum emission in blazars. In: E. Valtaoja, M. Valtonen (eds.) Variability of Blazars, pp. 85–101. Cambridge University Press, Cambridge (1992)
63. A.P. Marscher, S.G. Jorstad, F.D. D'Arcangelo et al.: Nature, **452**, 966 (2008)
64. A.P. Marscher, S.G. Jorstad, J.L. Gómez et al.: Astrophys. J., **665**, 232 (2007)
65. J.C. McKinney, R. Narayan: Mon. Not. R. Astron. Soc., **375**, 531 (2007)
66. D.L. Meier, S. Koide, Y. Uchida: Science, **291**, 84 (2000)
67. R. Mittal, R. Porcas, O. Wucknitz et al.: Astron. Astrophys., **447**, 515 (2006)
68. A. Pacholczyk: Radio Astrophysics. Freeman, San Francisco (1970)
69. R.A. Perley, J.W. Dreher, J.J. Cowan: Astrophys. J., **285**, L35 (1984)
70. B. Punsley: Astrophys. J., **473**, 178 (1996)
71. A.B. Pushkarev, D.C. Gabuzda, Yu.N. Vetukhnovskaya et al.: Mon. Not. R. Astron. Soc., **356** 859 (2005)
72. A.C.S. Readhead: Astrophys. J., **426**, 51 (1994)
73. C.S. Reynolds, A.C. Fabian, A.Celotti et al.: Mon. Not. Royal Astron. Soc., **283**, 873 (1996)

74. M. Ruszkowski, M.C. Begelman: Astrophys. J., **573**, 485 (2002)
75. M. Sikora, M.C. Begelman, G.M. Madejski et al.: Astrophys. J., **625**, 72 (2005)
76. M. Sikora, G. Madejski: Astrophys. J., **534**, 109 (2000)
77. H. Sol, G. Pelletier, E. Asséo: Mon. Not. R. Astron. Soc., **237**, 411 (1989)
78. M. Türler, T.J.-L. Courvoisier, S. Paltani: Astron. Astrophys., **349**, 45 (1999)
79. M. Türler, T.J.-L. Courvoisier, S. Paltani: Astron. Astrophys., **361**, 850 (2000)
80. N. Vlahakis: Disk-Jet connection. In: H.R. Miller, K. Marshall, J.R. Webb et al. (eds.) Blazar Variability Workshop II: Entering the GLAST Era, Astronomical Society of the Pacific Conf. Ser., vol. 350, pp. 169–177. Astron. Soc. Pacific, San Francisco (2006)
81. N. Vlahakis, A. Königl: Astrophys. J., **605**, 656 (2004)
82. R.C. Walker, J.M. Benson, S.C. Unwin: Astrophys. J., **316**, 546 (1987)
83. J.F.C. Wardle, D.C. Homan, R. Ojha et al.: Nature, **395**, 457 (1998)

Chapter 8
X-Ray Variability of AGN and Relationship to Galactic Black Hole Binary Systems

I. McHardy

Abstract Over the last 12 years, AGN monitoring by RXTE has revolutionised our understanding of the X-ray variability of AGN, of the relationship between AGN and Galactic black hole X-ray binaries (BHBs) and hence of the accretion process itself, which fuels the emission in AGN and BHBs and is the major source of power in the universe. In this chapter I review our current understanding of these topics.

I begin by considering whether AGN and BHBs show the same X-ray spectral-timing "states" (e.g. low flux, hard spectrum or "hard" and high flux, soft spectrum or "soft"). Observational selection effects mean that most of the AGN which we have monitored will probably be "soft-state" objects, but AGN are found in the other BHB states, although possibly with different critical transition accretion rates.

I examine timescale scaling relationships between AGN and BHBs. I show that characteristic power spectral "bend" timescales, T_B, scale approximately with black hole mass, M_{BH}, but inversely with accretion rate $\dot{m}_E$ (in units of the Eddington accretion rate), probably signifying that T_B arises at the inner edge of the accretion disc. The relationship $T_B \propto M_{BH}/\dot{m}_E$ is a good fit, implying that no other potential variable, e.g. black hole spin, varies significantly. Lags between hard and soft X-ray bands as a function of Fourier timescale follow similar patterns in AGN and BHBs.

I show how our improved understanding of X-ray variability enables us to understand larger scale properties of AGN. For example, the width of the H_β optical emission line, V, scales as $T_B^{1/4}$, providing a natural explanation of the observed small black hole masses in Narrow Line Seyfert Galaxies; if M_{BH} were large then, as $T_B \propto M_{BH}/\dot{m}_E$, we would require $\dot{m}_E > 1$ to obtain narrow lines.

I note that the rms X-ray variability scales linearly with flux in both AGN and BHBs, indicating that the amplitude of the shorter timescale variations is modulated by that of the longer timescale variations, ruling out simple shot-noise variability models. Blazars follow approximately the same pattern. The variations may therefore arise in the accretion disc and propagate inwards until they hit, and modulate, the X-ray emission region which, in the case of blazars, lies in a relativistic jet.

I. McHardy (✉)

School of Physics and Astronomy, University of Southampton, Southampton SO17 1BJ, UK,
imh@astro.soton.ac.uk

McHardy, I.: *X-Ray Variability of AGN and Relationship to Galactic Black Hole Binary Systems.*
Lect. Notes Phys. **794**, 203–232 (2010)
DOI 10.1007/978-3-540-76937-8_8

Short timescale (weeks) optical variability arises from reprocessing of X-rays in the accretion disc, providing a diagnostic of X-ray source geometry. On longer timescales, variations in the disc accretion rate may dominate optical variations.

AGN X-ray monitoring has greatly increased our understanding of the accretion process and there is a strong case for continued monitoring with future observatories.

8.1 Introduction

Understanding the relationship between AGN, which are powered by accretion onto supermassive black holes, and the much smaller Galactic black hole binary systems (BHBs) is currently one of the major research areas in high-energy astrophysics. Possible similarities between AGN and BHBs have been mooted ever since the late 1970s and early 1980s when it was first realised that they were both black hole systems (e.g. [86]). However comparison of their X-ray variability properties provided the first quantitative method for this comparison [49]. More recently considerable attention has been devoted to the jet properties of black hole systems and a strong scaling has been shown by means of comparing radio and X-ray luminosities [44, 23]. However, here I concentrate on the considerable insight which can be gained regarding the scaling between AGN and BHBs by studying their X-ray timing similarities.

I begin this review with a discussion of AGN "states", as we must be sure, when comparing AGN with BHBs, that we are comparing like with like. I then discuss characteristic X-ray variability timescales and how they scale with mass and accretion rate. I then show how larger scale AGN properties, e.g. AGN optical permitted line width, are related to the small-scale accretion properties. I discuss models for the origin of the variability and show how the variability of blazars fits the same pattern as for Seyfert galaxies, showing that the source of the X-rays is separated from the source of the variations, which may be the same in all accreting objects. The long AGN X-ray monitoring programmes which were set up to enable us to compare AGN and BHB X-ray variability properties have also allowed us, through correlated monitoring programmes in other wavebands, to understand a little about what drives the optical variability of AGN. I therefore conclude this review with an examination of that topic.

8.2 AGN X-Ray Variability and AGN "States"

8.2.1 States of Galactic Black Hole X-Ray Binary Systems

BHBs are found in a number of "states", originally defined in terms of their medium energy (2–10 keV) X-ray flux and spectral hardness. These states are discussed fully elsewhere in this volume (See "States and Transitions in Black-Hole Binaries" by T.M. Belloni and " 'Disc-Jet' Coupling in Black Hole X-ray Binaries and Active

Galactic Nuclei" by R. Fender, this volume) but, for completeness here, and at the risk of some oversimplification, I briefly summarise the properties of the most commonly found states, i.e. the "hard", "soft" and "VHS" states.

In the "hard" state, the medium energy X-ray flux is low and the spectrum is hard. In the "soft" state, the medium energy X-ray flux is high and the spectrum soft. The main spectral characteristic of the very high state (VHS), sometimes also known as the high intermediate state, is that the X-ray flux is very high. The spectrum is intermediate in hardness between the hard and soft states. The accretion rate rises as we go from the hard to soft to VHS states. The X-ray timing properties of the different spectral states are also very different and quite characteristic and may, in fact, provide a better state discriminant than the spectral properties.

X-ray variability is usually quantified in terms of the power spectral density (PSD) of the X-ray light curve. In Cyg X-1 there is a large amplitude of variability in the soft-intermediate and soft states and the PSD is well described by a simple bending powerlaw. At high frequencies, the power spectral slope is -2 or steeper (i.e. $P(\nu) \propto \nu^{-\alpha}$ with $\alpha \geq 2$) bending, below a characteristic frequency ν_B (or timescale T_B) to a slope $\alpha \simeq 1$ [16, 55]. In most other BHBs, e.g. GX 339-4 [7], which typically differ from Cyg X-1 in reaching the soft state only during very high accretion rate transient outbursts but then reaching a more pronounced soft state, the variable component is swamped by a quiescent thermal component and so is very hard to quantify. However, the PSD of the weak variable component, when measurable, is consistent with the shape seen for Cyg X-1.

In the hard state the PSD can also be approximated by bending powerlaws, and the high-frequency bend is also seen. In addition, a second bend to a slope of $\alpha = 0$ is seen approximately 1.5–2 decades below the high-frequency bend. However, in the hard state, BHB PSDs are better described by a combination of Lorentzian-shaped components [59], with prominent components being located at approximately the bends in the bending powerlaw approximation to the PSD. In the very high state and, indeed, in any state apart from the soft state, the PSD can be described by the sum of Lorentzian components, as in the hard state. However, in the very high state, the corresponding Lorentzian frequencies are higher than in the hard state.

8.2.2 Quantifying AGN States by Power Spectral Analysis

AGN typically have X-ray spectra similar to those of hard state BHBs (i.e. photon indices ~ 2), and much harder than soft-state BHBs. (Note that, unless specifically mentioned to the contrary, by "AGN" I mean an active galaxy, such as a Seyfert galaxy, where the emission does not come from a relativistic jet. Relativistically beamed AGN, i.e. blazars, are discussed in Sect. 8.7.) Simple spectra do not provide a particularly direct means of state comparison as, in soft-state BHBs, the very hot accretion disc leads to a large, soft spectrum, thermal component which dominates the medium energy emission. In AGN, the accretion disc is much cooler and so does not affect the medium energy flux, even at high accretion rates. However, if one can disentangle the relative contribution of the accretion disc from the total luminosity then AGN and BHB do seem to occupy broadly similar regions of disc-fraction/luminosity diagrams (e.g. [32], See " 'Disc-Jet' Coupling in Black Hole X-ray Binaries and Active Galactic Nuclei" by R. Fender, this volume). Nonetheless, timing properties may provide a cleaner and simpler state discriminant and means of comparing AGN with BHBs. However, there are considerable difficulties in measuring AGN PSDs over a wide enough frequency range that the equivalent of the bends seen in BHB PSDs may be found. Assuming to first order that system sizes, and hence most relevant timescales, will scale with black hole mass, M_{BH}, then as $\nu_B \sim 10\,\mathrm{Hz}$ for the BHB Cyg X-1 in the high state ($M_{BH} \sim 10 - 20 M_\odot$; [30, 88]), we might expect $\nu_B \sim 10^{-7}\,\mathrm{Hz}$ for an AGN with $M_{BH} = 10^8$ $M_\odot$. To measure such a bend requires a light curve stretching over a number of years.

The first really detailed observations of AGN X-ray variability were made by EXOSAT. In the case of NGC 4051 [35] and NGC 5506 [47], variability on timescales of less than $\sim$day was shown to be scale invariant, initially dashing hopes of finding some characteristic timescale from which black hole masses could be deduced. However, by combining the EXOSAT observation of NGC 5506 with archival observations from a variety of satellites, it was possible to make a PSD covering timescales from $\sim$years to $\sim$minutes. Although poorly determined on long timescales, the PSD did roughly resemble the PSD of a BHB and was sufficient to show that the scale invariance which prevailed on timescales shorter than a day broke down at timescales of around a few days to a week [49, 50]. The bend timescale was broadly consistent with linear scaling with mass. But the long timescale data were very poor. Subsequent archival searches produced a measurement of a bend frequency in NGC 4151 [64], but again the long timescale data were poorly sampled. However, the long timescale monitoring of AGN, on which the production of high-quality PSDs relies, was revolutionised in late 1995 with the launch of the RXTE which was specifically designed for rapid slewing, allowing for monitoring on a variety of timescales from hours to years. As a result a number of groups began monitoring AGN. Long-timescale PSDs of good quality were made and reliable bend frequencies were measured for a number of AGN (e.g. [54, 20]).

8.2.3 Power Spectral Analysis of Irregularly Sampled Data

Determining the shape of AGN PSDs from sampling that is usually discrete and irregular is not trivial. Fourier analysis of the raw observational data will result in the true underlying PSD being distorted by the window function of the sampling pattern. The non-continuous nature of most monitoring observations means that high-frequency power is not properly sampled, which results in that power being "aliased", or reflected, back to lower frequencies. Also, as the overall light curves have a finite length, low-frequency power is not properly measured and so additional spurious power leaks into the PSD from low frequencies. This latter effect is known as red noise leak. The shape of the true underlying PSD can, however, be estimated by modelling. The procedure which we use is called PSRESP [83]. This procedure builds on earlier work (the "response" method, [18]) and other previous authors have also used a similar method to analyse AGN X-ray variability [26]. In PSRESP we assume a model form for the underlying PSD and, using a now standard prescription [75], a light curve is then simulated. This light curve is folded through the real observational sampling pattern and a "dirty" simulated PSD is produced. This procedure is repeated many times to establish an average model dirty PSD and to establish its errors. Observations from a number of different observatories (e.g. RXTE XMM-Newton ASCA) can be modeled simultaneously. The real, observed, dirty PSD is then compared with the dirty model average and the model parameters are adjusted until the best fit is obtained.

Using PSRESP, we have determined the overall PSD shapes of a number of AGN, covering timescales from years to tens of seconds (e.g. [83, 55, 52, 82]). In only one case (Akn 564, described below) do we find evidence for more than one bend in the PSD. In all other cases we find only one bend, from $\alpha = 2$ at high frequencies to $\alpha = 1$ at lower frequencies. In NGC 4051 [55] and MCG-6-30-15 [52] the slope of $\alpha = 1$ continues for at least four decades below the upper bend and in a number of other AGN (e.g. NGC 3227, [82]) the slope can be traced for at least three decades. In such cases we can be reasonably sure that a hard state PSD is ruled out and a soft-state PSD is required. In other cases there may not be sufficient low-frequency data to distinguish between hard and soft states but in all cases a soft-state PSD is consistent with the data.

The lack of hard-state AGN PSDs may, however, just be a selection effect. Most of the AGN that have recently been monitored are X-ray bright and have moderate accretion rates (typically above a few per cent of the Eddington accretion rate, $\dot{m}_E$). In BHBs, whenever the accretion rate is below $\sim$0.02 $\dot{m}_E$, the sources are found in the hard state. Although hard states are also found above $\sim$0.02 $\dot{m}_E$, no soft states are found below that accretion rate. Thus the finding that almost all AGN PSDs may be soft-state PSDs is consistent with the expectation based on their accretion rates. It is also possible that the transition accretion rate between the hard and soft states is different in AGN. Soft states are associated with an optically thick accretion disc reaching close in to the black hole. In the lower temperature discs around more massive black holes (i.e. AGN) it is possible that the optically thick part of the accretion disc might survive without evaporation, at the same accretion rate, at smaller radii than in the hotter discs around BHBs. Thus the transition accretion rate might be lower around AGN. However, the data do not yet allow us to test this possibility.

There had been a tentative suggestion that there might be a second bend, at low frequencies, in the PSD of NGC 3783 and hence that it might be a hard-state system [41]. As NGC 3783 does not have a particularly low accretion rate ($\sim$0.07 $\dot{m}_E$), this suggestion was a little puzzling. However, our further observations [75] show that, in fact, NGC 3783 is another soft-state AGN.

8.2.4 The Unusual Case of Akn 564 – A Very High-State AGN

The finding of a VHS PSD in a high-accretion rate AGN would significantly strengthen the growing link between AGN and BHBs. Arakelian 564 is one of the highest accretion rate AGN known ($\dot{m}_E \sim 1$) and is the only AGN for which there is presently good evidence that its PSD contains more than one bend. From observations with RXTE, evidence for a low-frequency bend was presented [68] and evidence for a second bend, at high frequencies, was also found from analysis of ASCA observations [62]. The combined RXTE and ASCA PSD was interpreted as a hard-state PSD [62], although it was noted that the frequency of the high-frequency bend did not scale linearly with mass to that of the hard state of the archetypal BHB Cyg X-1; the frequency difference is too small. However, proper modelling of the combined long- and short-timescale PSD was hampered by the gaps which occur in the ASCA light curves, and hence in the PSD, at the orbital period (~5460 s). We therefore made a 100 ks continuous observation with XMM-Newton to fill that gap in the PSD and the resulting overall XMM-Newton ASCA and RXTE PSD is shown in Fig. 8.1, bottom panel. Although the PSD of Akn 564 can be fitted with a doubly bending powerlaw as well as with two Lorentzian components, strong support for the Lorentzian interpretation is given by the spectrum of the lags between the hard and soft X-ray bands. It has long been noted in both BHBs (e.g. [60]) and AGN (e.g. [65, 55]) that the X-ray emission in the harder energy bands usually lags behind that in the softer bands and that the lag increases with increasing energy separation between the bands. However, if we split the light curves into components of different Fourier frequency we can measure the lag as a function of Fourier frequency, i.e. we can determine the lag spectrum. Although I do not discuss it here, the degree of correlation between light curves is known as the coherence which, together with the lags, can provide a very useful diagnostic of the source geometry and emission mechanism. The calculation of lags and coherence is known as cross-spectral analysis and a full description of the technique and its applications is given elsewhere [85].

In BHBs where the PSD is well described by the sum of Lorentzian components, i.e. in every state apart from the soft state, it is found that the lags are reasonably constant over the frequency range where the PSD is dominated by any one Lorentzian component, but the lags change abruptly at the frequencies at which the PSD changes from one Lorentzian component to the next. In most AGN the PSD and lag spectra are not defined well enough to distinguish different Lorentzian components but for Akn 564 we can measure the lag spectrum (Fig. 8.1, top panel). We see that the lag changes rapidly from one fairly constant level to another at the frequency where the PSD changes from one Lorentzian to another. Given the extremely high accretion rate of Akn 564, we interpret these observations as strongly supporting the VHS, rather than hard-state interpretation. We also note that, at the highest frequencies (above 10^{-3} Hz), the lags become slightly negative, i.e. the soft band lags the hard band. One possible explanation is that, at these frequencies, which may come from the very innermost part of the accretion disc, the soft

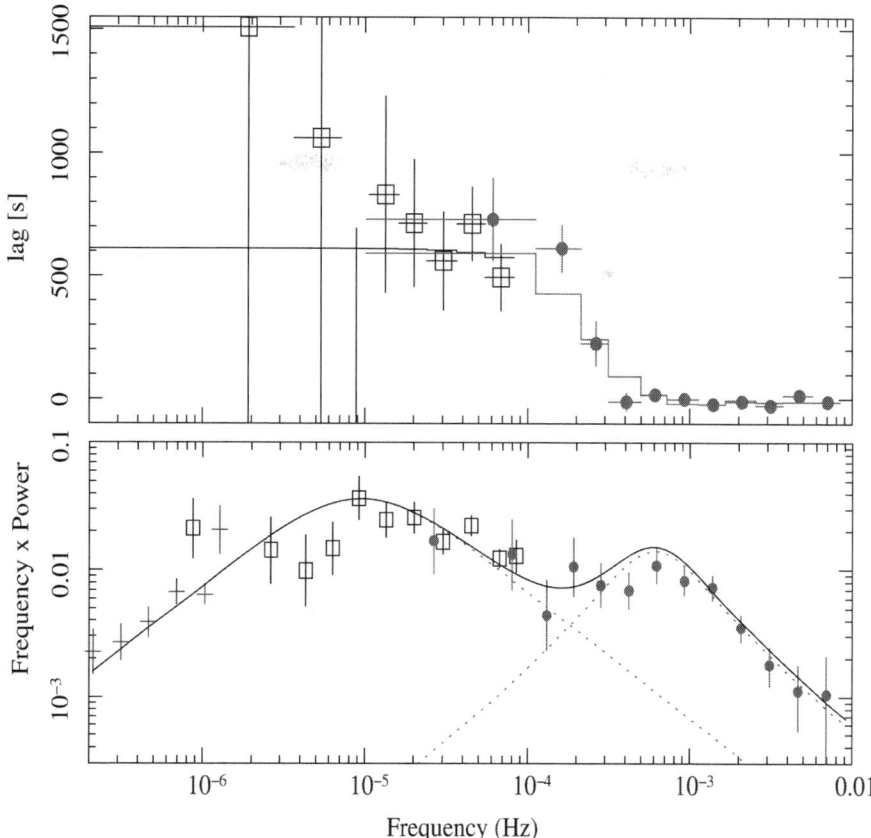

Fig. 8.1 *Top panel*: Lag between the hard (2–10 keV) and soft (0.5–2 keV) X-ray bands as a function of Fourier frequency for Akn 564. A positive lag means that the hard band lags the soft. Above 10^{-3} Hz the lags become slightly negative (-11 ± 4.3 s), i.e. the soft band lags the hard band. *Bottom panel*: PSD of Akn 564 showing a fit to the two-Lorentzian model [51]. Note how the cross-over frequency between the two Lorentzians corresponds to the frequency at which the lags rapidly change

X-rays arise mainly from reprocessing of harder X-rays by the accretion disc (see Sect. 8.6.3).

Ton S180 In our recent PSD survey of 32 well-observed AGN (Summons et al., in prep.) the only AGN besides Akn 564 for which there are reasonable hints that a double-Lorentzian, rather than a single-bending powerlaw model, is required is Ton S180 (Fig. 8.2). Interestingly, Ton S180 has, like Akn 564, an accretion rate close to the Eddington limit and so may very well be another VHS system. As with Ark 564, we have calculated the lag spectrum but the data are not as good as for Ark 564. The lags are, however, consistent with a VHS interpretation, as in Ark 564.

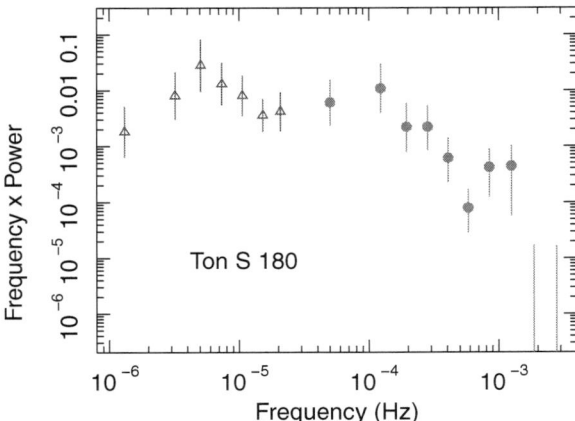

Fig. 8.2 Combined RXTE and ASCA PSD of Ton S180, unfolded from the distortions introduced by the sampling pattern, assuming a simple model of a powerlaw with a bend at high frequencies. Two large bumps are seen in the residuals which can be fitted equally well with two Lorentzian-shaped components (Summons et al., in prep.)

8.2.5 Periodicities in AGN

Although broad Lorentzian features can be fitted to the PSDs of Akn 564 and Ton S180 there have been, until very recently, no believable detections of highly coherent periodicities in AGN. However, one periodicity of high-quality factor ($Q > 16$) has now been found in the NLS1 RE J1034+396 (Fig. 8.3, from [24]). The periodicity does change in character during the observation with slight shifts in times of minimum and variations in amplitude so it is very similar to the quasi-periodic oscillations (QPOs) seen in BHBs (e.g. [70]).

The reason that a narrow QPO is found in RE J1034+396 is not known. It has a high accretion rate, but so too does Akn 564 and Ton S180. It has been suggested that perhaps the phenomenon is transient as in BHBs [24] and the observers were just very lucky. Alternatively, it may have something to do with the overall spectral shape of RE J1034+396 which is unique, even for an NLS1, peaking in the far UV. Whatever the reason, this observation is very important as it shows that the X-ray variability similarities between AGN and BHB extend to almost all known phenomena. It also enables us to estimate the black hole mass. In BHBs, there is a tentative relationship between the frequency of the high-frequency QPOs and the mass, $f_0 = 931(M/M_\odot)$Hz (e.g. [70]), where f_0 is the fundamental frequency of the pair of high-frequency QPOs whose frequency ratio is $2f{:}3f$. In RE J1034+396 then (and depending a little on which of the two high-frequency QPOs we think the observed periodicity is) we can estimate a mass around $10^7 M_\odot$.

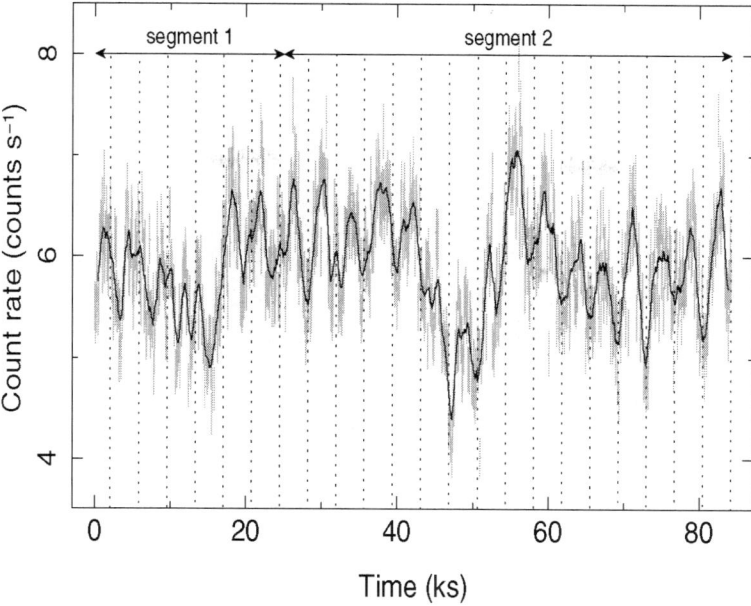

Fig. 8.3 XMM-Newton 0.3–10 keV light curve of RE J1034+396 [24]. The *thick black line* represents the running average over nine bins. The *vertical dotted lines* show the expected times of minima obtained from folding segment 2 with the period 3733 s

8.3 Scaling Characteristic Timescales with Mass and Accretion Rate

The early suggestions of a scaling of T_B with M_{BH} [52, 66] were supported by the first results from the RXTE monitoring programmes (e.g. [56, 21, 85, 44]). However, once T_B had been measured in more than just a handful of AGN it was clear that there was much more spread in the T_B vs. M_{BH} relationship than could be accounted for just by observational error. Interestingly it was noted that for a given M_{BH}, the narrow line Seyfert 1 galaxies (NLS1s) had shorter values of T_B than broad line AGN and so it was suggested that accretion rate, as well as M_{BH}, might also affect the value of T_B [55, 52, 82]. The reason for the dependence on accretion rate was not clear but one possibility was that, if T_B is somehow associated with the inner edge of the optically thick part of the accretion disc, and if that edge moves closer to the black hole when the accretion rate rises [21], then T_B would decrease. Also, the increase in accretion rate would lead to an increase in luminosity and hence in an increase in the inner radius of the broad line region (R_{BLR}), and therefore in a decrease of the typical velocities, or widths, associated with the emission lines [52].

8.3.1 Fitting the Timing Scaling Relationship for Soft-State Objects

In order to properly quantify the dependence of T_B on M_{BH} and $\dot{m}_E$, and motivated by the approximate linear relationship between T_B and $\dot{m}_E$ and the approximate inverse relationship between T_B and $\dot{m}_E$ we hypothesised that

$$\log T_B = A \log M_{BH} - B \log L_{bol} + C,$$

where L_{bol} is the bolometric luminosity, and performed a simple 3D parameter grid search to determine the values of the parameters A, B and C. As the AGN under consideration were, in all cases where the PSD was well-defined, soft-state objects, then $\dot{m}_E \sim L_{bol}/L_{Edd}$. However, we preferred to fit to L_{bol} rather than $\dot{m}_E$, as L_{bol} is an observable, rather than a derived, quantity. The best fit to a sample of 10 AGN was $A = 2.17^{+0.32}_{-0.25}$, $B = 0.90^{+0.3}_{-0.2}$ and $C = -2.42^{+0.22}_{-0.25}$ [53].

In order to determine whether the same scaling extends down to BHBs we included two bright BHBs (Cyg X-1 and GRS 1915+105) in radio-quiet states where, for proper comparison with the AGN, their high-frequency PSDs are well described by the same cut-off, or bending powerlaw model which best describes AGN and where broad band X-ray flux provides a good measurement of bolometric luminosity. For Cyg X-1, we combined measurements of T_B [5] with simultaneous measurement of the bolometric luminosity [87] over a range of luminosities. For GRS 1915+105, we measured an average T_B from the original X-ray data and determined L_{bol} from the published fluxes [77] and generally accepted distance (11 kpc). For the combined fit to the 10 AGN and the 2 BHBs, we find complete consistency with the fit to the AGN on their own, i.e. $A = 2.1 \pm 0.15$, $B = 0.98 \pm 0.15$ and $C = -2.32 \pm 0.2$. The confidence contours for the fit to the AGN on their own and to the combined AGN and BHB sample are shown in Fig. 8.4. We can see that, even at the 1σ (i.e. 68% confidence) level, the contours for the AGN on their own, and for the combined AGN and BHB sample, completely overlap (as do the offset constants, C). Thus we answer a long-standing question and show, using a self-consistently derived set of AGN and BHB timing data, that over a range of $\sim 10^8$ in mass and $\sim 10^3$ in accretion rate, AGN behave just like scaled-up BHBs. Assuming $\dot{m}_E = L_{bol}/L_{Edd}$, then $T_B \propto M_{BH}^{1.12}/\dot{m}_E^{0.98}$.

We can see how well the scaling relationship fits the data by comparing the observed bend timescales with the values predicted by the best fit parameters (Fig. 8.5). The only noticeable outlier is NGC 5506 (at $\log T_{\text{predicted}} = +1.3$). However, we note that the black hole mass estimate for NGC 5506 ($10^8 M_\odot$) which we used [53] is based on the width of the [OIII] lines, which is only a secondary mass estimator. The resultant implied accretion rate (2.5% $\dot{m}_E$) is surprisingly low given its recent classification, from the width of the IR P_β line, as an obscured NLS1 [56]. However, using the recently measured width of its stellar absorption lines [28], a smaller mass (few $\times 10^6 M_\odot$) is derived. With that lower mass, NGC 5506 lies much closer to the best-fit line.

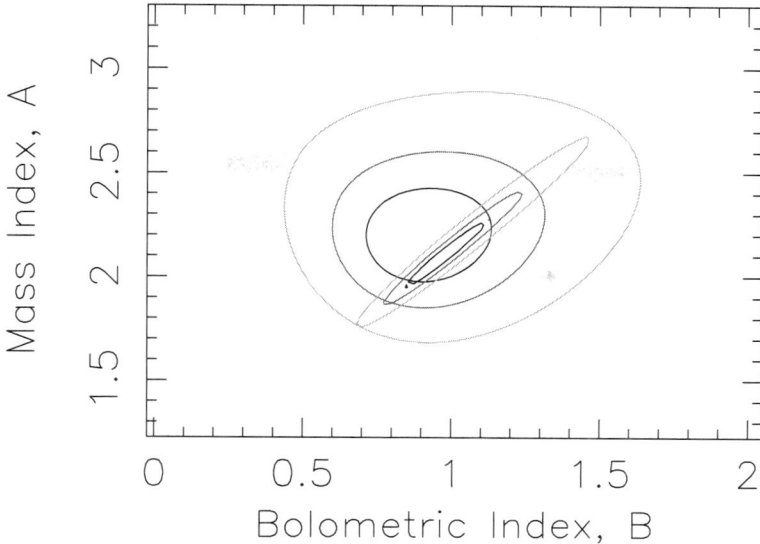

Fig. 8.4 Assuming a relationship between the PSD bend timescale T_B, the black hole mass M_{BH} and the bolometric luminosity L_{bol} of the form $\text{Log}\,T_B = A\text{Log}\,M_{BH} - B\text{Log}\,L_{bol} + C$, where C is a constant, we present here the 68 (*black*), 90 (*dark grey*) and 95% (*light grey*) confidence contours for A and B. The thick contours are derived from a fit to 10 AGN only. The thin contours also include the BHBs Cyg X-1 and GRS 1915+105 in their soft (radio quiet) states. Note how the contours overlap completely, even at the 68% confidence level

8.3.2 Constraint on Black Hole Spin

In the fits discussed above, the best-fit reduced χ^2 was very close to unity. These fits were performed without introducing any additional systematic error into the fit, but purely by using the best estimates of the observational errors. In many such fits an unknown additional error has to be introduced to bring the reduced χ^2 down to unity, but we do not require any additional source of uncertainty. The implication is that no other parameter, which we do not include in the fit, varies greatly from object to object. One parameter which could conceivably affect T_B is black hole spin. In faster spinning black holes, the innermost stable orbit is closer to the black hole. Thus if T_B is related to the radius of the inner edge of the accretion disc and if that edge reaches right up to the last stable orbit, then T_B might be lower in faster spinning black holes. The implication of the good quality of the present fit is that spin does not vary greatly from object to object. Thus if some X-ray emitting black holes are considered, from the width of their Fe K_α emission lines, to be maximally spinning [22], then probably they all are.

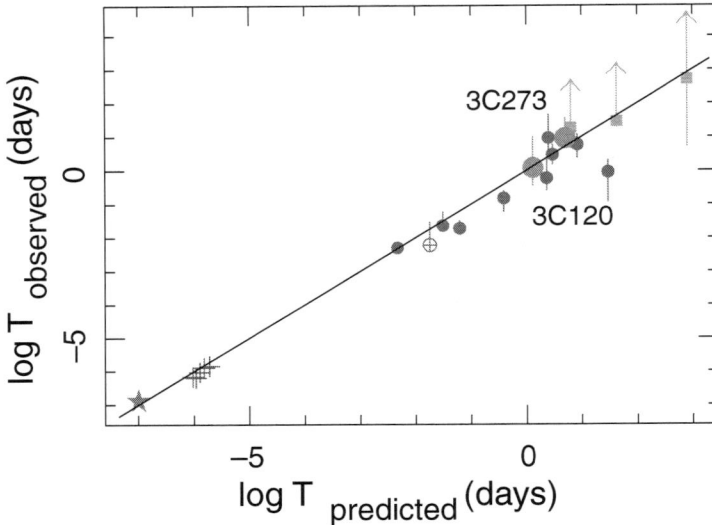

Fig. 8.5 Log of the observed vs. the predicted PSD bend timescales (in units of days), using the best-fit relationship derived from Fig. 8.4. The star on the *lower left* is GRS 1915+105, the *crosses* are Cyg X-1 and the *circles* are the AGN which were included in the fit. The three squares, whose upper bend timescales are unbounded, are AGN which were not included in the fit. However, it can be seen that the measured bend values are, nonetheless, consistent with the fit from the AGN with well-defined PSD bend timescales. Similarly we also plot the blazar 3C 273 (McHardy et al., in prep., and Sect. 8.7 here) and the radio galaxy 3C 120 ([74] and Marshall et al., in prep.), which were also not included in the fit

8.3.3 Scaling Relationship for Hard-State Objects

An inverse dependence of a characteristic timescale on accretion rate had been noted elsewhere [45]. There it was noted that the frequency of the highest frequency Lorentzian component, ν_{high}, in the hard-state PSD of the BHB GX 339-4, varied with radio luminosity, L_R, approximately as $\nu_{high} \propto L_R^{1.4}$. In standard jet models (e.g. [9]), the total power in the jet is $L_J \propto L_R^{1.4}$. Assuming that $L_J \propto \dot{m}_E$, then it is expected that $\nu_{high} \propto \dot{m}_E$ [45].

A similar scaling of characteristic timescale with accretion rate is seen in hard-state observations of the BHB XTE J1550-564. In their Fig. 8.3, Done et al. [15] plot ν_{high} against a parameter which they label " L_{bol}". However, this parameter is derived by simple linear transformation from X-ray flux, S_X. From that same Fig. 8.3 [15] it can be seen that, in 2002, when the source was in a steady hard state, $\nu_{high} \propto S_X^{0.45\pm0.05}$. In hard-state, jet-dominated systems, $S_X \propto \dot{m}_E^2$ [40]. Thus for the 2002, hard state, $\nu_{high} \propto \dot{m}_E^{0.9\pm0.1}$, i.e. the same dependence of timescale on accretion rate in an individual object that was previously found for the sample of AGN and BHBs.

In a comprehensive study it has been shown [33], using radio luminosity as a tracer of accretion rate in hard-state objects [31], that hard-state black hole systems and neutron stars also follow the same dependence of characteristic timescale with M_{BH} and $\dot{m}_E$ that was derived earlier [53], which thus seems to be characteristic of the majority of accreting objects.

8.3.4 RMS Variability as a Mass Diagnostic

From observations which are too sparse or too short to allow a well-defined PSD to be produced, it is possible to estimate the bend timescale if we assume a particular, fixed, power spectral shape. The variance in a light curve, σ^2, equals the integral of the power spectrum over the frequency range sampled by the light curve. Thus if we assume a fixed PSD shape, i.e. typically a fixed normalisation, low-frequency slope and high-frequency slope, then we can deduce the bend frequency from the observed variance in the light curve (e.g. [61]). This technique has been used to demonstrate a strong relationship between T_B and black hole mass but the dependence of T_B on $\dot{m}_E$ has not been clear [61, 58, 57]. The reason that the dependence on $\dot{m}_E$ has not been clear in these analyses is probably that the samples under consideration did not contain a great range of accretion rates, with few NLS1s involved, and also that the assumption of a universal, fixed, PSD shape is not absolutely correct. Small differences in normalisation (e.g. compare NGC 3227 and NGC 5506 [82]) or in PSD slopes (the low-frequency slope in MCG-6-30-15 is -0.8 whereas in NGC 4051 it is -1.0 [52, 55]) can easily move the derived bend frequency by half a decade, or more, which is sufficient to blur any dependence on accretion rate. High-frequency PSD slope is also a function of energy, being flatter at higher energies (e.g. [55]), and thus great care must be taken to ensure uniform energy coverage. Also, in the translation from σ^2 to T_B, a hard-state PSD shape has generally been assumed whereas we now know that a soft state is generally more likely.

Nonetheless, σ^2 does still provide a useful diagnostic of variability and has been used to good effect in the study of the evolution of variability properties with cosmic epoch. In an analysis of a deep ROSAT observation it has been shown [1] that, for a particular luminosity, the variance of high-redshift ($z > 0.5$) QSOs was greater than that of low-redshift QSOs. A similar, though more detailed, analysis has been performed [63] using the extensive XMM-Newton observations of the Lockman hole which show that, for the same luminosity, black hole mass is lower, but accretion rate is higher, for the higher redshift QSOs, indicative of the growth of black holes with cosmic epoch.

8.3.5 High-Frequency Scaling Relationships

Measurements of rms variability are usually carried out on individual observations lasting for maybe a day or so. These observations thus mainly sample the high-frequency part of the PSD. There have been related methods of quantifying variability which also use the high-frequency part of the PSD.

In the late 1980s we used the amplitude of the normalised PSD (after removing the Poisson noise level) at a standard frequency of 10^{-4} Hz (the "normalised variability amplitude", NVA, [49] and [26]). This frequency was chosen to be as high as possible whilst avoiding the region dominated by Poisson noise in typical EXOSAT observations. At that time very few AGN black hole masses were available and so NVA had to be plotted against luminosity as a proxy, and NVA was seen to decrease with increasing luminosity. Subsequently, broadly similar techniques were used [29, 17] to show that high-frequency variability timescales scaled approximately with mass.

Recently the PSDs of a number of hard-state BHBs have been studied and it is found that the high-frequency part of the PSD does not change greatly with source flux [25], although the lower frequency part does change. It is surmised that the relatively unchanged high-frequency tail part of the PSD may represent some limit, such as the last stable orbit, for the black hole in question. Nonetheless, a plot of the amplitude of the PSD tail extrapolated to 1 Hz (C_M – very similar to the NVA discussed above) against mass reveals no correlation with mass within the BHB sample alone. However, if a sample of AGN [57] is included, estimating C_M from the variance by model fitting assuming a hard-state PSD, there is a correlation with mass. As many AGN are probably soft-state systems, and as the range of accretion rates in that AGN sample is small, further work is required to determine whether the high-frequency part of the PSD gives a mass measurement, independent of any accretion rate variations.

Overall, measures of variability timescale based solely on the high-frequency part of the PSD can provide a reasonable mass estimate for many AGN samples where there is not a great range in accretion rate. Future work may show that they give an accretion rate independent measure of mass, but that is not yet confirmed. These methods have the advantage of being applicable to short observations. However, measurement of the PSD bend timescale is more robust to changes of PSD slope and normalisation and provides a more sensitive diagnostic of mass and accretion rate.

8.4 Relationship Between Nuclear Variability Properties and Larger Scale AGN Properties

One of the more important observable AGN parameters is the width of their permitted optical emission lines, V. This linewidth is often used as a means of classifying different types of AGN. It has been shown that the most variable X-ray emitting AGN (where variability has typically been quantified in terms of σ_{rms}) tend to have the narrowest optical emission lines (e.g. [78]). However, there is a good deal of spread in the relationship between σ_{rms} and V, and no quantitative physical explanation of the relationship has yet been forthcoming.

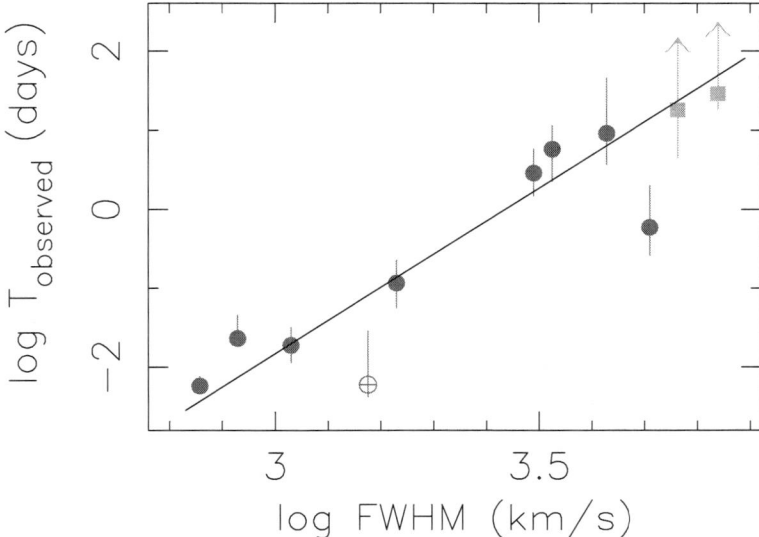

Fig. 8.6 Correlation of the observed PSD bend timescale vs. the FWHM of the H_β optical emission line

In Fig. 8.6 we show a plot of linewidth, V, against T_B [53]. Here we note a remarkably tight relationship. Parameterising $\log T_B = D \log V + E$, we found that $D = 4.2^{+0.71}_{-0.56}$ [53]. Such a tight relationship should have a simple explanation and, indeed, such an explanation can be derived from simple scaling relationships.

We assume that the linewidth arises from the Doppler motions of neutral gas at the inner edge of the broad line region (BLR) and that the BLR is excited by radiation from the central black hole. If the gas is in virial motion then $V^2 \sim GM_{BH}/R_{BLR}$, where R_{BLR} is the radius of that inner edge. We assume that the ionising luminosity $L \sim M_{BH}\dot{m}_E$. For the locally optimised condition for the production of emission lines by gas at the same optimum density and ionisation state we expect $R_{BLR} \propto L^{0.5}$ and measurements of the radius of the BLR confirm that expectation, finding that $R_{BLR} \propto L^{0.518\pm0.039}$ [8]. Putting these simple scaling relationships together we then expect that $V^4 \sim M_{BH}/\dot{m}_E$. As it has already been shown that $T_B \sim M_{BH}/\dot{m}_E$, we naturally explain the observation that $T_B \sim V^4$ and, by self-consistency, we show that the basic assumptions that we have made above about the BLR are probably correct.

We therefore have a strong link between small-scale nuclear accretion properties (X-ray variability) and larger scale properties of the AGN (linewidth). The relationship shown in Fig. 8.6 is particularly useful as it is purely a plot of two observational quantities and does not rely on any assumptions or derivations. Thus from a simple measurement of an optical linewidth, we can make at least a first-order prediction of how the X-ray emission from a Seyfert galaxy should vary.

8.5 Origin of T_B

We have now measured, reasonably well, the way in which T_B varies with M_{BH} and $\dot{m}_E$ in all types of accreting systems. But what is the physical origin of T_B? On timescales shorter than T_B, the variability power drops rapidly and so it is natural to associate T_B with some cut-off, or edge, in the accreting system. One obvious possibility, therefore, is the inner edge of the optically thick accretion disc. However, unless we push the inner edge of the disc out to much larger radii ($\gtrsim 20R_G$) than are considered reasonable for soft-state systems [12], the simple dynamical timescale, T_d, at the edge of the disc is much shorter than the observed PSD bend timescale. Other possible characteristic timescales are the thermal timescale, $T_{\text{therm}} = T_d/\alpha$, where α is the viscosity parameter, typically taken to be ~ 0.1, and the viscous timescale, $T_{\text{visc}} = T_{\text{therm}}/(H/R)^2$, where H/R is the ratio of the scale height to radius of the disc [76].

For any fixed black hole mass, the viscous timescale, as well as the dynamical timescale, can be varied only by varying the inner disc radius (assuming that the viscosity parameter does not change). Measurements of the disc radius, usually by means of spectral fitting to the disc spectrum (e.g. [12]), show that the inner disc radius is large at low accretion rates and decreases as the accretion rate rises. However, once the accretion rate exceeds a few percent of the Eddington accretion rate (i.e. typically in soft states), the radius has typically decreased to a few gravitational radii and it is hard to measure further changes with increasing accretion rate. Although such measurements are prone to large uncertainties, nonetheless there does not yet seem to be evidence for the large changes of radius with accretion rate that would be required to explain the spread of timescales, for a given black hole mass, that are observed (e.g. [52]).

The viscous timescale can, however, be changed, for the same radius, by altering the scale height of the disc. If we assume that, when in the soft state, the inner edge of the accretion disc reaches the innermost stable orbit at, say $1.23R_G$ (the limit for a maximally spinning Kerr black hole), and that the observed bend timescale is actually a viscous timescale at that radius, then we can derive the scale height of the disc (H/R). We then find that H/R varies between about 0.1 for $\dot{m}_E \sim 0.01$ and about unity at $\dot{m}_E \sim 1$, which is a sensible range, broadly in line with theoretical expectations. Thus T_B might well be associated with the viscous timescale at the inner edge of the accretion disc: that edge first moves in as the accretion rate rises in the hard state and then, once the edge has reached the innermost stable orbit, further rises in accretion rate increase H/R.

8.6 Origin of the Variations

We have seen that AGN and BHBs have similar "states" and display similar patterns of variability. We have found scaling relationships which enable us to link the timing properties of accreting objects of all masses. We are beginning to have some idea of the origin of the main characteristic timescale in AGN. But what is the underlying origin of the variability?

8.6.1 Shot-Noise Models

A model commonly used to describe variability is the random shot-noise model. In this model, light curves are made up of the superposition of pulses, or shots, randomly distributed in time. If all of the shots have the same decay timescale then we have white noise. However, if we allow for a range of decay timescales then we can reproduce the bending powerlaw PSDs seen in BHBs and AGN [6, 36, 50]. Also, if more luminous sources simply contain more shots, then the fractional variability of more luminous sources will be less than that of less luminous sources. This behaviour is generally seen but from the fact that the normalised variability amplitude, NVA, decreases with increasing luminosity less slowly than would be expected for random, similarly sized, shots (i.e. we see $NVA \propto L^{-0.34}$ rather than the expected $NVA \propto L^{-0.5}$), it was concluded that if the variability does arise from the summed emission from a number of separate regions, then there must be some correlation between those regions [26].

8.6.2 The Rms–Flux Relationship

Although shot-noise models can explain the shape of the power spectra, they cannot explain the way in which the rms variability of the light curves varies with flux. Following some discussions regarding the benefits of calculating PSDs in absolute or rms units, it was noted that the rms variability of the light curves is directly proportional to the flux [81]. This proportionality is now known as the rms–flux relationship. In Fig. 8.7 we show the rms–flux relationship for the BHB Cyg X-1 [81] and for the Seyfert galaxy NGC 4051 [55]. This linear relationship has now been found in all BHBs and AGN for which the observations were sufficient to

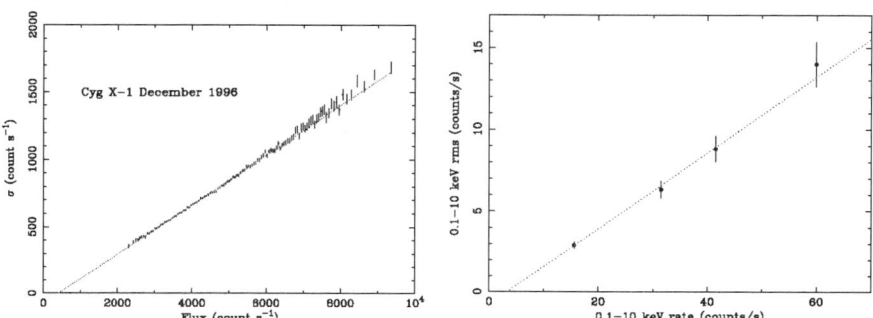

Fig. 8.7 *Left panel*: rms–flux relationship for Cyg X-1 [81]. *Right panel*: rms–flux relationship for the Seyfert galaxy NGC 4051 [55]

define it. We incidentally note that the rms–flux relationship usually does not go through the origin but has a positive intercept on the flux axis at zero rms, indicative of a constant (e.g. thermal) component to the X-ray flux.

The important implication of the rms–flux relationship is that when the long timescale Fourier components of the light curve, which determine the overall average flux level, are large, the short timescale components, which determine the rms variability, are also large, and vice versa. Thus different timescales know about each other.

In simple additive shot-noise models the shots are independent and so the rms will be constant and independent of flux. Simulations show that a random distribution of similar shots will not reproduce the extended low-flux periods seen in NGC 4051 [28, 80], i.e. the short timescale variability knowing that the average long-timescale flux is very low. In simple shot-noise models at least one large shot will always appear in the low-flux periods. As the rms/flux relationship applies, at least to BHBs, on whichever timescale one measures the rms (e.g. 1 s, 10 s bins), then to make shot-noise models produce the rms–flux relationship, we would have to introduce some mechanism which alters the shot amplitude on a large variety of timescales. What that mechanism would be is unknown and hence we are really no further forward in our understanding.

8.6.3 Propagating Fluctuation Models

An alternative, and simpler, explanation of the rms–flux relationship is provided by the propagating fluctuation model, initially proposed by Lyubarskii [37]. In this model fractional variations in mass accretion rate are produced on timescales longer than the local viscous timescale so that longer timescale variations are produced further out in the accretion disc. These variations then propagate inwards. If the low-frequency variations then modulate the amplitude of higher frequency variations produced closer in (as in amplitude modulation in radio communications) then we produce a linear rms–flux relationship. Note that this relationship between frequencies is multiplicative, not additive. If each inner frequency is modulated then by the product all lower frequency modulations produced further out, then the rms–flux relationship will apply no matter what frequency range one chooses to measure the rms over, or at which lower frequency one chooses to measure the flux.

There are a number of implications when the time series is produced by multiplying together a set of variations of different frequencies. First let us consider the distribution of fluxes. If a light curve is produced by the sum of many random sub-processes then, by the central limit theorem, the fluxes will be normally distributed, i.e. they will follow a Gaussian distribution. Summative processes are linear (e.g. [69]), multiplying the input (e.g. the accretion rate) by some constant will multiply the output (e.g. the luminosity) by the same factor. However, if a light curve is produced by the product of many random sub-processes, then the flux distribution will be lognormal (i.e. the distribution of the log of the fluxes will be Gaussian). The flux

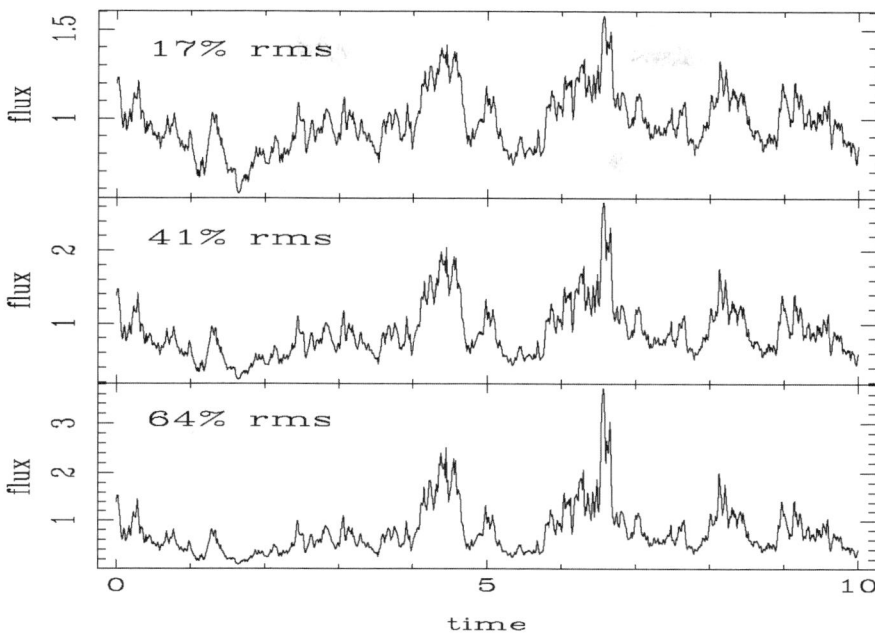

Fig. 8.8 Simulated exponential light curves [84] showing increasing nonlinearity as the rms variability increases

distributions of the well-measured BHBs such as Cyg X-1 do indeed follow such a lognormal distribution. (A full discussion of these topics is presented elsewhere [84].)

If the observed light curve is actually the product of variations on a variety of timescales, then it is straightforward to show [84] that the observed light curve is the exponential of an underlying linear light curve. The observed light curve is therefore nonlinear. As the nonlinearity arises from the coupling between variations on different timescales, being ultimately driven by the lowest frequency variations (from the outer part of the disc, in the Lyubarskii model), then the larger the driving variations are, the greater will be the nonlinearity. In Fig. 8.8 I show examples of light curves simulated using the exponential formulation, with increasing rms variability (from top to bottom) but with similar random number sequence used in their generation. It can clearly be seen how the light curves become more nonlinear as the rms variability increases. The lower rms variability light curves are typical of many ordinary Seyfert galaxies whilst the higher rms, more nonlinear, light curves are typical of the NLS1s.

A qualitative model which can explain many of the spectral timing observations of AGN and BHBs has been proposed [15, 34]. In this model the $1/f$ power spectrum of variations which is produced in the Lyubarskii model propagates inwards, perhaps in an optically thin corona over the surface of the disc, until it hits the X-ray emitting region, whose emission it modulates. A numerical version of this model has been developed [3] which can reproduce many observed aspects of X-ray spectral variability. The critical aspect of these models is that the source of the variations is not the same as the source of the X-rays.

8.6.4 Lags and Coherence

As well as providing an indication of X-ray state (Sect. 8.2), lags and coherence are potentially very powerful diagnostics of emission models and geometry. A standard interpretation of the lag of the soft band by the hard band is given by Comptonisation models. In these models the low-energy seed photons from the accretion disc are Compton-scattered up to higher energies by a surrounding corona of very hot electrons. As more scatterings are required to reach the higher energies, the higher energy X-rays will be delayed relative to the lower energy photons.

However, there are other observed aspects of X-ray variability which are not so easily explained by simple Comptonisation scenarios, e.g. the observation that the slope of the PSD above the bend is flatter for most AGN at higher energies (e.g. Fig. 8.9). If the variability is driven by variations in the seed photons, one would expect high-frequency variability to be washed out by many scatterings and so would expect the PSD to be steeper above the bend at higher frequencies. We might retain the Comptonisation model if the high-frequency variability is driven by variations in the corona rather than in variations in the seed photon flux. However, the temperature and optical depth of the corona would have to be carefully tuned so that seed photons are raised to high energies by very few scatterings and are then immediately able to leave the corona.

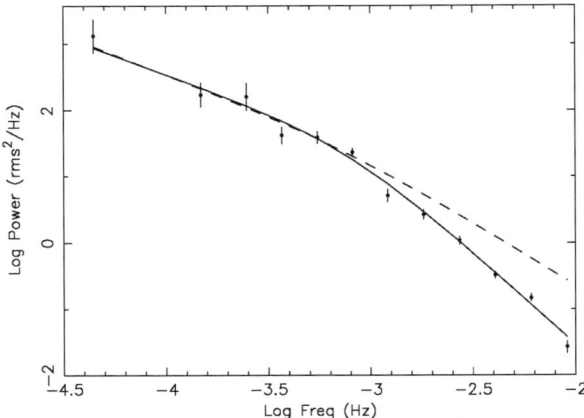

Fig. 8.9 PSDs of NGC 4051 from XMM-Newton observations. The data are from the 0.2–2 keV band and the *solid line* is the best fit to these data. The *dashed line* is the fit to the 2–10 keV PSD. Note that this PSD is flatter than that of the 0.2–2 keV PSD at higher frequencies. (The normalisation of the 2–10 keV PSD has been slightly increased to make it the same as that of the 0.2–2 keV PSD at low frequencies.)

A perhaps more natural explanation of the lags, the coherence and the energy-dependent PSD shapes can be provided by the propagating fluctuations scenario if we introduce an X-ray emitting region with a hardness gradient, such that higher energy X-rays are emitted, preferentially, closer towards the black hole [34]. The geometry of the emission region is not critically defined but might extend over the inner region of the accretion disc. The overall emissivity at all energies would still rise with decreasing radius, perhaps $\propto r^{-3}$, in accordance with the expectation for the release of gravitational potential energy, but the proportion of higher energy photons released would increase with decreasing radius. If viewed from far out in the accretion disc, there would thus be an emission-weighted centroid for each energy, and this centroid would move to larger radius as the energy decreases. The observed lag between the hard and soft bands would therefore simply represent the time taken for the incoming perturbations to travel between the centroids of the particular soft and hard energy bands under consideration.

This scenario can also explain why the lags vary with the Fourier frequency of the perturbations. The lowest frequency perturbations will originate far beyond any X-ray emission region and so will produce the largest possible lag. The lag will remain the same as we move to higher frequencies until we reach frequencies which are produced within the X-ray emission region. Once we move to frequencies produced at radii inside the X-ray emission region, then the centroids of the emission produced inside those radii will move closer to the black hole. As the radial profile of the X-ray emission will be more extended at lower energies, the centroid of the lower energy emission will move inwards more than the centroid of the higher energy emission. Therefore, the lag will decrease as we move to higher frequencies (e.g. NGC 4051, [55]).

In Akn 564 we note (Fig. 8.1) that, at the very highest frequencies, the lags are slightly negative (-11 ± 4.3s [51]), i.e. the hard band leads. In the model discussed here the very high frequencies are produced in the region where the X-ray source is very hard. The main source of soft photons in this region may therefore not be the intrinsic source spectrum but may come from reprocessing of the intrinsic very hard spectrum by the accretion disc. The lag may then represent approximately twice the light travel time between the hard X-ray corona and the reprocessing accretion disc.

By altering the emissivity profile of the disc it is possible to produce lags which vary as a function of frequency either smoothly (as in NGC 4051 or Mkn 335, e.g. Fig. 8.10 [2]) or in a stepwise manner (e.g. Akn 564, [51]). The propagating fluctuation model can simultaneously explain the energy-dependent shape of the PSDs, thereby providing a simple and self-consistent explanation of many spectral-timing properties of accreting systems.

We note that the truncation of the optically thick disc leads to a rapid drop in the lags at a frequency close to the bend in the PSD. Although we do not yet have accurate lag measurements for many AGN, the indications are that the lag drops follow the same mass (and possibly accretion rate) scaling that is found for the PSD bend timescales.

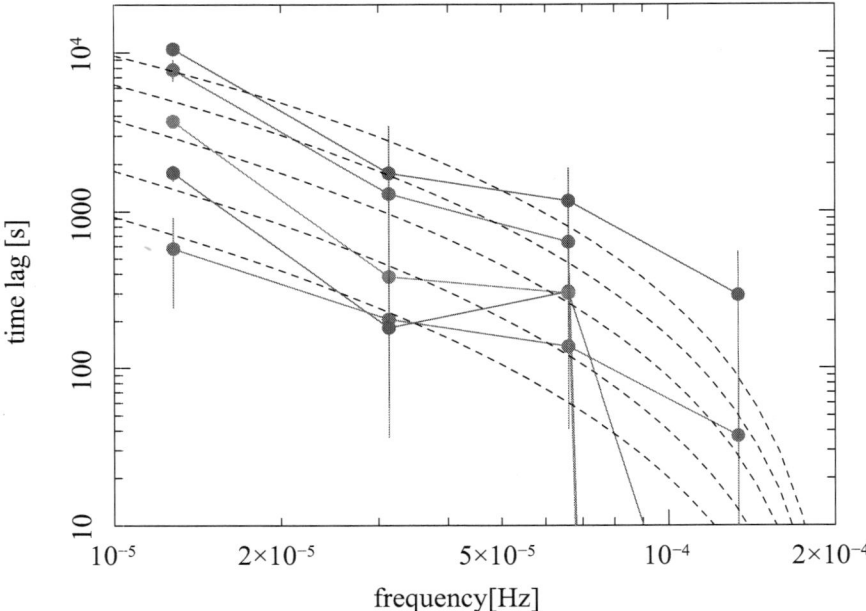

Fig. 8.10 The data points show the time lag, as a function of Fourier frequency, between the 0.2–0.4 keV band and the 0.4–0.6, 0.6–1, 1–2, 2–5 and 5–10 keV bands for Mkn 335. The lags increase as we move to larger energy separations and in all cases the hard band lags behind the soft band. The *faint dashed lines* represent model fits to these data based upon the propagating fluctuation scenario with a disc emissivity profile (at least in the 0.2–0.4 keV band) $\propto r^{-3}$ and the inner edge of the optically thick disc being truncated at $6R_g$ [2]

8.7 Variability of Blazars

The radio through optical emission in blazars like 3C 273 and 3C 279, i.e. AGN with apparent superluminal radio components, is synchrotron emission from a relativistic jet oriented at a small angle to the line of sight. In 3C 273 the tight correlation between the synchrotron component (e.g. IR) and X-rays, with the X-rays lagging the synchrotron by ∼day (e.g. [48]), indicates that the X-ray emission mechanism is almost certainly synchrotron self-Compton (SSC) emission from the jet (e.g. [72, 48]). A similar mechanism probably applies in all other blazars. Thus the X-ray emission region (in a relativistic jet) and emission process are different to those of Seyfert galaxies where the X-ray emission is not relativistically beamed and the emission process is probably a thermal Compton emission process. It is therefore not at all clear that we should expect similar X-ray variability characteristics.

The 3C 273 is one of the brightest AGN in the sky and, as such, has been observed by all major observatories since the 1970s. We have collected these data, including extensive observations with RXTE, and the long-term light curve is shown in Fig. 8.11a. Using PSRESP, in our normal manner, I have made the PSD which is shown, unfolded from the distortions introduced by the sampling pattern, in

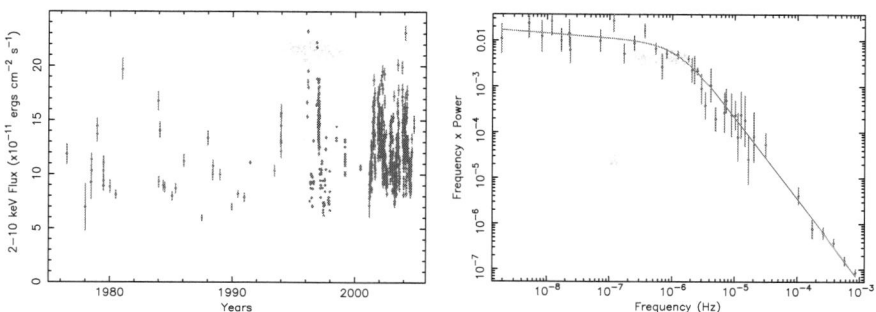

Fig. 8.11 (**a**) *Left panel*: Long-term 2–10 keV X-ray lightcurve of 3C 273. The intensive monitoring after 1996 is from RXTE; the earlier data are from the Ariel V SSI, Exosat and Ginga. (**b**) *Right panel*: Unfolded PSD of the blazar 3C 273, showing an excellent fit to the same bending powerlaw model that fits the Seyfert galaxies (McHardy et al., in prep.)

Fig. 8.11b. It can be seen that this PSD is exactly like that of a soft-state Seyfert galaxy PSD. The bend timescale is 10 days. I have placed this observed timescale together with the predicted bend timescale in Fig. 8.5. For comparison with most of the other non-beamed AGN on that figure we use the black hole mass derived from reverberation measurements by the same observers [66]. It can be seen that 3C 273 fits the scaling relationship between M_{BH}, $\dot{m}_E$ and T_B as well as all the Seyfert galaxies.

We note that there is no necessity to alter the observed bend timescale by any relativistic time dilation factor to make the observed bend timescale fit with the relationship derived for Seyfert galaxies and BHBs. For a clock moving with the jet of 3C 273, that factor would be about 10. The implication, therefore, is that the source of the variations – note, not the source of the X-rays – lies outside the jet and is simply modulating the emission from the jet. Coupled with the fact that the PSD of 3C 273 looks exactly like that of soft-state Seyfert galaxies and BHBs, this implication is then entirely consistent with Seyfert galaxies, BHBs and blazars all suffering the same sort of variations, e.g. variations propagating inwards through the disc. The jet then would be seen as an extension of the corona which dominates the emission in Seyfert galaxies.

8.7.1 Rms–Flux Relationship for 3C 279

The 3C 279 is an even more relativistically beamed blazar than 3C 273 (e.g. [14]). For both of these objects we have calculated the rms–flux relationship from the RXTE monitoring observations. Both objects show a strong linear relationship (e.g. Fig. 8.12). This relationship provides strong confirmation that the underlying process driving the variability in blazars has the same multiplicative relationship between different frequencies that is seen in BHBs and Seyfert galaxies. Coupled with our observation that the PSD shape of 3C 273 is identical to that of Seyfert galaxies, the simplest conclusion is that the process driving the variability in all systems is the same, i.e. fluctuations propagating inwards through the disc.

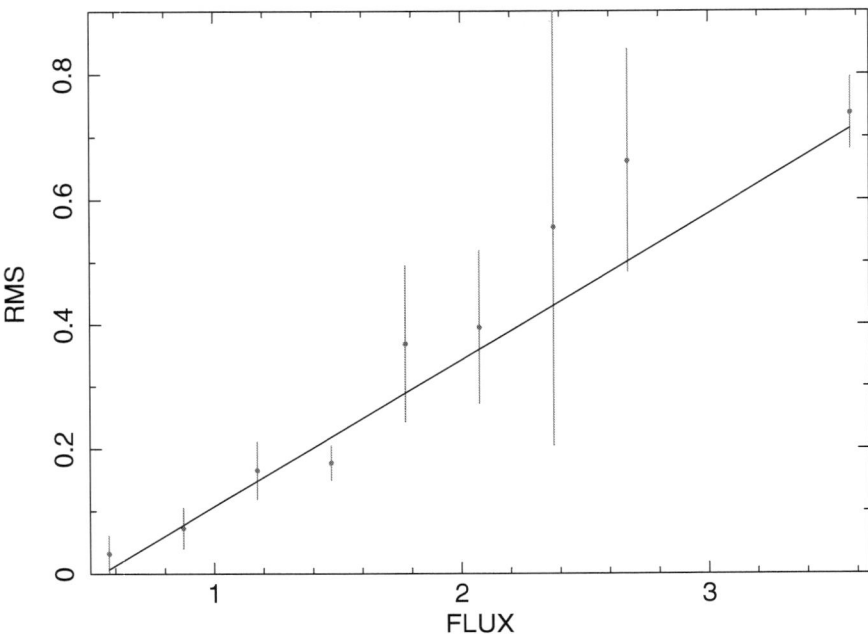

Fig. 8.12 Rms–flux relationship for 3C 279 with flux and rms both being measured in the 2–10 keV band in units of 10^{-11} ergs cm^{-2} s^{-1}

8.8 X-Ray/Optical Variability

For many years the origin of the optical variability of AGN has been, at best, a mystery and, at worst, a complete muddle. It has often been suggested that optical variability arises from reprocessing of X-ray variations but most early X-ray/optical studies found either weak [67] or non-existent [19] X-ray/optical correlations. In the case of NGC 3516 a strong correlation appeared to exist in the first set of observations, with the optical leading the X-rays by 100 d [38], but this correlation was not confirmed by subsequent observations [39]. However, during a 100 ks XMM-Newton observation of NGC 4051, the UV (2900Å) emission was observed to lag behind the X-rays by $\sim$0.2 d [43]. The optical variability amplitude was much less than that of the X-rays, consistent with reprocessing.

Our understanding of X-ray/optical variability has, however, improved considerably recently due largely to the long timescale RXTE X-ray monitoring programmes of ourselves and others, coupled with extensive optical monitoring programmes from groups such as AGNwatch and from robotic optical telescopes. In the case of NGC 5548 a strong X-ray/optical correlation has also been found [79] but here the optical band varies at least as much as the X-rays, which is energetically difficult to achieve with reprocessing. NGC 5548 has a more massive black hole ($10^8 M_\odot$) than NGC 4051 ($10^6 M_\odot$) or NGC 3516 ($4.3 \times 10^7 M_\odot$), prompting thoughts that the correlation might be mass dependent. In higher mass (or lower accretion rate) black hole systems the disc temperature is lower (for a given radius, in terms of gravitational radii). Thus the optical emitting region will be closer to, and will subtend a greater solid angle at, the central X-ray emitting source. This geometry would increase the efficiency of reprocessing but would also mean that both optical and X-ray emitting regions would be subject to more similar variations than if they were widely separated. In recent X-ray/optical monitoring of another high mass ($10^8 M_0$) AGN, MR2251-178 [4], a similar pattern to that of NGC5548 is seen. However, in the case of Mkn 79 [10] ($10^8 M_\odot$) we see correlated X-ray and optical variations on short timescales (tens of days, Fig. 8.13) but large amplitude changes in the optical band on ∼years timescales which are not seen in the X-ray band.

If we remove the long-term trend in the optical light curve of Mkn 79, we find an extremely strong (>99.5% significance) correlation between the X-ray and optical variations with a lag very close to zero days (±∼2d). Similar behaviour is seen in other AGN of medium or low black hole mass [11]. This correlation provides strong support for the model in which the short timescale optical variability in AGN is dominated by reprocessing of X-ray photons. This result is consistent with the observations of wavelength-dependent lags between different optical bands in AGN [13, 71] which are also best explained by reprocessing.

However, the slower optical variations in both MR 2251-178 [4] and Mkn 79 [10] cannot be explained by reprocessing from an X-ray emission region on an accretion disc both of constant geometry. As reprocessing from the accretion disc occurs on

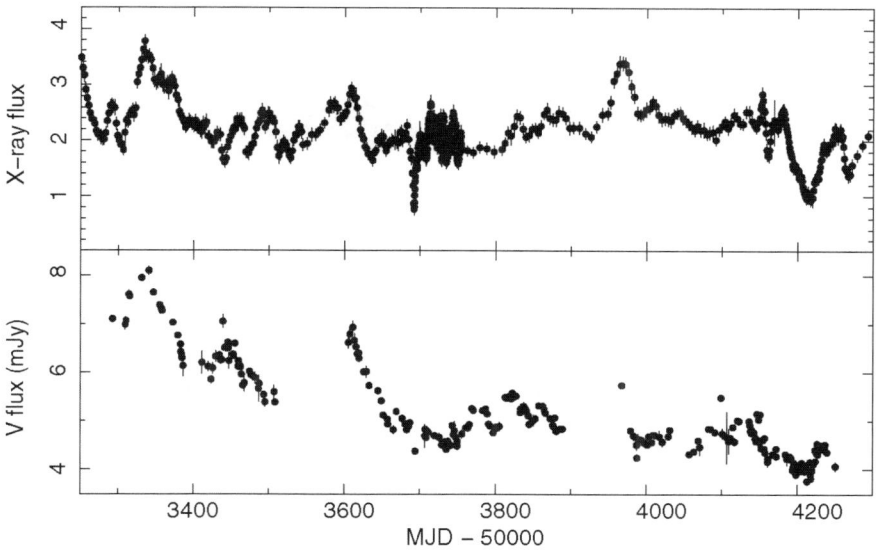

Fig. 8.13 Optical and X-ray variations in Mkn 79 [10]. Note how the variations are correlated on short timescales (weeks/month), probably showing that the optical variations arise from reprocessing of the optical emission. However, the optical emission varies independently on longer timescales, possibly indicating changes in the intrinsic emission from the disc due to accretion rate changes or possibly indicating changes in the geometry of the X-ray source

the light travel time which is, at most, a few days, it is impossible to remove all of the fast variations seen in the X-ray band. It might be possible to produce very smooth reprocessed light curves if the reprocessor is much larger than the accretion disc e.g. if it is in the broad line region or torus. Another possibility is that the slow optical variations can be explained either by varying the accretion rate in the disc, or by varying the geometry of the X-ray source or accretion disc. The radius within which half of the thermal optical emission from the accretion disc in Mkn 79 is produced is $70R_G$. The viscous timescale at this radius is about 4.7 years. Thus variations in accretion rate, propagating inwards, would be expected to vary the intrinsic thermal emission from the disc on that sort of timescale, which is consistent with the optical long-term trends seen in Fig. 8.13. Alternatively, if we approximate the X-ray emitting corona as a point source at its centroid, above the accretion disc, we can change the strength of the reprocessed optical signal by altering the height of that point source. We can also alter the strength of the reprocessed signal by changing the inner radius of the accretion disc. Further work is underway to determine which of these possibilities is the most likely.

We note that a variable height lamppost model has been invoked to explain continuum variability coupled with much lower variability of the iron line in MCG-6-30-15 [46]. When the source is close to the disc, a greater fraction of the continuum is gravitationally bent back to the disc or into the black hole, thereby reducing the flux as seen by a distant observer but not altering the reprocessed iron line flux greatly. Although a reasonable explanation of spectral variability, this model cannot explain the rms–flux relationship where the observed variability is least at the lowest continuum fluxes.

8.9 Conclusions

Our understanding of AGN variability has improved enormously in the last 20 years and the analogy with BHBs is now very strong. Almost all of the bright AGN with well-measured PSDs have PSDs similar to those of Cyg X-1 in the soft-state. These AGN typically have accretion rates similar to those of soft-state BHBs. Akn 564, which has a very high accretion rate, perhaps even exceeding the Eddington limit, has a PSD similar to that of BHBs in the very high state where accretion rates are similarly high. Ton S180, another very high accretion rate AGN, has a less well-measured PSD which is also similar to that of a VHS BHB. As yet, we have no firm evidence, from PSD shapes, for a hard-state AGN, but this result is probably just due to two selection effects. Hard-state BHBs have low accretion rates ($<2\%$ $\dot{m}_E$) and so have low X-ray luminosities. Thus, unless they are very nearby, hard-state AGN will be too faint to detect with dedicated monitoring satellites such as RXTE. Second, in order to confirm a hard PSD state, one must detect the second bend, at low frequency, to a slope of zero and, as bend frequencies scale inversely with accretion rate, typical bend timescales will be far too long to be measured within a human lifetime.

Measurement of the lags between different energy bands as a function of Fourier timescale provides another strong state diagnostic and in the very high accretion rate AGN Akn 564 we see a lag spectrum very similar to that seen in very high accretion rate BHBs. In particular, we see a rapid change in lag at the frequency at which the PSD changes from being dominated by one Lorentzian shaped component to another, again, just as in BHBs.

Characteristic timescales in accreting systems have now been shown to vary almost exactly linearly with compact object mass and inversely with accretion rate (in Eddington units). This scaling relationship applies over a huge range of over 10 decades in timescale. The physical origin of the high-frequency bend in the PSD is not yet entirely clear but is very likely related to a timescale, probably viscous, at the inner radius of the optically thick part of the accretion disc. This radius probably moves inwards with rises in accretion rate during the hard state but, in the soft state, it may hit the ISCO where an increase in disc scale height with increasing accretion rate could explain the decreasing characteristic timescales.

The fact that the relationship $T_B \propto M_{BH}^{1.12}/\dot{m}_E^{0.98}$ is a good fit within the measurement errors without having to invoke an additional unknown source of error indicates that no other parameters besides M_{BH} and $\dot{m}_E$ have a major effect on determining T_B. If T_B is associated with the inner edge of the accretion disc then an obvious additional parameter would be spin, which alters the radius of the last stable orbit and hence of all associated timescales. The implication of the good fit is therefore that the spin of all black holes whose accretion rate is high enough for them to be powerful X-ray sources is broadly similar.

The discovery of the rms–flux relationship in both AGN and BHBs shows that the process underlying the variability is multiplicative, rather than additive. Thus the amplitude of the shorter timescale variations is modulated by the amplitude of the longer timescale variations, leading to occasions (e.g. in NGC 4051) where the X-ray flux almost completely turns off. Standard shot-noise models, with independent shots, are thus ruled out. The rms–flux relationship predicts a lognormal flux distribution, which is seen in BHBs where the count rate is high enough for reliable measurement. The link between variations on different timescales implies that the light curves will be nonlinear, with the greatest nonlinearity in the sources with the highest rms variability, exactly as is seen, e.g. in the narrow line Seyfert 1 galaxies.

The physical models which best explain the overall patterns of X-ray variability, i.e. the PSD shape, the lag spectra and the coherence between different energy bands, are those in which the production of the variations is separated from the production of the X-rays. Thus accretion rate variations in the accretion disc (e.g. [38]) may propagate inwards and modulate an X-ray emission region with an energy-stratified emission profile (e.g. [34, 15]).

The origin of optical variability in Seyferts, and its relationship to the X-ray variability, is also partly explained by propagating fluctuations. Although short-timescale (few days/weeks) optical variability is well explained by X-ray reprocessing, longer timescale optical variations, which have no parallel in the X-ray band, may result from perturbations propagating inwards through the accretion disc which may take many years to travel from the optical to X-ray emission region.

There is some evidence that the same general pattern of variability that is seen in the non-relativistically beamed AGN and BHBs is also seen in blazars. The X-ray emission mechanism in blazars (e.g. synchrotron self-Compton emission) is almost certainly different to that in Seyfert galaxies (probably thermal Comptonisation), but blazars also demonstrate the rms–flux relationship and, in at least one well-observed case (3C 273) the PSD shape is identical to that of soft-state Seyfert galaxies. Moreover, the characteristic PSD timescale in 3C 273 is entirely consistent with that expected, based on its black hole mass and accretion rate, from Seyfert galaxies, without any requirement for modification to take account of the relativistic motion of the emitting region. The conclusion, therefore, is that the source of the variations lies outside the jet, whose emission it modulates, and that that source is probably the same as in BHBs and Seyfert galaxies.

A major task for future observatories is to extend the variability studies of Seyfert galaxies to galaxies of very low accretion rate (i.e. $<1\%$ $\dot{m}_E$). The aims are to determine whether such AGN are hard-state systems, whether the transition accretion rate in AGN is the same as, or different to, that in BHBs and whether such AGN behave in exactly the same way as scaled-up hard-state BHBs. However, such AGN are likely to be faint and the required monitoring timescales will be long. Thus I encourage the building of a sensitive (sub-mCrab on $\sim$day timescales) all-sky X-ray monitor, with few arcmin resolution to avoid confusion problems and with a lifetime of $\sim$10 years rather than the $\sim$2 years typical of many missions.

References

1. O. Almaini, A. Lawrence, T. Shanks et al.: MNRAS, **315**, 325 (2000)
2. P. Arévalo, I.M. McHardy, D.P. Summons: MNRAS, **388**, 211 (2008)
3. P. Arévalo, P. Uttley: MNRAS, **367**, 801 (2006)
4. P. Arévalo, P. Uttley, S. Kaspi et al.: MNRAS, **389**, 1479 (2008)
5. M. Axelsson, L. Borgonovo, S. Larsson: A&A, **438**, 999 (2005)
6. T. Belloni, G. Hasinger: A&A ,**227**, L33 (1990)
7. T. Belloni, J. Homan, P. Casella et al.: A&A, **440**, 207 (2005)
8. M.C. Bentz, B.M. Peterson, R.W. Pogge et al.: ApJ, **644**, 133 (2006)
9. R.D. Blandford, A. Konigl: ApJ, **232**, 34 (1979)
10. E. Breedt, P. Arévalo, I.M. McHardy et al.: MNRAS, **394**, 427 (2008)
11. E. Breedt, I. McHardy, P. Arevalo et al.: MNRAS, in press (2009)
12. C. Cabanac, R.P. Fender, R.J.H. Dunn et al.: MNRAS, **396**, 1415 (2009)
13. E.M. Cackett, K. Horne: MNRAS, **365**, 1180 (2006)
14. R. Chatterjee, S.G. Jorstad, A.P. Marscher et al.: ApJ, **689**, 79 (2008)
15. E. Churazov, M. Gilfanov, M. Revnivtsev: MNRAS, **321**, 759 (2001)
16. W. Cui, W.A. Heindl, R.E. Rothschild et al.: ApJ, **474**, L57 (1997)
17. C. Done, M. Gierliński: MNRAS, **364**, 208 (2005)
18. C. Done, G.M. Madejski, R.F. Mushotzky et al.: ApJ, **400**, 138 (1992)
19. C. Done, M.J. Ward, A.C. Fabian et al.: MNRAS, **243**, 713 (1990)
20. R. Edelson, K. Nandra: ApJ, **514**, 682 (1999)
21. A.A. Esin, R. Narayan, W. Cui et al.: ApJ, **505**, 854 (1998)
22. A.C. Fabian: ApS&S, **300**, 97 (2005)
23. H. Falcke, E. Körding, S. Markoff: A&A, **414**, 895 (2004)

24. M. Gierliński, M. Middleton, M. Ward et al.: Nature, **455**, 369 (2008)
25. M. Gierliński, M. Nikołajuk, B. Czerny: MNRAS, **383**, 741 (2008)
26. A.R. Green, I.M. McHardy, H.J. Lehto: MNRAS, **265**, 664 (1993)
27. Q. Gu, J. Melnick, R.C. Fernandes et al.: MNRAS, **366**, 480 (2006)
28. M. Guainazzi, F. Nicastro, F. Fiore et al.: MNRAS, **301**, L1 (1998)
29. K. Hayashida, S. Miyamoto, S. Kitamoto et al.: ApJ, **500**, 642 (1998)
30. A. Herrero, R.P. Kudritzki, R. Gabler et al.: A&A **297**, 556 (1995)
31. E.G. Körding, R.P. Fender, S. Migliari: MNRAS, **369**, 1451 (2006)
32. E.G. Körding, S. Jester, R.P. Fender: MNRAS, **372**, 1366 (2006)
33. E.G. Körding, S. Migliari, R. Fender et al.: MNRAS, **380**, 301 (2007)
34. O. Kotov, E. Churazov, M. Gilfanov: MNRAS, **327**, 799 (2001)
35. A. Lawrence, M.G. Watson, K.A. Pounds et al.: Nature, **325**, 694 (1987)
36. H.J. Lehto: In: J. Hunt, B. Battrick (eds.), Two Topics in X-Ray Astronomy, vol. 1, volume 296 of *ESA Special Publication*, pp. 499–503 (1989)
37. Y.E. Lyubarskii: MNRAS, **292**, 679 (1997)
38. D. Maoz, R. Edelson, K. Nandra: AJ, **119**, 119 (2000)
39. D. Maoz, A. Markowitz, R. Edelson et al.: AJ, **124**, 1988 (2002)
40. S. Markoff, H. Falcke, R. Fender: A&A, **372**, L25 (2001)
41. A. Markowitz, R. Edelson, S. Vaughan et al.: ApJ, **593**, 96 (2003)
42. K. Marshall, W.T. Ryle, H.R. Miller et al.: ApJ, **696**, 601 (2009)
43. K.O. Mason, I.M. McHardy, M.J. Page et al.: ApJ, **580**, L117 (2002)
44. A. Merloni, S. Heinz, T. di Matteo: MNRAS, **345**, 1057 (2003)
45. S. Migliari, R.P. Fender, M. van der Klis: MNRAS, **363**, 112 (2005)
46. G. Miniutti, A.C. Fabian, R. Goyder et al.: MNRAS, **344**, L22 (2003)
47. I. McHardy, B. Czerny: Nature, **325**, 696 (1987)
48. I. McHardy, A. Lawson, A. Newsam et al.: MNRAS, **375**, 1521 (2007)
49. I.M. McHardy: Mem S. A. It., **59**, 239 (1988)
50. I.M. McHardy: In: J. Hunt, B. Battrick (eds.), Two Topics in X-Ray Astronomy, vol. 1, volume 296 of *ESA Special Publication*, pp. 1111–1124, (1989)
51. I.M. McHardy, P. Arévalo, P. Uttley et al.: MNRAS, **382**, 985 (2007)
52. I.M. McHardy, K.F. Gunn, P. Uttley et al.: MNRAS, **359**, 1469 (2005)
53. I.M. McHardy, E. Koerding, C. Knigge et al.: Nature, **444**. 730 (2006)
54. I.M. McHardy, I.E. Papadakis, P. Uttley In: The Active X-ray Sky: Results from BeppoSAX and RXTE, pp. 509–514 (1998)
55. I.M. McHardy, I.E. Papadakis, P. Uttley et al.: MNRAS, **348**, 783 (2004)
56. N.M. Nagar, E. Oliva, A. Marconi et al.: A&A, **391**, L21 (2002)
57. M. Nikołajuk, B. Czerny, J. Ziółkowski et al.: MNRAS, **370**, 1534 (2006)
58. M. Nikolajuk, I.E. Papadakis, B. Czerny: MNRAS, **350**, L26 (2004)
59. M.A. Nowak: MNRAS, **318**, 361 (2000)
60. M.A. Nowak, B.A. Vaughan, J. Wilms et al.: ApJ, **510**, 874 (1999)
61. I.E. Papadakis: MNRAS, **348**, 207 (2004)
62. I.E. Papadakis, W. Brinkmann, H. Negoro et al.: A&A, **382**, L1 (2002)
63. I.E. Papadakis, E. Chatzopoulos, D. Athanasiadis et al.: A&A, **487**, 475 (2008)
64. I.E. Papadakis, I.M. McHardy: MNRAS, **273**, 923 (1995)
65. I.E. Papadakis, K. Nandra, D. Kazanas: ApJ, **554**, L133 (2001)
66. B.M. Peterson, L. Ferrarese, K.M. Gilbert et al.: ApJ, **613**, 682 (2004)
67. B.M. Peterson, I.M. McHardy, B.J. Wilkes et al.: ApJ, **542**, 161 (2000)
68. K. Pounds, R. Edelson, A. Markowitz et al.: ApJ, **550**, L15 (2001)
69. M.B. Priestley: Spectral Analysis and Time Series, Academic Press, London (1982)
70. R.A. Remillard, J.E. McClintock: ARA&A, **44**, 49 (2006)
71. S.G. Sergeev, V.T. Doroshenko, Y.V. Golubinskiy et al.: ApJ, **622**, 129 (2005)
72. A. Sokolov, A.P. Marscher, I.M. McHardy: ApJ, **613**, 725 (2004)
73. D.P. Summons, P. Arevalo, I.M. McHardy et al.: MNRAS, **378**, 649 (2007)

74. D.P. Summons: PhD Thesis, University of Southampton (2007)
75. J. Timmer, M. Koenig: A&A, **300**, 707 (1995)
76. A. Treves, L. Maraschi, M. Abramowicz: PASP, **100**, 427 (1988)
77. S.P. Trudolyubov: ApJ, **558**, 276 (2001)
78. T.J. Turner, I.M. George, K. Nandra et al.: ApJ, **524**, 667 (1999)
79. P. Uttley, R. Edelson, I.M. McHardy et al.: ApJ, **584**, L53 (2003)
80. P. Uttley, I.M. McHardy, I.E. Papadakis et al.: MNRAS, **307**, L6 (1999)
81. P. Uttley, I.M. McHardy: MNRAS, **323**, L26 (2001)
82. P. Uttley, I.M. McHardy: MNRAS, **363**, 586 (2005)
83. P. Uttley, I.M. McHardy, I.E. Papadakis: MNRAS, **332**, 231 (2002)
84. P. Uttley, I.M. McHardy, S. Vaughan: MNRAS, **359**, 345 (2005)
85. B.A. Vaughan, M.A. Nowak: ApJ, **474**, L43 (1997)
86. N.E. White, A.C. Fabian, R.F. Mushotzky: A&A, **133**, L9 (1984)
87. J. Wilms, M.A. Nowak, K. Pottschmidt et al.: A&A, **447**, 245 (2006)
88. J. Ziółkowski: MNRAS, **358**, 851 (2005)

Chapter 9
Theory of Magnetically Powered Jets

H.C. Spruit

Abstract The magnetic theory for the production of jets by accreting objects is reviewed with emphasis on outstanding problem areas. An effort is made to show the connections behind the occasionally diverging nomenclature in the literature, to contrast the different points of view about basic mechanisms and to highlight concepts for interpreting the results of numerical simulations. The role of dissipation of magnetic energy in accelerating the flow is discussed and its importance for explaining high Lorentz factors. The collimation of jets to the observed narrow angles is discussed, including a critical discussion of the role of "hoop stress." The transition between disk and outflow is one of the least understood parts of the magnetic theory; its role in setting the mass flux in the wind, in possible modulations of the mass flux, and the uncertainties in treating it realistically are discussed. Current views on most of these problems are still strongly influenced by the restriction to two dimensions (axisymmetry) in previous analytical and numerical work; 3-D effects likely to be important are suggested. An interesting problem area is the nature and origin of the strong, preferably highly ordered magnetic fields known to work best for jet production. The observational evidence for such fields and their behavior in numerical simulations is discussed. I argue that the presence or absence of such fields may well be the "second parameter" governing not only the presence of jets but also the X-ray spectra and timing behavior of X-ray binaries.

9.1 The Standard Magnetic Acceleration Model

The magnetic model has become the de facto standard for explaining (relativistic) jets, that is, collimated outflows. In part this has been a process of elimination of alternatives, in part it is due to analytic and numerical work which has provided a sound theoretical basis for some essential aspects of the mechanism. It should be remembered that a key observational test of the model is still largely missing. Evidence for magnetic fields of the configuration and strength required by the model

H.C. Spruit (✉)

Max-Planck-Institut für Astrophysik, Postfach 1317, D-85741 Garching, Germany,
henk@mpa-garching.mpg.de

Spruit, H.C.: *Theory of Magnetically Powered Jets*. Lect. Notes Phys. **794**, 233–263 (2010)
DOI 10.1007/978-3-540-76937-8_9   © Springer-Verlag Berlin Heidelberg 2010

is indirect at best. Magnetic fields are detected indirectly through synchrotron radiation (such as the radio emission of extragalactic jets), and in some cases directly through the Zeeman effect in spectral lines (OH or H_2O masers) in young stellar objects and protoplanetary nebulae (e.g., [35, 3, 79]). Most of these detections, however, do not refer to the inner regions of the flow where much of the magnetic action is expected to take place. On the theoretical side, the acceleration process itself is the best studied aspect. Problems such as the precise conditions leading to the launching of a flow from a magnetic object or the collimation of the flow to a jet-like state are still under debate.

In the magnetic model, outflows are produced by magnetic fields of a (rapidly) rotating object. These objects include rapidly rotating magnetically active stars, young pulsars, or accretion disks such as those in young stellar objects, X-ray binaries, cataclysmic variables and active galactic nuclei (AGN). In these cases, the magnetic fields are "anchored" in the material of the rotating object. A related kind of process is the Blandford–Znajek mechanism [10], in which a rotating black hole with an externally imposed magnetic field is the energy source of a flow. In the following I limit the discussion to the illustrative case of flows from accretion disks, for which much observational data are available. For more on the Blandford–Znajek mechanism, see [39].

9.1.1 Flow Regions

In the standard magnetocentrifugal acceleration model for jets produced by an accretion disk [9, 11] there are three distinct regions. The first is the accretion disk; here the kinetic energy of rotation (perhaps also the gas pressure) dominates over the magnetic energy density. As a result, the field lines corotate with the disk in this region: they are "anchored" in the disk.

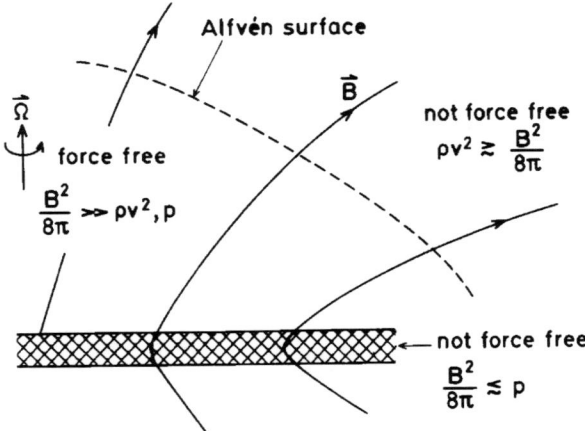

Fig. 9.1 Regions in a magnetically accelerated flow from an accretion disk (central object is assumed at the left of the sketch). In the atmosphere of the disk up to the Alfvén surface the magnetic field dominates over gas pressure and kinetic energy of the flow. This is the region of centrifugal acceleration

The second is a region extending above and below the disk. Assuming the disk to be cool, the atmosphere of the disk has a low density and gas pressure. In this region, the magnetic pressure dominates over gas pressure, so that the field must be approximately force free $[(\nabla \times \mathbf{B}) \times \mathbf{B} = 0]$, like the magnetic field in much of the solar atmosphere. It forces the flow of gas into corotation with the disk, with only the velocity component along the field unrestricted by magnetic forces. The flow experiences a centrifugal force accelerating it along the field lines, much as if it were carried in a set of rotating rigid tubes anchored in the disk.

This acceleration depends on the inclination of the field lines: there is a net upward force along the field lines only if they are inclined outward at a sufficient angle. Field lines more parallel to the axis do not accelerate a flow. The conditions for collimation and acceleration thus conflict somewhat with each other. Explanation of the very high degree of collimation observed in some jets thus requires additional arguments in the magnetic acceleration model (see Sect. 9.6).

Finally, as the flow accelerates and the field strength decreases with distance from the disk, the approximation of rigid corotation of the gas with the field lines stops being valid. This happens roughly at the *Alfvén radius*: the point where the flow speed equals the Alfvén speed (for exceptions see Sect. 9.3.5). At this point, the flow has reached a significant fraction of its terminal value. The field lines start lagging behind, with the consequence that they get "wound up" into a spiral. Beyond the Alfvén radius, the rotation rate of the flow gradually vanishes by the tendency to conserve angular momentum, as the flow continues to expand away from the axis. If nothing else were happening, the field in this region would thus be almost purely azimuthal, with one loop of azimuthal field being added to the flow for each orbit of the anchoring point, see Fig. 9.2. In fact, this state is not likely to survive for much of a distance beyond r_A, because of other (3-D) things actually happening (see Sect. 9.5).

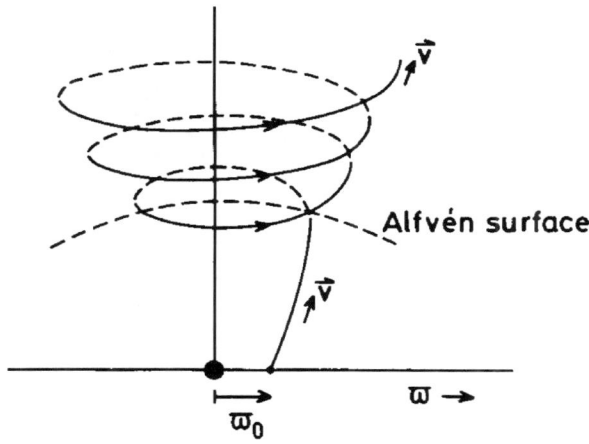

Fig. 9.2 Beyond the Alfvén distance the field lines lag behind the rotation of their foot-points and are coiled into a spiral (very schematic: the Alfvén surface actually has a more complicated shape)

9.1.2 Launching, Acceleration, Collimation

The three regions in Fig. 9.1 play different roles in the formation of the jet. At the surface of the disk, a transition takes place from the high-β interior to the magnetically dominated atmosphere of the disk. This is also the region which determines the amount of mass flowing into the jet: it is the *launching* region (Sect. 9.7). At some height in the atmosphere the flow reaches the sound speed (more accurately, the slow magnetosonic cusp speed v_c given by $v_c^2 = c_s^2 v_A^2/(c_s^2 + v_A^2)$, cf. [32]). If the gas density at this point is ρ_0, the mass flow rate is $\dot{m} \approx c_s \rho_0$, just as in standard stellar wind theory (cf. [51, 68]).

When the temperature in this region is high, for example, due to the presence of a hot corona, the atmosphere extends higher above the disk surface and it is easier to get a mass flow started. If the disk atmosphere is cool (temperature much less than the virial temperature), the gas density declines rapidly with height and the mass flow rate becomes a sensitive function of physical conditions near the disk surface. Since these cannot yet be calculated in sufficient detail for realistic disks, the mass flow rate is usually treated as an external parameter of the problem. This is discussed further in Sect. 9.7.

After launching, the flow is first accelerated by the centrifugal effect, up to a distance of the order of the Alfvén radius. The flow velocity increases approximately linearly with distance from the rotation axis (Fig. 9.3).

For acceleration by the centrifugal mechanism to be effective, the field lines have to be inclined outward: the centrifugal effect does not work on field lines parallel to the rotation axis. In the magnetic acceleration model, the high degree of collimation[1] observed in some of the most spectacular jets must be due to an additional process

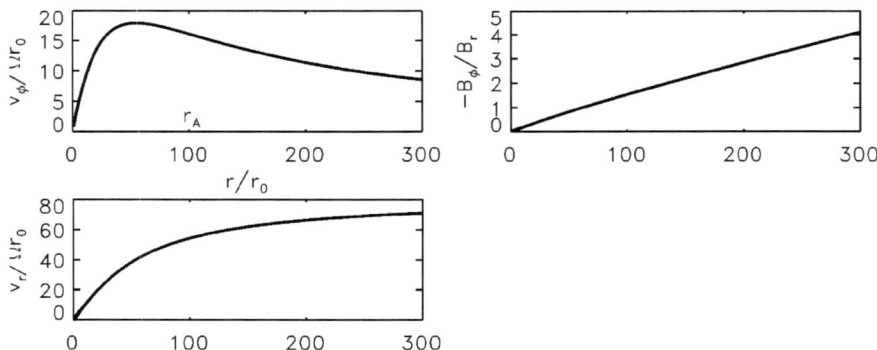

Fig. 9.3 Properties of a magnetocentrifugally accelerated flow. Rotation rate, flow speed, and azimuthal field angle as functions of distance from the rotation axis (cold Weber–Davis model). The Alfvén distance in this example is at 100 times the foot-point distance r_0

---

[1] Collimation is meant here in the same sense as in optics: the angle measuring the degree to which the flow lines in the jet are parallel to each other. This is different from the *width* of the jet (a length scale).

beyond the Alfvén surface (Sect. 9.6). It is conceivable that this does not happen in all cases: less collimated flows may also exist. They would be harder to detect, but have already been invoked for observations such as the "equatorial outflows" in SS 433 [67] and inferred from the rapidly varying optical emission in the accreting black holes GX 339-4 and KV UMa [37].

The transfer of energy powering the outflow is thus from gravitational energy to kinetic energy of rotation and from there to kinetic energy of outflow via the magnetic field. Note that in the centrifugal picture the magnetic field plays an energetically passive role: it serves as a conduit for energy of rotation, but does not itself act as a source of energy. The function of the magnetic field in the acceleration process can also be viewed in a number of different ways, however; this is discussed further below.

9.2 Length Scales

The energy release powering a relativistic outflow happens near the black hole, say 10^7 cm in the case of a microquasar. The narrow jets of microquasars seen at radio-wavelengths appear on scales of the order 10^{17} cm. In other words, on scales some ten orders of magnitude larger. It is quite possible that some of the jet properties are determined on length scales intermediate between these extremes, at least in some cases. In [70], for example, we have argued that collimation of the flow may actually take place on scales large compared with the Alfvén radius, at least in very narrow jets. In Sect. 9.5, it is shown that such intermediate length scales can also be crucial for acceleration to high Lorentz factors, besides the region around the Alfvén radius that plays the main role in the axisymmetric centrifugal acceleration process.

Much of the current thinking about the processes of launching, acceleration, and collimation of the jet is based on previous analytical models (Fig. 9.4). Numerical simulations of magnetic jets are now becoming increasingly realistic and useful. They are, however, quite restricted in the range of length scales and timescales they can cover. This leads to a bias in the interpretation of such simulations: the tendency is to assume that all steps relevant to the final jet properties happen within the com-

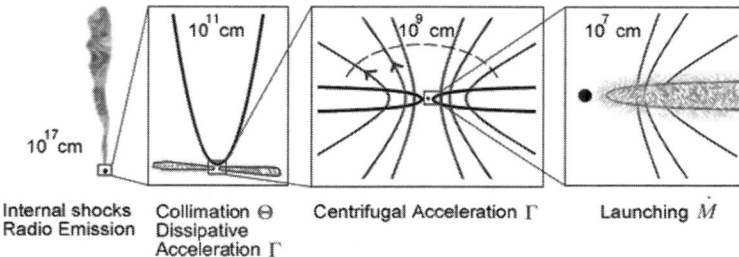

Fig. 9.4 Length scales in a microquasar jet. Processes determining the mass flow $\dot{M}$ in the jet take place in the "launching region" close to the black hole ($\sim 10^7$ cm), but the processes determining the final Lorentz factor (Γ) and opening angle (θ) may take place on much larger scales

putational box (cf. Sect. 9.7.1). This bias is likely to persist as long as simulations covering realistic ranges in length and timescale are impossible.

9.3 Magnetic Jets

9.3.1 Power Sources, Composition of the Jet

Jets powered by the rotation of a black hole (Blandford–Znajek mechanism) are often assumed to consist of electron–positron pair plasmas, while outflows from rotating disks are regarded as consisting of a normal ion–electron plasma. These associations are not exclusive, however. Since isolated black holes cannot hold a magnetic field, a field threading the hole requires the presence of an accretion disk holding it in place. Hence it is quite likely that (part of) the jet accelerated by the hole is actually fed with mass from the disk, rather than a pair plasma generated in situ. The simulations by [22, 47] are examples of this.

The opposite may also happen. A strong field threading a thin (cool) disk will not be easily loaded with mass from the disk unless the field lines are sufficiently inclined outward, away from the vertical [11]. In addition, the mass loading decreases with increasing field strength, for a given field line geometry [56]. If too little mass is loaded onto the field lines, the MHD approximation may not hold. The field lines rotating in a (near) vacuum may then produce a pair plasma above the disk, like the relativistic pair plasma outflows from pulsars. This case has not received much attention so far, perhaps because it would be as difficult to calculate as pulsar winds.

In the literature, the phrase 'Poynting flux' is sometimes associated specifically with relativistic and/or pair-dominated flows. It applies quite generally, however, equally to relativistic and nonrelativistic flows and independent of their composition. See also Sect. 9.3.2.

9.3.2 "Centrifugal" vs. "Magnetic" vs. "Poynting Flux"

The physical description of the flow-acceleration process has been a source of confusion. There are alternatives to the centrifugal picture sketched above: descriptions in terms of magnetic forces or in terms of "Poynting flux conversion." The different descriptions are largely equivalent, however; which one to take is a matter of personal taste, or the particular aspect of the problem to be highlighted. A term like "Poynting jet" for example, does not refer to a separate mechanism, but rather to a particular point of view of the process.

In a frame of reference corotating with the anchoring point of a field line, the flow is everywhere parallel to the magnetic field (e.g., [51]). The component of the Lorentz force parallel to the flow vanishes in this frame. There is no magnetic force accelerating the flow: the role of the Lorentz force is taken over by the centrifugal force. This is sometimes viewed as a contradiction for a magnetic model of acceleration: Can one still call the acceleration magnetic if there is no work done by magnetic forces?

If the same process is evaluated in an inertial frame, the centrifugal force is absent. Instead, one finds that in this frame the flow is accelerated by a force associated with the azimuthal component of the magnetic field: $\mathbf{F} = -\nabla B_\phi^2/8\pi - \mathbf{e}_\varpi B_\phi^2/(4\pi\varpi)$, where ϖ is the distance from the axis.[2]

The two descriptions, magnetic and centrifugal, are thus mathematically equivalent: they are related by a simple frame transformation. The centrifugal picture is an elegant way to visualize the acceleration as long as the magnetic field lines corotate with their anchoring point. Where they do not corotate, the field gets wound up into a predominantly azimuthal field, the centrifugal picture loses its meaning, and the acceleration is described most simply in terms of the forces exerted by the azimuthal field component B_ϕ.

Finally, the same process can also be viewed as the conversion of a Poynting flux of electromagnetic energy into kinetic energy. To see this, recall that in magneto-hydrodynamics the electric field $\mathbf{E}$ is given by $E = \mathbf{v} \times \mathbf{B}/c$, so that the Poynting flux

$$\mathbf{S} = \frac{c}{4\pi}\mathbf{E} \times \mathbf{B} \tag{9.1}$$

can be written as

$$\mathbf{S} = \mathbf{v}_\perp B^2/(4\pi), \tag{9.2}$$

where $\mathbf{v}_\perp$ is the component of the flow velocity perpendicular to the magnetic field. This shows that it is not necessary to think of Poynting flux as EM waves in vacuum. It applies equally well in MHD and not only to waves but also to stationary magnetic flows.

Expression (9.2) can be interpreted as a flux of magnetic energy, advected with the fluid, in a direction perpendicular to the field lines.[3] Borrowing a useful analogy from hydrodynamic flows, the Poynting flux in MHD plays the role of a "magnetic enthalpy flux." The centrifugal acceleration process is equivalent to the gradual (and incomplete) conversion of Poynting flux into a flux of kinetic energy, much like the conversion of enthalpy into kinetic energy in an expanding hydrodynamic flow. Near the base of the flow (for example, the surface of the accretion disk which supplies the mass flux into the wind), the enthalpy flows almost entirely in the form of a Poynting flux $\mathbf{S}$. $\mathbf{S}$ declines gradually with distance and the kinetic energy increases

---

[2] The poloidal field component $\mathbf{B}_p$ is absent from the accelerating force. Explanations invoking acceleration of the flow by $\mathbf{B}_p$ are erroneous since the poloidal velocity is parallel to $\mathbf{B}_p$ in a steady flow.

[3] The actual flux of magnetic energy would of course be $\mathbf{v}_\perp B^2/8\pi$. The "missing" $B^2/8\pi$ represents the "PdV work" done by the source of the flow against the magnetic pressure at the base of the flow. See also Sect. 9.5.

correspondingly, most of the energy transfer taking place around the Alfvén radius (in the axisymmetric case, see, however, Sect. 9.5).

9.3.3 Poynting Flux Conversion Efficiency: Axisymmetric

Since the flow is magnetic everywhere, at least some of the energy is carried in the form of a magnetic energy flux. The work done by the central engine is not converted completely into kinetic energy, and one may wonder what determines the efficiency of conversion of Poynting flux.

It turns out that the answer depends critically on the symmetry conditions imposed on the flow. When 3-D, nonaxisymmetric processes are allowed, conversion can be more efficient than in axisymmetric flows. This is discussed further in Sect. 9.5. Since much of the current views are still based on axisymmetric models, however, consider these first.

If S_0 is the Poynting flux at the base of the flow ($\sim$the power output of the jet) and F_K the kinetic energy flux, we can define this efficiency f as

$$f = F_{K\infty}/S_0 = F_{K\infty}/(F_{K\infty} + S_\infty), \tag{9.3}$$

where $_\infty$ denotes the asymptotic values at large distance. A simple model for which this can be calculated is the cold Weber–Davis model (for introductions see [51, 68]; for a concise and elegant mathematical treatment see [62]). In this model the poloidal field lines are straight and radial, so that the poloidal field strength varies as $1/r^2$ (a "split monopole"). In this approximation the azimuthal field and the flow can be calculated exactly, but the force balance in the latitudinal (θ) direction is neglected. In the "cold" version of the model, the gas pressure is also neglected. In the nonrelativistic limit, the conversion efficiency in this model is of order unity, so a significant fraction of the power delivered by the central engine remains in the flow as magnetic energy.

In the relativistic case, i.e., when the flow reaches large Lorentz factors (Γ), a smaller fraction of the Poynting flux is converted to kinetic energy than in the nonrelativistic case. This can be illustrated with the relativistic extension of the Weber–Davis model, given already by Michel's model [52, 29]. This model gives an exact solution for a relativistic, magnetically accelerated flow, in the approximation of a purely radial geometry for the poloidal field.

If (r, θ, ϕ) are spherical coordinates centered on the source of the flow, the simplest case to visualize is a flow near the equatorial plane, $\theta = \pi/2$. The split monopole assumption for the poloidal field components implies $v_\theta = B_\theta = 0$. As in the nonrelativistic case, the field at large distances is nearly exactly azimuthal. The radial component of the Lorentz force is then

$$F_r = -\partial_r B_\phi^2/(8\pi) - B_\phi^2/(4\pi r). \tag{9.4}$$

From the induction equation one finds that

$$B_\phi r v_r = \text{cst}, \tag{9.5}$$

i.e., the "flow of azimuthal field lines" is constant. Asymptotically for $\Gamma \to \infty$, $v \approx c$, so $B_\phi \sim 1/r$. The two terms in the Lorentz force then cancel. Equation (9.4) holds at all latitudes in this split monopole configuration.

The consequence of this cancellation is that a flow in a purely radial poloidal field stops being accelerated as soon as it develops a significant Lorentz factor. From then on acceleration and conversion of Poynting flux slow down. Moderately efficient conversion of Poynting flux to kinetic energy is possible, but only up to modest Lorentz factors. High Lorentz factors are also possible, but at the price of converting only a small fraction of the energy flux. This is seen in the expression for the terminal Lorentz factor Γ_∞ in Michel's model:

$$\Gamma_\infty \approx m^{1/3}, \tag{9.6}$$

where m is Michel's magnetization parameter,

$$m = B_0^2/(4\pi \rho_0 c^2), \tag{9.7}$$

and B_0 and ρ_0 are the magnetic field strength and mass density, respectively, at the base of the flow (where it is still nonrelativistic). If conversion of Poynting flux into kinetic energy were complete, it would produce a flow with Lorentz factor Γ_c,

$$\Gamma_c = m. \tag{9.8}$$

The actual efficiency of conversion is thus

$$f \equiv \Gamma_\infty/\Gamma_c \approx m^{-2/3} \approx 1/\Gamma_\infty^2. \tag{9.9}$$

This is a small number if large Lorentz factors are to be achieved. One gets either good conversion of Poynting flux into kinetic energy or large terminal speeds but not both.

9.3.3.1 Conversion in Diverging Flows

The conclusion from the previous section holds under the "split monopole" assumption that (apart from its azimuthal component) the flow expands exactly radially. If this is not the case, the cancellation is not exact and continued acceleration possible. To achieve this, the magnetic pressure gradient term in (9.4) has to be larger, relative to the second term, than it is in the split monopole geometry. This is the case when the azimuthal field *decreases more rapidly* with distance, in some region of the flow (cf. [58, 6]). This is perhaps the opposite of what intuition would tell, but similar to the acceleration of supersonic flows in expanding nozzles.

To make this a bit more precise, consider the magnetic forces in a steady axisymmetric flow. The flow can be considered separately along each poloidal field line. Let

z be the distance along the axis of the jet, and $R(z)$ the cylindrical radial coordinate
of a field line. Let $d(z)$ be the distance, in a meridional plane, between two neighboring poloidal field lines. Consider distances far enough from the Alfvén radius
that the rotation rate of the flow has become negligible. The field has then become
purely azimuthal. If the flow is steady, it follows from the induction equation that

$$vB_\phi d = cst, \tag{9.10}$$

i.e., the "flow rate of azimuthal field loops" is constant along a flow line. To investigate under which conditions the flow is accelerated, consider the component F_s
of the Lorentz force along the flow line. It is the sum of a curvature force and a
magnetic pressure gradient. Instead of calling the equation of motion into action,
it is sufficient to evaluate the forces under the assumption that the flow speed is
constant. Acceleration is then indicated if there is a net outward force under this
assumption.

If θ is the angle $\mathrm{atan}(dR/dz)$ of the field line with the axis, the component of
the curvature force along the flow line is $-\sin\theta\, B_\phi^2/(4\pi R)$ and the magnetic pressure gradient along the flow is $-\cos\theta\, dB_\phi^2/dz/8\pi$. Summing up while using (9.10)
yields

$$F_s = \frac{B_\phi^2}{4\pi}\frac{\mathrm{d}}{\mathrm{d}s}\ln(d/R), \tag{9.11}$$

where $\mathrm{d}/\mathrm{d}s = \cos\theta\, \mathrm{d}/\mathrm{d}z$ is the derivative along the flow line. For a purely radial
flow, $d \sim R$, and the force vanishes as expected. The result is more general, how-
ever: it applies to any "homologous" flow, with d/R independent of z. That is, it
holds if the distance d between poloidal field lines varies in the same way as the
distance R of the field line from the axis.

For net acceleration it is thus not sufficient that the overall expansion of the
flow is more rapid than radial. Expansion must also be non-self-similar: acceleration takes place only on field lines that diverge from their neighbors faster than
the overall expansion of the flow. This is unlike the case of the "nozzle effect" in
supersonic hydrodynamical flows.

The result can also be derived by considering the Poynting flux $\mathbf{S}$ itself. In ideal
MHD it is given by Eq. (9.2): $\mathbf{S} = \mathbf{v}_\perp B^2/4\pi$ where $\mathbf{v}_\perp$ is the velocity component
perpendicular to the flow. Again assuming that the field is already purely azimuthal,
$\mathbf{S}$ is parallel to the flow as well, $\mathbf{S} = vB_\phi^2/4\pi$. The total Poynting flux flowing
between two poloidal surfaces at distance $R(z)$ from the axis and separated by a
distance d as before is then

$$S_\mathrm{p} = S\, 2\pi Rd. \tag{9.12}$$

Poynting flux is converted into kinetic energy if this quantity decreases with distance
along the flow. Again using (9.10) to eliminate B_ϕ

$$S_{\mathrm{p}} = \frac{k^2}{2v} \frac{R}{d},\tag{9.13}$$

where $k = vB_\phi d$, the flow rate of the azimuthal field. It follows that for Poynting flux conversion it is necessary for d to increase with distance z faster than R, in agreement with the derivation above.

The conclusion is that at distances where the field has already become azimuthal, efficient Poynting flux conversion is not possible by the simplest expanding-nozzle effect. The expansion must be non-homologous, and acceleration takes place only in parts of the flow where the separation between neighboring flow lines increases faster than average. See [55, 75] for further discussion.

The assumption of a purely azimuthal field does not hold closer to the source and up to several times the Alfvén distance, and the above reasoning does not apply there. The region around the fast mode critical point of the flow appears to be conducive to Poynting fux conversion even in homologously expanding flows ($d \sim R$). A detailed study has been given by [20], for the general relativistic case and including the contributions from thermal pressure. Reasonable conversion efficiencies, of order 50%, are achieved if the opening angle of the flow increases by a factor of a few in the region around the fast mode point (see also the numerical simulation by [4]).

The limitations of this process of acceleration by divergence of the opening angle become more severe when actual physical conditions leading to divergence are considered. A limiting factor is causality. In a flow of Lorentz factor Γ, parts of the fluid moving at angles differing by more than $\theta_{\mathrm{max}} = 1/\Gamma$ are causally disconnected: there is no physical mechanism that can exchange information between them. Hence there are no physically realizable processes that can cause them to either converge to or diverge from each other. Jets accelerated by flow divergence therefore must satisfy

$$\theta\Gamma < 1,\tag{9.14}$$

where θ is the asymptotic opening angle of the jet. Jets in gamma-ray bursts, with inferred opening angles of a few degrees and minimum Lorentz factors of ≈ 100, can thus not be accelerated by homologous expansion.

Condition (9.14) is less constraining in the case of AGN jets, with inferred Lorentz factors in the range 3–30. Since efficient Poynting flux conversion by this process requires an expansion in opening angle by a factor of several; however, the initial opening angle of an AGN jet would have to be several times smaller than the observed angles. It is not clear if this is compatible with observations. In Sect. 9.5 an efficient conversion process is presented that does not require divergence of the flow (in fact it works best in a converging flow geometry) and is applicable to both AGN and GRB.

9.3.3.2 Conversion Efficiency: Possible Artifacts

Because of the near cancellation of terms in the magnetic acceleration, some cau-
tion is needed when interpreting results of magnetically driven flow calculations,
whether they are analytic or numerical. In analytic models, simplifying assumptions
can tip the balance in favor of one of the two terms, leading to a spurious acceleration
or deceleration.

In numerical work, the unavoidable presence of numerical diffusion of field lines
(e.g., lower order discretization schemes) can cause artificial acceleration. If such
diffusion is effective, i.e., the numerical resolution low, the toroidal field can potentially decay by annihilation across the axis. This would result in a decrease of the
magnetic pressure with distance along the axis, increasing the pressure gradient term
in (9.4), also resulting in acceleration. The signature of such an artifact would be that
the acceleration found decreases as numerical resolution is improved.

Acceleration by a decrease in magnetic energy along the flow is itself a real
effect, however, if a mechanism for dissipating magnetic energy within the jet is
present. I return to this in Sect. 9.5.

9.3.4 "Magnetic Towers"

Another picture of magnetic jets produced by a rotating, axisymmetric source is
that of a "magnetic tower," sometimes presented as an intrinsically separate mechanism with its own desirable properties. It is a simplified picture of the magnetic
acceleration process, in which the magnetic field is depicted as a cylindrical column
of wound-up magnetic field. One loop of toroidal field is added to the column for
each rotation of the foot-points (cf. Fig. 9.2). The tower is assumed to be in pressure
balance with an external confining medium. The attraction of this model is that it
is easily visualizable. In addition, some of the numerical simulations look much
like this picture. This is the case in particular for simulations done in a cylindrical
numerical grid.

The description of a magnetic jet as a "magnetic tower" does not address the
acceleration of the flow, nor does it address how a jet is collimated. It is a kinematic model describing the shape of the field lines once a flow has been assumed.
To explain acceleration of the flow, the dynamics of the centrifugal and dissipative
acceleration mechanisms described above and below have to be included.

9.3.5 Flows with High Mass Flux

At low mass flux in the wind, the Alfvén surface is at a large distance from the
disk,[4] and it moves inward with increasing mass flux. One may wonder what hap-

---

[4] Except in relativistic disks: in this case the Alfvén surface approaches the light surface ("cylinder") corresponding to the rotation rates of the foot-points on the disk.

pens when so much mass is loaded onto the field lines that the field is too weak to enforce corotation. When conditions in the wind-launching zone (Sect. 9.7) produce such a high mass flux, the "centrifugal" acceleration picture does not apply any more. Cases like this are likely to be encountered in numerical simulations, since the opposite case of low mass flux is much harder to handle numerically. Low mass fluxes cause problems associated with the high Alfvén speeds in the accelerating region and the larger computational domain needed, so the characteristic behavior of a centrifugal wind, with Alfvén surface far from the source, is not the first one expects to encounter in simulations.

The high mass flux case can be illustrated with an analytical model: the "cold Weber–Davis" model mentioned above. Consider for this a flow with the poloidal component of the B-field purely radial ($B_\theta = 0$) near the equatorial plane of a rotating source (i.e., the plane of the disk). Define the mass-loading parameter μ as

$$\mu = \rho_0 \frac{4\pi v_0 \Omega r_0}{B_0^2} = \frac{v_0 \Omega r_0}{v_A^2 0},$$ (9.15)

where Ω is the rotation rate of the field line with foot-point at distance r_0 from the axis, v_0 and v_{A0} the flow velocity and Alfvén speed at r_0, respectively. The solution of the model then yields for the Alfvén distance r_A

$$r_A = r_0 \left[\frac{3}{2}(1 + \mu^{-2/3}) \right]^{1/2}.$$ (9.16)

Various other properties of the flow can be derived (see the summary in Sect. 9.7 of Spruit 1996). An example is the asymptotic flow speed:

$$\frac{v_\infty}{\Omega r_0} = \mu^{-1/3}.$$ (9.17)

Figure 9.5 shows how the field lines are wound up for a low mass flux and a high mass flux case. These scalings have been derived only for the rather restrictive assumptions of the cold Weber–Davis model. It turns out, however, that they actually hold more generally, at least qualitatively; they have been reproduced in 2-D numerical simulations [2].

Equation (9.17) shows that at high mass flux, the outflow speed goes down, as expected, and for $\mu > 1$ actually drops below the orbital velocity at the foot-point. At such low velocities, the travel time of the flow, say to the Alfvén radius, becomes longer than the crossing time t_A of an Alfvén wave, $t_A = r/v_A$. There is then enough time for Alfvénic instabilities to develop, such as buoyant (Parker) instability and/or kink modes before the flow reaches substantial speed, interfering with the acceleration process. Since these are nonaxisymmetric, such effects do not show up in the typical 2-D calculations done so far. It is quite possible that the high mass flux case will turn out to be highly time dependent and poorly represented by the above

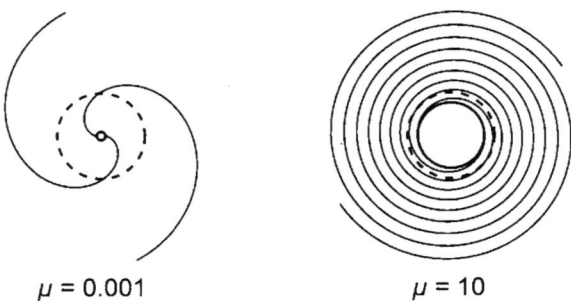

$\mu = 0.001$ $\mu = 10$

Fig. 9.5 Shape of the magnetic field lines of the cold Weber–Davis wind model, for two values of the mass-loading parameter μ. At low mass loading, the Alfvén radius r_A (*dashed*) is far from the source surface (*solid circle*), the angle of the field lines at r_A is of order unity. At high mass flux, the field lines are already wound up into a tight spiral before the flow reaches r_A. The flow is slow in this case, and it is likely to be subject to various nonaxisymmetric instabilities

scalings derived for steady flow. Three-dimensional simulations of high mass flow cases would therefore be interesting, but likely to be challenging.

High mass flux conditions may well occur in astrophysical objects, but they probably will not produce the familiar highly collimated high-speed jets. They might be involved in slower 'equatorial' outflows inferred in some objects.

9.4 Ordered Magnetic Fields

9.4.1 Impossibility of Generation by Local Processes

An *ordered* magnetic field such as sketched in Fig. 9.1 is usually assumed in work on magnetocentrifugal acceleration: a field of uniform polarity threading the (inner regions of a) disk. This is sometimes chosen as a representative idealization of more complicated configurations such as might result from magnetic fields generated in the disk. The ordered configuration has the advantage of simplicity: all field lines anchored in the disk extend to infinity, and the flow can be a smooth function of distance from the axis. If the field is not of uniform polarity, some field lines form closed loops connecting parts of the disk surface instead of extending to infinity, and the outflow will be patchy (cf. Fig. 9.5 in [11]).

A field of uniform polarity, however, is subject to an important constraint: it cannot be created in situ by local processes in the disk. It can only exist as a consequence of either the initial conditions or of magnetic flux entering or leaving the disk through its outer (radial) boundary. To see this formally, consider a circle at $r = R, z = 0$ (in cylindrical coordinates r, ϕ, z centered on the disk), where R could be the outer radius of the disk, or the radial outer boundary of the computational domain. Let S be a surface with this circle as its boundary, with normal vector $\mathbf{n}$, and let $\Phi = \int dS\, \mathbf{B} \cdot \mathbf{n}$ the magnetic flux through this surface. On account of div $\mathbf{B} = 0$, Φ is independent of the choice of S, as long as the boundary at $r = R$ is fixed, and we can take S to be in the midplane $z = 0$ of the disk. With the induction equation, we have

$$\partial_t \Phi = \int dr d\phi \, r [\nabla \times (\mathbf{u} \times \mathbf{B})]_z. \tag{9.18}$$

With $u_r(0, \phi, z) = B_r(0, \phi, z) = 0$ by symmetry of the coordinate system, this yields

$$\partial_t \Phi = -\int d\phi \, R[u_z B_r - u_r B_z], \tag{9.19}$$

where the integrand is evaluated at $r = R$. The square bracket can be written as $u_\perp B_p$, where $\mathbf{B}_p$ is the poloidal field (B_r, B_z) and $\mathbf{u}_\perp$ the velocity component perpendicular to it. The RHS of (9.19) can thus be identified as the net advection of poloidal field lines across the outer boundary.

If this flow of field lines across the outer boundary vanishes, the net magnetic flux Φ through the disk is constant. If it vanishes at $t = 0$, it remains zero: it cannot be created by local processes in the disk, including large-scale dynamos (even if these were to exist in accretion disks).

The magnetic flux through a disk is therefore a *global* quantity rather than a local function of conditions near the center. It depends, if not on initial conditions, on the way in which magnetic flux is transported through the disk as a whole. Since it is not just a function of local conditions in the disk, it acts as a *second parameter* in addition to the main global parameter, the accretion rate. This has an interesting observational connection: not all disks produce jets, and the ones that do, do not do it all the time. The possibility suggests itself that this variation is related to variations in the magnetic flux parameter of the disk [72].

9.4.2 Field Strengths

A particular attraction of ordered fields is that they can be significantly stronger than the fields produced by magnetorotational (MRI) turbulence. The energy density in MRI fields is limited to some (smallish) fraction of the gas pressure at the midplane of the disk. The exact fraction achievable still appears to depend on details such as the numerical resolution, with optimistic values of order 0.1 commonly quoted, while values as low as 0.001 are being reported from some of the highest resolution simulations [27].

The strength of ordered fields can be significantly higher, limited in principle only by equipartitition of magnetic energy with orbital kinetic energy, or equivalently by the balance of magnetic forces with gravity. In practice, interchange instabilities already set in when the fractional support against gravity reaches a few percent, as shown by the numerical simulations of [74]. For a thin disk, however, this can still be substantially larger than equipartition with gas pressure, since the orbital kinetic energy density is a factor $(r/H)^2$ larger than the gas pressure at the midplane. Magnetic fields of this strength actually suppress magnetorotational instability. Instead, their strength is limited by new instabilities-driven magnetic energy rather than the shear in the orbital motion (see discussion and results in [74]).

Strong fields are also indicated by the observations of rapidly varying optical emission in some accreting black holes, in particular GX 339-4 and KV UMa. As argued in [25], the only realistic interpretation of this radiation is thermal synchrotron emission from a compact region near the black hole. The inferred optical depth requires very strong magnetic fields [37], probably larger than can be provided by MRI turbulence.

9.4.3 Ordered Magnetic Fields in Numerical Simulations

Equation (9.19) says that the net magnetic flux Φ through the disk does not change unless there is a net advection of field lines into or out of the disk boundary. This implies that the velocity field inside the disk cannot create a net poloidal flux, no matter how complex or carefully construed the velocities.

A bundle of ordered magnetic flux threading a black hole, such as seen in simulations (e.g., [22, 47, 31]), can only have appeared "in situ" by violation of div $\mathbf{B} = 0$. Since this is unlikely with the codes used, the flux bundles seen must have developed from flux that was already present at the start of the simulation.

The way this happens has been pointed out in [36] and [73]; it can be illustrated with the simulations of [47]. The initial state used there is a torus of mass, with an initially axisymmetric field consisting of closed poloidal loops. The net poloidal flux through the midplane of the calculation thus vanishes: downward flux in the inner half of the torus is compensated by upward flux in the outer part (Fig. 9.6). The differential rotation in the torus generates MRI turbulence, causing the torus to spread quasi-viscously. The inner parts spread toward the hole while the outer parts spread outward. The magnetic loops share this spreading: the downward flux in the inner part spreads onto the hole, the upward flux spreads outward.

This explains the formation of a flux bundle centered on the hole in the simulations, but also makes clear that the result is a function of the initial conditions. The process as simulated in this way, starting with a torus close to the hole, is not really representative for conditions in an extended long-lived accretion disk. If a poloidal loop as in the initial conditions of present simulations were present in an actual

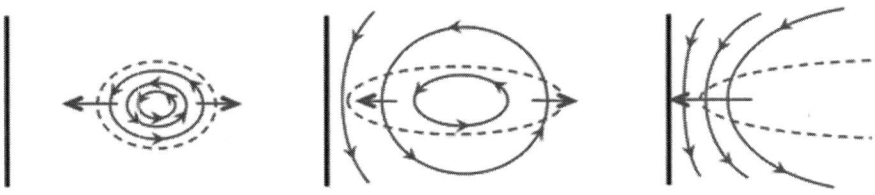

Fig. 9.6 Formation of a central flux bundle by a spreading torus containing loops of poloidal field

accretion flow, the downward part of the flux would accrete onto the hole as well, canceling the flux threading the hole.

The dependence on initial conditions is demonstrated more explicitly by some of the results in [46] and [22]. These results show that different initial conditions (a toroidal instead of a poloidal field) result in very similar magnetic turbulence in the disk, but without an ordered Poynting flux jet. See also the recent discussion in [5].

This disagrees with earlier suggestions [22, 40] that such jets would be a natural generic result of MRI-generated magnetic fields, or even the claim [49] that a net magnetic flux through the disk would appear from MRI turbulence.

Existing simulations thus show that a flux bundle at the center of a disk can form from appropriate initial conditions, but leave open the question which physics would lead to such conditions. As I argue below, this need not be seen only as an inconvenience. The same question may well be related to the puzzling phenomenology of X-ray states in X-ray binaries. At the same time, the simulation results are important since they appear to demonstrate that there is a flaw in the analytical models previously used for the accretion of net magnetic flux through a disk. This is discussed further in the next section.

9.4.4 Accretion of Ordered Magnetic Fields?

If magnetic fields of net polarity cannot be created internally in a disk, but a net polarity at the center of the disk is still considered desirable, there are two possibilities:

- The field is accreted from the outside (a companion, or the interstellar medium),
- It forms by systematic separation of polarities somewhere in the disk.

The first of these has been a subject of several studies, the conclusions of which are discouraging. The model used is that of a diffusing disk, where angular momentum transport is mediated by a viscosity ν, and the magnetic field diffuses with diffusivity η. If both result from some quasi-isotropic turbulence, they are expected to be of similar magnitude. Numerical simulations addressing this question [30] show that the ratio ν/η (the magnetic Prandtl number) is close to unity.

Vertical field lines are accreted through the disk at a rate $\sim \nu/r$, while diffusing outward at a rate $\sim \eta/r$. In a steady state the balance between the two would yield a strong increase of field strength toward the center of the disk. The assumption of a vertical field being accreted is very unrealistic, however, since accreted field lines cannot stay vertical. In the vacuum above the disk the field lines bend away from the regions of strong field, so that the field lines make a sharp bend on passing through the disk. As shown first by [77], the result is that magnetic flux is accreted very inefficiently. Because of the sharp bend, the length scale relevant for magnetic diffusion is the disk *thickness* H, rather than r, and diffusion correspondingly faster. The result is that accretion in a disk with $\nu \approx \eta$ cannot bend field lines by more than an angle $\sim H/r$ [42], and the increase of field strength toward the disk center is negligible.

This result would seem to exclude the accretion of magnetic flux to the center of a disk by amounts significant enough to create a strong ordered field around the central mass. The observational indications for the existence of such fields are nevertheless rather compelling. An attempt to circumvent the diffusion argument above was made by [72]. We appealed there to the fact that in addition to the external field to be accreted, the disk also has its own, magnetorotationally generated small-scale magnetic field. This field is likely to be highly inhomogeneous, with patches of strong field separated by regions of low strength, as seen in recent numerical simulations [45, 27, 30]. An external field accreted by the disk then does not cross the disk uniformly, but through the patches of strong field. Such patches can effectively lose angular momentum through a wind, thereby beating the diffusion argument and causing the external field to be accreted.

The ability of an accretion flow to maintain a bundle of strong field at its center, first proposed by [8], is suggested by the simulations of [36], [47], and others, at least for the geometrically thick flows that are accessible with numerical simulations.

9.4.5 Spontaneous Separation

The alternative possibility of a spontaneous separation of magnetic polarities from a mixture (as generated by MRI turbulence, for example) is also possible in principle, if a process of "coordinated small-scale action" exists. Assume that some form of magnetic turbulence in the disk contains small-scale ($\Delta r \sim H$) loops of poloidal field (in addition to a toroidal field component). If there is a reason why the loops are of the same sign throughout the disk, as sketched in Fig. 9.7, reconnection between them can create an ordered radial field. Allowing this field to escape through the upper and lower surfaces of the disk leaves two bundles of field lines of opposite sign crossing the disk, near its inner and outer boundaries. Though this still has not produced a net flux through the disk, it has separated polarities enough that the canceling flux in the outer disk need not influence the flux bundle in the inner disk much. The scenario is unrealistic, however.

9.4.6 Dynamos

The assumption of coordinated action on which the scenario Fig. 9.7 is based is unrealistic. For the loops to know the orientation needed for the process to work, at a minimum the information communicating this knowledge, the Alfvén speed, has to travel through the disk sufficiently fast. The lifetime of poloidal loops like those in Fig. 9.7 is of the order of the local orbital period, however. An Alfvén wave travels only of the order of a disk thickness in this time. By the time the loop's orientation

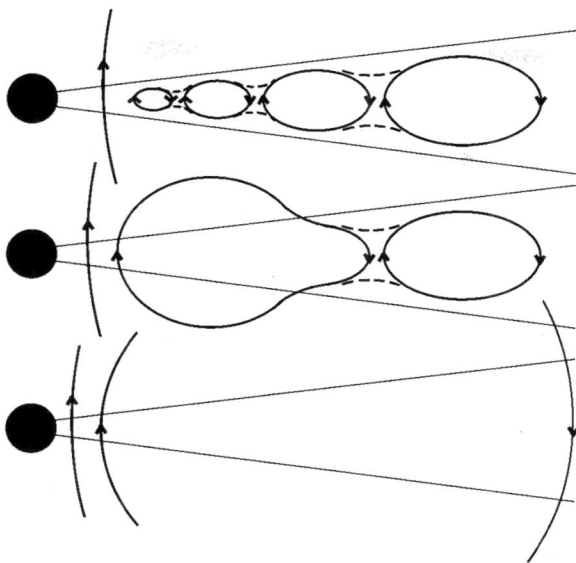

Fig. 9.7 Separation of magnetic polarities by coordinated small-scale action. If small-scale loops created in a disk have the same orientation (*arrows*), reconnection (*dashed lines* and *middle panel*) with subsequent escape from the disk would leave a net flux bundle in the inner disk

has been communicated to any larger distance, it has been replaced randomly by another loop. Larger scales thus will be acausal, governed by the statistics of random superposition. Simulations of magnetic turbulence in disks confirm this [30]. This calls into question the literature on large-scale fields produced with "disk dynamo" equations, which have the assumption of large-scale field generation already built into them. The well-established practice of using such mean field equations does not replace justification of their applicability.

9.4.7 The "Second Parameter" in Accreting X-Ray Sources

Previous analytic theory, as well as the recent numerical simulations, shows the advantage of an ordered magnetic field near the central object for creating powerful outflows; though less effective forms of outflow associated with random magnetorotationally generated magnetic fields appear possible as well [36, 46].

Ordered fields also make the puzzling behavior of X-ray binaries easier to understand. The phenomenology of X-ray binaries (black hole and neutron star systems) includes changes in the X-ray spectrum and the time variability of the X-ray emission (see contributions elsewhere in this volume). For many years the prevailing view in the interpretation of X-ray binaries has been that this phenomenology is governed by a single parameter: the accretion rate (not counting the system parameters of the binary itself).

This view was supported by the fact that X-ray binaries with neutron star primaries showed a systematic behavior, with their spectral and timing properties ordered approximately along a single track in color–color or color–intensity diagrams. It did not agree with the observations of X-ray transients (mostly black hole systems) including the "canonical" black hole transient GU Mus. These do not conform to the movement back and forth along a single track in color–intensity diagrams expected from a single-parameter system, showing instead motion around wide and/or irregular "loops".

In spite of this, the phenomenology of these transients has traditionally been interpreted as a single sequence of states with declining accretion rate, the apparent anomalies blamed on, for example, the transient nature of the sources. The X-ray spectrum and the properties of the time variability in these sources show strong similarities in states of very different brightness (e.g., the "very high" and "intermediate" states [61, 7, 33]). This provides a compelling clue that the phenomenology is not simply a function of the instantaneous accretion rate alone. Anomalies in the neutron star binaries ("parallel tracks," e.g., [78]), though smaller in magnitude, point in the same direction.

Instead of just the instantaneous accretion rate, one could imagine that the state of the system depends also on the *history* of the accretion rate. This would be the case if there is a physical property of the disk causing hysteresis, such that the state is different during increasing and decreasing accretion rates, for example. Such a mechanism might be an evaporation process depleting the inner regions of the disk, such that the size of the depleted zone depends on the history of the accretion rate (see review by [34]).

A more radical idea is that a true "second parameter" is involved in the state of the accretion flow and its X-ray and timing properties. Apart from binary parameters such as the masses and orbital separation, the mass transfer rate from the companion is the only parameter determining a (steady) hydrodynamic accretion disk. In the influential standard theory of disks based on local viscosity prescriptions, the physical state at a given point of the disk is only a function of the *local* mass flux. This makes the theory much more deterministic than the properties of X-ray transients seem to indicate.

A useful second parameter would therefore preferably be a *global* quantity: a property of the disk as a whole, independent of the accretion rate, which varies between disks or with time in a given disk. It is hard to come up with plausible candidates. As argued in [72] a promising one, however, is the *net flux* Φ of field lines crossing the disk. As shown above, this is a truly global parameter: its value is determined only by inheritance from the initial conditions and the boundary conditions at its outer edge; it cannot be changed by local processes in the disk. One could imagine, for example, that a large amount of flux would interfere with the accretion process in the inner parts of the disk [8], so flux could get concentrated there. This could be related to the nature of the poorly understood hard X-ray state, and the indications for 'truncation' of the inner disk (cf. [19, 21] and references therein).

Another useful property of the global magnetic flux as a second parameter is the observed relation between X-ray states and the occurrence of jets from X-ray binaries. If the hard X-ray state is indeed one with a high magnetic flux in the inner disk, its connection with jets would be natural since current theory strongly suggests them to be magnetically driven phenomena.

9.5 Flow Acceleration by Magnetic Dissipation

In three dimensions, the energy carried in the form of the wound-up magnetic field can decay by internal dissipation, something which is excluded in axisymmetry. This turns out to be a very efficient way of accelerating the flow to high Lorentz factors [23, 28]. The fact that this has not been recognized much may be due to the at first sight anti-intuitive nature of the effect and the emphasis on axisymmetric models in previous work. It may well be the actual mechanism of "Poynting flux conversion" in jets, replacing the mechanisms seen in 2-D [55].

An initially axisymmetric flow of nearly azimuthal magnetic field is bound to be highly unstable to kinking modes. This is to be expected specially when it is highly collimated: in a frame comoving with such a flow, the field is close to a static, almost azimuthal configuration. The details of such configurations are well known since the early days of controlled fusion research. Purely azimuthal fields like this are found to be unconditionally unstable. Instability reduces the energy B_ϕ^2 of the magnetic field, and small length scales developing under the instability can lead to reconnection, further reducing the magnetic energy. The growth time of the instability is of the order of the Alfvén travel time around a loop of azimuthal field. Instability is thus more destructive in highly collimated jets than in wider outflows. Dissipation of magnetic field energy into radiation by such instability has been proposed by [44] as a mechanism to power the prompt emission of gamma-ray bursts.

Another way of dissipating magnetic energy is to generate the flow from a *non*axisymmetric rotating magnetic field. A classic example is the pulsar wind generated by a rotating neutron star with a magnetic field inclined with respect to the rotation axis. If the asymmetry is strong enough, the azimuthal field in the outflow will change sign on a length scale $L \sim \pi v/\Omega$ (a "striped wind" [38, 12]), where Ω is the rotation rate of the source and v is the speed of the outflow. For a relativistic outflow, L is of the order of the light cylinder radius of the rotator. This is generally quite small compared with the distances traveled by the flow. Dissipation of magnetic energy by reconnection of field lines can be very efficient on such short length scales.

The effect of dissipation of magnetic energy on the flow is dramatic. In the absence of dissipation the balance between pressure gradient and curvature force tends to impede the acceleration and the conversion of Poynting flux. In axisymmetry, the balance between these forces can shift in favor of acceleration by the magnetic pressure gradient if (parts of) the flow expand more rapidly than in a radial ("conical") flow (see Sect. 9.3.3). In 3-D, diffusion caused by instabilities reduces this effect by making the flow more uniform across its cross section. But at the same time, the decay of magnetic energy by instabilities reduces the magnetic pressure along the path of the flow, causing the balance to shift in favor of the pressure gradient (see Eq. (9.4)), again resulting in acceleration of the flow. See Sect. 9.3.3 and [55]. This effect is efficient in particular in GRB outflows, where observations indicate the largest Lorentz factors [23, 28].

It may seem strange that one can convert Poynting flux into kinetic energy by "throwing away magnetic energy." The magnetic energy carried by the flow accounts for only half of the Poynting flux, however. The Poynting flux in MHD is $v_\perp B^2/4\pi$, where $v_\perp$ is the velocity component perpendicular to $\mathbf{B}$. This is twice the rate of advection of magnetic energy $B^2/8\pi$. The other half is accounted for by the work done by the central engine on the magnetic pressure of the outflow.

This is entirely analogous to the case of a steady hydrodynamic flow, where the energy balance is expressed by the Bernoulli function. The relevant thermal energy in that case is the enthalpy, the sum of the internal energy (equivalent to $B^2/8\pi$ in our case), and the pressure (also equivalent to $B^2/8\pi$). If internal energy is taken away along the flow (for example, by radiation), a pressure gradient develops which accelerates the flow. The energy for this acceleration is accounted for by the PdV work done at the source of the flow. For a more extended discussion of this point see [71].

This mechanism does not require an increasing opening angle of the flow lines. If the dissipation is due to magnetic instabilities, it works best at high degrees of collimation. If it is due to reconnection in an intrinsically nonaxisymmetric flow it works independent of the degree of collimation [23]. It may well be one of the main factors determining the asymptotic flow speed in many jets [24, 28]. However, the mechanism becomes effective mostly at distances significantly beyond the Alfvén radius of the flow. Simulations which focus on the region around the black hole do not usually cover these distances very well (cf. the discussion on length scales in Sect. 9.2). Recent 3-D simulations covering a large range in distance show that dissipation by kink instabilities can effectively become complete after 10–30 Alfvén radii [54, 55].

9.5.1 Observational Evidence

The development of the magnetic acceleration model has raised the question how observations can tell if jets are actually magnetically powered. On the parsec-and-larger scales in AGN jets, there does not appear to be strong evidence for magnetic fields being a major component of the energy content of the flow [66] (for a recent observation see [50]). Rather than interpreting this as a failure of magnetic models, it can be understood as evidence for the effectiveness of dissipation of magnetic energy in the flow. Parsec scales in AGN are large compared with the expected Alfvén radius in magnetic models, and there is ample opportunity for dissipation by internal instabilities of the helical highly coiled magnetic fields found in axisymmetric models. The 3-D instabilities allow the flow to shed the magnetic field that powered it, so that it ends as a ballistic, essentially nonmagnetic flow (cf. Sect. 13.9 in [68]).

The consequences of this interpretation are significant. First, it implies that there is no point in interpreting magnetic field observations on these large scales in terms of the simple winding-up process happening near the Alfvén distance (Fig. 9.2). Instabilities will have destroyed the original organized helical fields long before the flow reaches these scales. This questions a popular interpretation of observations of radio polarization in AGN jets. Second, flow acceleration is an automatic consequence of internal dissipation of magnetic energy. The effectiveness of such dissipation as indicated by the observations implied that it can be a significant, perhaps even the dominant factor determining the observed jet speeds [28].

9.6 Jet Collimation

In this section, the reader is reminded of the problems with the idea of collimating a jet by magnetic "hoop stress." The notion that a coiled magnetic field, as in the outflow from a magnetic rotator, will confine itself by hoop stresses is incorrect. Accommodation to this intuition has led to confused discussions in the observational and numerical literature (even where technically correct, e.g., [48]). Collimation of jets as observed must be due to some external agent; suggestions are discussed at the end of this section.

To see this, recall that a magnetic field is globally expansive, corresponding to the fact that it represents a positive energy density. That is, a magnetic field can only exist if there is an external agent to take up the stress it exerts. In the laboratory, this agent is a current-carrying coil or the solid-state forces in a bar magnet. The stress exerted by the magnetic field on the coils generating them is what limits the strengths of the magnets in fusion devices or particle accelerators.

A well-known theorem, particularly useful in the astrophysical context, is the "vanishing force-free field theorem." A magnetic field on its own, i.e., without other forces in the equation of motion, must be force free, $(\nabla \times \mathbf{B}) \times \mathbf{B} = 0$. The theorem says that if a field is force free everywhere and finite (i.e., vanishing sufficiently fast at infinity), it vanishes identically (for proofs see, e.g., [60, 51, 41]).

In the case of magnetic jets, this means that they can exist only by virtue of a surface that takes up the stress in the magnetic field. In most numerical simulations this is the external medium surrounding the flow. Its presence is assumed as part of the physical model (as in the "magnetic towers") or simply to ease numerical problems with low gas densities. Consider the boundary between the magnetic field ("jet") and the field-free region around it in such a calculation. Pressure balance at this boundary is expressed by

$$B_{\mathrm{p}}^2 + B_{\phi}^2 = P_{\mathrm{e}}, \tag{9.20}$$

where $\mathbf{B}_{\mathrm{p}} = (B_r, B_z)$ is the poloidal field and B_{ϕ} the azimuthal component, P_{e} the external pressure, while the internal pressure has been neglected without loss of generality. For a given poloidal field configuration (i.e., shape and magnetic flux of the jet), addition of an azimuthal field component *increases* the pressure at the boundary. Everything else being equal, this will cause the jet to *expand*, in spite of curvature forces acting in the interior of the jet.

Confusion about the role of curvature force is old and reappears regularly in the astrophysical literature. For a discussion with detailed examples see [57] (Chap. 8, esp. pp. 205–234). The azimuthal component *can* cause constriction to the axis, but only in a *part* of the volume around the axis itself in particular. This part cannot live on its own. It is surrounded by a continuation of the field out to a boundary where the stress of the entire configuration can be taken up. These facts are easily recognizable in existing numerical simulations (cf. [76]).

9.6.1 Collimation of Large-Scale Relativistic Jets

Since magnetic jets do not collimate themselves, an external agent has to be involved. A constraint can be derived from the observed opening angle θ_∞. Once the flow speed has a Lorentz factor $\Gamma > 1/\theta_\infty$, the different directions in the flow are out of causal contact, and the opening angle does not change any more (at least not until the jet slows down again, for example, by interaction with its environment as in the case of a GRB). Turning this around, collimation must have taken place at a distance where the Lorentz factor was still less than $1/\theta_\infty$.

Once on its way with a narrow opening angle, a relativistic jet needs no external forces to keep it collimated. Relativistic kinematics guarantees that it can just continue ballistically, with unconstrained sideways expansion. This can be seen in a number of different ways. One of them is the causality argument above, alternatively with a Lorentz transformation. In a frame comoving with the jet the sideways expansion is limited by the maximum sound speed of a relativistic plasma, $c_{s,m} = c/\sqrt{3}$. Since it is transversal to the flow, the apparent expansion rate in a lab frame (a frame comoving with the central engine, say) is reduced by a factor Γ: the time dilatation effect. In the comoving frame, the same effect appears as Lorentz contraction: the jet expands as quickly as it can, but distances to points a long its path are reduced by a factor Γ (for example, the distance to the lobes: the place where jet is stopped by the interstellar medium). In AGN jets, with Lorentz factors of order ~20, the jet cannot expand to an angle of more than about a degree. This holds if the flow was initially collimated: it still requires that a sufficiently effective collimating agent is present in the region where the jet is accelerated. Comparisons of Lorentz factors and opening angles of AGN jets might provide possible clues on this agent.

9.6.2 External Collimating Agents

The agent responsible for collimation somehow must be connected with the accretion disk (especially in microquasars where there is essentially nothing else around). One suggestion [13] is that the observed jet is confined by a slower outflow from the accretion disk. In AGN, the "broad line region" outflow might serve this role.

Something similar may be the case in protostellar outflows ([64] and references therein). In microquasars, such flows are not observed.

Another possibility [70] is that the collimation is due to an ordered magnetic field kept in place by the disk: the field that launches the jet from the center may may be part of a larger field configuration that extends, with declining strength, to larger distances in disk. If the strength of this field scales with the gas pressure in the disk, one finds that the field lines above the disk naturally have a nearly perfectly collimating shape (see analytical examples given by [63, 15]). The presence and absence of well-defined jets at certain X-ray states would then be related to the details of how ordered magnetic fields are accreted through the disk (cf. Sect. 9.4.4).

Near the compact object, the accretion can be in the form of an ion-supported flow (with ion temperatures near virial) which is geometrically thick ($H/r \approx 1$). Jets launched in the central "funnel" of such a disk are confined by the surrounding thick accretion flow. As shown by current numerical simulations, this can lead to a fair degree of collimation, though collimation to angles of a few degrees and less as observed in some sources will probably require an additional mechanism.

9.6.3 Occurrence of Instabilities, Relation to Collimation

In a cylindrical (i.e., perfectly collimated) jet, the wound-up, azimuthal component of the field will always be unstable, whether by external or internal kink instabilities. In a rapidly expanding jet, on the other hand, the Alfvén speed drops rapidly with distance, and an Alfvénic instability may get "frozen out" before it can develop a significant amplitude. For which types of collimated jet should we expect instability to be most effective in destroying the azimuthal magnetic field?

An estimate can be made by comparing the instability timescale with the expansion timescale of the jet radius. If the jet expands faster than the Alfvén speed based on B_ϕ, there is no time for an Alfvénic instability to communicate its information across the jet, and instability will be suppressed.

To see how this works out [54], let the distance along the jet be z, the jet radius $R(z)$ a function of z. As a reference point take the Alfvén distance, the distance where the flow speed v first exceeds the Alfvén speed v_{Ap} based on the poloidal field strength (cf. Sect. 9.1). Call this point z_0, and denote quantities evaluated at this point with an index $_0$. We then have the approximate equalities:

$$B_{p0} \approx B_{\phi 0}, \qquad v_0 \approx v_{A0}, \qquad (9.21)$$

while the flow speed reaches some modest multiple k of its value at z_0:

$$v = kv_0. \qquad (9.22)$$

In the following, it is assumed that the jet has reached this constant asymptotic speed v. The shape of the jet $R(z)$ depends on external factors such as an external collimating agent. Assume for the dependence on distance

$$R = \epsilon z_0 (\frac{z}{z_0})^\alpha. \tag{9.23}$$

i.e., ϵ is the opening angle of the jet at the Alfvén distance. The mass flux $\dot{m}$ is constant (steady flow is assumed), so

$$\dot{m} = \rho R^2 v = \rho_0 R_0^2 v = \rho_0 \epsilon^2 z_0^2 v, \tag{9.24}$$

At z_0 the azimuthal field component is of the same order as the poloidal component B_p. In the absence of dissipation by instability, the azimuthal field strength thus varies with jet radius as

$$B_\phi \approx B_{p0} \left(\frac{R}{R_0}\right)^{-1} = B_{p0} \left(\frac{z}{z_0}\right)^{-\alpha}. \tag{9.25}$$

The (azimuthal) Alfvén frequency is

$$\omega_A(z) = v_{A\phi}/R, \tag{9.26}$$

where $v_{A\phi} = B_\phi/(4\pi\rho)^{1/2}$. The instability rate η is some fraction of this:

$$\eta = \gamma \omega_A(z). \tag{9.27}$$

With the expressions for R and ρ this is

$$\eta = \gamma \frac{v_{A0}}{\epsilon z_0}(\frac{z}{z_0})^{-\alpha}. \tag{9.28}$$

The expansion rate ω_e of the jet is:

$$\omega_e = \frac{d \ln R}{dt} = v \frac{d \ln R}{dz} = \alpha \frac{v}{z}. \tag{9.29}$$

The ratio is

$$\frac{\eta}{\omega_e} = \frac{\gamma}{k\epsilon\alpha}(\frac{z}{z_0})^{1-\alpha}. \tag{9.30}$$

For an increasingly collimated jet ($\alpha < 1$) the instability rate will become larger than the expansion rate at some distance, and kink instability will become important. Decollimating jets ($\alpha > 1$) do not become very unstable since the instability soon "freezes out" due to the decreasing Alfvén speed. For the in-between case of a constant opening angle, a conical jet $\alpha = 1$, the ratio stays constant and it depends on the combination of factors of order unity $\gamma/(k\epsilon)$ whether instability is to be expected. A numerical study of these effects is given in [54].

In unstable cases it may take some distance before the effects of instability become noticeable, depending on the level of perturbations present at the source of the flow. Then again, as noted above (cf. Sect. 9.2), in most observed jets the range of length scales is quite large. Even a slowly growing instability can have dramatic effects that do not become evident in, for example, numerical simulations covering a limited range in length scale.

When instability is present, it reduces the azimuthal field strength (since this is what drives the instability) until the growth rate of the instability has settled to a value around the expansion rate ω_e.

Since the decay of magnetic internal energy has an accelerating effect on the flow, a relation between acceleration and collimation is to be expected. Jets which go through an effective (re)collimation stage should achieve a better "Poynting flux conversion" efficiency by dissipation of magnetic energy. This is the opposite of the (axisymmetric) process of acceleration by decollimation discussed in Sect. 9.3.3, which yields the best conversion in strongly, in particular nonuniformly, diverging flows.

9.7 The Launching Region

As *launching region* we define the transition between the high-β disk interior and the flow region above the disk. It contains the base of the flow, defined here as the point (called the *sonic point*) where the flow speed reaches the slow magnetosonic cusp speed. The mass flux in the jet is determined by the conditions at this point; these are visualized most easily in the centrifugal picture of acceleration. If Ω is the rotation rate of the foot-point of the field line, r the distance from the axis, and Φ the gravitational potential, the accelerating force can be derived from an effective potential $\Phi_e = \Phi - \frac{1}{2}\Omega^2 r^2$. As in other hydrodynamic problems, the sonic point lies close to the peak of the potential barrier, the maximum of Φ_e. Its height and location depend on Ω and the strength and inclination of the field. As in the case of hydrodynamic stellar wind theory, the mass flux is then approximately the product of gas density and sound speed at the top of the potential barrier.

9.7.1 Models for the Disk-Flow Transition

If the Alfvén surface is not very close to the disk surface, the magnetic field in the disk atmosphere is approximately force free since the gas pressure declines rapidly with height. As with any force-free or potential field, the shape of the field lines in this region is a *global* problem. The field at any point above the disk is determined by the balance of forces inside the field: field lines sense the pressure of their neighbors. At points where the field strength at the underlying disk surface is high, the field lines above it spread away from each other, like the field lines at the pole of a bar magnet.

The inclination of the field lines at the base of the flow is thus determined in a global way by the distribution of field lines at the disk surface, i.e., the vertical component $B_z(r)$ (assuming axisymmetry for this argument). Most of the physics inside a thin disk can be treated by a local approximation, that is, only a region with a radial extent similar to the disk thickness needs to be considered. The field inclination at the base of the flow, however, a key factor in the launching problem, cannot be computed in this way.

Several more mathematically inclined studies have nevertheless attempted to find "self-consistent" field configurations in a local approximation, by extrapolating field configurations along individual field lines from inside the disk into the flow region ([80], [26] and references therein, [65, 14]). By ignoring magnetic forces in the low-β region, these results do not yield the correct field configuration above the disk except in singular cases. The high-β disk interior and the low-β disk atmosphere are regions of different physics and so are the factors determining the field line shape in these regions.

The transition from the disk to the flow regime can still be studied in a local approximation provided one gives up the ambition of at the same time determining the field configuration above the disk. Since the field inclination is determined also by conditions at distances that are not part of the local region studied, the inclination at some height above the disk then has to be kept as an *external parameter* in such a local study.

This has been done in the detailed study by [56]. Their results show how the mass flux depends on the strength of the field and its asymptotic inclination. If the magnetic stress $f_r = B_z B_r / 4\pi$ is kept fixed, the mass flow increases with increasing inclination of the field lines with respect to the vertical as expected. The flow rate *decreases* with increasing field strength, however. This is due to the fact that the curvature of magnetic field lines shaped as in Fig. 9.1 exerts the outward force f_r (against gravity) on the disk. The rotation rate is therefore a bit lower than the Keplerian value. This is equivalent to an increase of the potential barrier in Φ_e for mass leaving the disk along field lines.

This complicates the conditions for launching a flow from the disk, compared with the simple estimate based on the field line inclination alone. Ignoring the slight deviation from Kepler rotation, a cool disk would launch a flow only if the inclination of the field lines with respect to the vertical is greater than 60° [11]. This condition is significantly modified by the sub-Keplerian rotation of the field lines, especially for the high magnetic field strengths that may be the most relevant for the generation of jets.

9.7.2 Limitations in Numerical Simulation

For a convincing numerical treatment of the launching and initial acceleration of the flow, the calculations would have to cover both the high-β disk interior and the low-β atmosphere. This implies a large range in characteristic velocities to be covered, with the region of large Alfvén speeds limiting the time step. To circumvent this *timescale problem*, modifications of the MHD equations have been explored in which the speed of magnetic waves is artificially reduced, cf. [53]. The problem is alleviated somewhat in relativistic calculations of flows near a black hole, where the various characteristic velocities of the problem converge on the speed of light.

Another way to reduce the timescale problem is by choosing conditions corresponding to a high mass flux, thus increasing the density and decreasing the Alfvén speeds in the wind. The Alfvén surface then decreases in size. This has the added benefit that the conceptually different regions in the flow (cf. Fig. 9.1) fit more comfortably inside an affordable computational volume.

It has to be realized, however, that this choice also limits the results to a specific corner of parameter space that may or may not be the relevant one for observed jets. It limits the jet speeds reached since the asymptotic flow speed decreases with mass loading. It strongly limits the generality of quantitative conclusions (in particular, about the mass flux in the jet, cf. [17]). It also tends to bias interpretations of jet physics to ones that are most meaningful in the high-mass flux corner of parameter space (cf. discussion in Sect. 9.3.5).

9.7.3 Ion-Supported Flows

At low disk temperature, the conditions for outflow are sensitive to the field strength and inclination near the disk surface, raising the question why the right conditions would be satisfied in any given jet-producing object. This sensitivity is much less in an ion-supported flow [59], where the (ion-)temperature of the flow is near the virial temperature and the flow is only weakly bound in the gravitational potential of the accreting object. This may, in part, be the reason why powerful jets tend to be associated with the hard states in X-ray binaries for which the ion-supported flow (also called ADAF [18]) is a promising model.

9.7.4 Instability of the Disk-Wind Connection, Knots

The same sensitivity of the mass flux to configuration and strength of the field can cause the connection between disk and outflow to become *unstable*. Since the wind carries angular momentum with it, an increase in mass loss in the wind causes an increase in the inward drift speed of the disk at the foot-points of the flow. This drift carries the vertical component of the magnetic field with it. Since the field configuration in the wind zone is determined by the distribution of foot-points on the disk, this feeds back on the wind properties. Linear stability analysis by [16] showed that this feedback leads to inward propagating unstable disturbances, with associated variations in mass flux in the wind. This had been assumed before in a model by [1]. These authors found that this kind of feedback causes the disk to become dramatically time dependent in a manner suggestive of the FU Ori outbursts in protostellar disks.

Strong ordered magnetic fields in disks have their own forms of instability, independent of the coupling to a wind [43, 69], driven instead by the energy in the field itself. The nonlinear evolution of such instabilities was studied numerically by [74] (see also Sect. 9.4 above). Their effect appeared to be similar to an enhancement of the rate of diffusion of the magnetic field through the disk.

Since both these kinds of instability cause changes in the vertical component of the field, which is the same on both sides of the disk, they produce symmetric variations in mass flow in jet and counterjet. They are thus good candidates for the time dependence often observed in the form of symmetric knot patterns in protostellar jets.

References

1. V. Agapitou: Ph.D. Thesis, Queen Mary and Westfield College (2007)
2. J.M. Anderson, Z.-Y. Li, R. Krasnopolsky, R.D. Blandford: ApJ, **630**, 945 (2005)
3. I. Bains, A.M.A. Richards, T.M. Gledhill et al.: MNRAS, **354**, 529 (2004)
4. M. Barkov, S. Komissarov: arXiv:0801.4861v1 [astro-ph] (2008)
5. K. Beckwith, J.F. Hawley, J.H. Krolik: ApJ, **678**, 1180 (2008)
6. M.C. Begelman, Z.-Y. Li: ApJ, **426**, 269 (1994)
7. T. Belloni, J. Homan, P. Casella et al.: A&A, **440**, 207 (2005)
8. G.S. Bisnovatyi-Kogan, A.A. Ruzmaikin: Ap&SS, **28**, 45 (1974)
9. G.S. Bisnovatyi-Kogan, A.A. Ruzmaikin: Ap&SS, **42**, 401 (1976)
10. R.D. Blandford, R.L. Znajek: MNRAS, **179**, 433 (1977)
11. R.D. Blandford, D.G. Payne: MNRAS, **199**, 883 (1982)
12. S.V. Bogovalov: A&A, **349**, 1017 (1999)
13. S.V. Bogovalov, K. Tsinganos: MNRAS, **357**, 918 (2005)
14. C.G. Campbell: MNRAS, **345**, 123 (2003)
15. X. Cao, H.C. Spruit: A&A, **287**, 80 (1994)
16. X. Cao, H.C. Spruit: A&A, **385**, 289 (2002)
17. F. Casse, R. Keppens: ApJ, **601**, 90 (2004)
18. X. Chen, M.A. Abramowicz, J-.P. Lasota et al: ApJ, **443**, L61 (1995)
19. E. Churazov, M. Gilfanov, M. Revnivtsev: MNRAS, **321**, 759 (2001)
20. F. Daigne, G. Drenkhahn: A&A, **381**, 1066 (2002)
21. C. D'Angelo, D. Giannios, C. Dullemond, H.C. Spruit: A&A, 488, 441(2008)
22. J.-P. De Villiers, J.F. Hawley, J.H. Krolik et al.: ApJ, **620**, 878 (2005)
23. G. Drenkhahn: A&A, **387**, 714 (2002)
24. G. Drenkhahn, H.C. Spruit: A&A, **391**, 1141 (2002)
25. A.C. Fabian, P.W. Guilbert, C. Motch et al.: A&A, **111**, L9 (1982)
26. J. Ferreira, G. Pelletier: A&A, **295**, 807 (1995)
27. S. Fromang, J.C.B. Papaloizou, G. Lesur et al.: A&A, **476**, 1123 (2007)
28. D. Giannios, H.C. Spruit: A&A, **450**, 887 (2006)
29. P. Goldreich, W.H. Julian: ApJ, **160**, 971 (1970)
30. X. Guan, C.F. Gammie, J.B. Simon, B.M. Johnson: ApJ, **694**, 1010 (2009)
31. J.F. Hawley, J.H. Krolik: ApJ, **641**, 103 (2006)
32. M. Heinemann, S. Olbert: J. Geophys. Res., **83**, 2457 (erratum in J. Geophys. Res., **84**, 2142) (1978)
33. J. Homan, T. Belloni: Ap&SS, **300**, 107 (2005)
34. J. Homan VIth Microquasar Workshop: Microquasars and Beyond, Proceedings of Science (http://pos.sissa.it), PoS(MQW6)093 (2006)

35. B. Hutawarakorn, T.J. Cohen, G.C. Brebner: MNRAS, **330**, 349 (2002)
36. I.V. Igumenshchev, R. Narayan, M.A. Abramowicz: ApJ, **592**, 1042 (2003)
37. G. Kanbach, C. Straubmeier, H.C. Spruit et al.: Nature, **414**, 180 (2001)
38. C.F. Kennel, F.V. Coroniti: ApJ, **283**, 710 (1984)
39. S.S. Komissarov: arXiv:0804.1912 (2008)
40. J.H. Krolik, J. Hawley: VIth Microquasar Workshop: Microquasars and Beyond, Proceedings of Science (http://pos.sissa.it), PoS(MQW6)046 (2006)
41. R. Kulsrud: Plasma Physics for Astrophysics, Princeton University Press, Princeton, NJ (2006)
42. S.H. Lubow, J.C.B. Papaloizou, J.E. Pringle: MNRAS, **267**, 235 (1994)
43. S.H. Lubow, H.C. Spruit: ApJ, **445**, 337 (1995)
44. M. Lyutikov, R. Blandford: arXiv:astro-ph/0312347 (2003)
45. M. Machida, M.R. Hayashi, R. Matsumoto: ApJ, **532**, L67 (2000)
46. M. Machida, R. Matsumoto, S. Mineshige: ArXiv e-prints, arXiv:astro-ph/0009004v1 (2000)
47. J.C. McKinney, C.F. Gammie: ApJ, **611**, 977 (2004)
48. J.C. McKinney, R. Narayan: arXiv:astro-ph/0607575v1 (2006)
49. J.C. McKinney, R. Narayan: MNRAS, **375**, 513 (2007)
50. K.T. Mehta, M. Georganopoulos, E.S. Perlman, C.A. Padgett, G. Chartas: ApJ, **690**, 1706 (2009)
51. L. Mestel: Stellar magnetism (International series of monographs on physics) Clarendon, Oxford, (1999)
52. F.C. Michel: ApJ, **158**, 727 (1969)
53. K.A. Miller, J.M. Stone: ApJ, **534**, 398 (2000)
54. R. Moll, H.C. Spruit, M. Obergaulinger: A&A, **492**, 621 (2008)
55. R. Moll: A&A, in press (2009)
56. G.I. Ogilvie, M. Livio: ApJ, **553**, 158 (2001)
57. E.N. Parker: Cosmical Magnetic Fields, Clarendon Press, Oxford (1979)
58. E.S. Phinney: Ph.D. Thesis, University of Cambridge, Cambridge (1983)
59. M.J. Rees, M.C. Begelman, R.D. Blandford et al.: Nature, **295**, 17 (1982)
60. P.H. Roberts: Magnetohydrodynamics, Longmans, London, Chap. 4.4 (1967)
61. R.E. Rutledge et al.: ApJS, **124**, 265 (1999)
62. T. Sakurai: A&A, **152**, 121 (1985)
63. T. Sakurai: PASJ, **39**, 821 (1987)
64. J. Santiago-García, M. Tafalla, D. Johnstone, R. Bachiller: ArXiv e-prints, arXiv:0810.2790 (2008)
65. D. Shalybkov, G. Rüdiger: MNRAS, **315**, 762 (2000)
66. M. Sikora, M.C. Begelman, G.M. Madejski, J.-P. Lasota: ApJ, **625**, 72 (2005)
67. R. Spencer: In VIth Microquasar Workshop: Microquasars and Beyond, Proceedings of Science (http://pos.sissa.it), PoS(MQW6)053 (2006)
68. H.C. Spruit: NATO ASI Proc. C477: Evolutionary Processes in Binary Stars, p. 249 (arXiv:astro-ph/9602022v1) (1996)
69. H.C. Spruit, R. Stehle, J.C.B. Papaloizou: MNRAS, **275**, 1223 (1995)
70. H.C. Spruit, T. Foglizzo, R. Stehle: MNRAS, **288**, 333 (1997)
71. H.C. Spruit, G.D. Drenkhahn: Astronomical Society of the Pacific Conference Series, **312**, 357 (2004)
72. H.C. Spruit, D.A. Uzdensky: ApJ, **629**, 960 (2005)
73. H.C. Spruit: In VIth Microquasar Workshop: Microquasars and Beyond, Proceedings of Science (http://pos.sissa.it), PoS(MQW6)044 (2006)
74. R. Stehle, H.C. Spruit: MNRAS, **323**, 587 (2001)
75. A. Tchekhovskoy, J.C. McKinney, R. Narayan: arXiv:0901.4776 (2009)
76. D.A. Uzdensky, A.I. MacFadyen: ApJ, **647**, 1192 (2006)
77. A.A. van Ballegooijen: Astrophys. Space Sci. Library, Kluwer, **156**, 99 (1989)
78. M. van der Klis: ApJ, **561**, 943 (2001)
79. W.H.T. Vlemmings, H.J. van Langevelde, P.J. Diamond: A&A, **434**, 1029 (2005)
80. M. Wardle, A. Königl: ApJ, **410**, 218 (1993)

Chapter 10
General Relativistic MHD Jets

J.H. Krolik and J.F. Hawley

Abstract Magnetic fields connecting the immediate environs of rotating black holes to large distances appear to be the most promising mechanism for launching relativistic jets, an idea first developed by Blandford and Znajek in the mid-1970s. To enable an understanding of this process, we provide a brief introduction to dynamics and electromagnetism in the space–time near black holes. We then present a brief summary of the classical Blandford–Znajek mechanism and its conceptual foundations. Recently, it has become possible to study these effects in much greater detail using numerical simulation. After discussing which aspects of the problem can be handled well by numerical means and which aspects remain beyond the grasp of such methods, we summarize their results so far. Simulations have confirmed that processes akin to the classical Blandford–Znajek mechanism can launch powerful electromagnetically dominated jets and have shown how the jet luminosity can be related to black hole spin and concurrent accretion rate. However, they have also shown that the luminosity and variability of jets can depend strongly on magnetic field geometry. We close with a discussion of several important open questions.

10.1 The Black Hole Connection

Jets can be found in association with many astronomical objects: everything from Saturn's moon Enceladus to proto-stars to symbiotic stars to planetary nebulae to neutron stars and the subject of this chapter, black holes. What distinguishes the black hole jets, whether they are attached to stellar mass black holes or supermassive black holes, is that they are the only ones whose velocities are relativistic. That this should be so is not terribly surprising because, of course, the immediate vicinity of a black hole is the most thoroughly relativistic environment one could imagine.

J.H. Krolik (✉)

Department of Physics and Astronomy, Johns Hopkins University, Baltimore, MD 21218, USA,
jhk@pha.jhu.edu

J.F. Hawley

Department of Astronomy, University of Virginia, Charlottesville, VA 22904-4325, USA,
jh8h@virginia.edu

Krolik, J.H., Hawley, J.F.: *General Relativistic MHD Jets*. Lect. Notes Phys. **794**, 265–287 (2010)
DOI 10.1007/978-3-540-76937-8_10   © Springer-Verlag Berlin Heidelberg 2010

Although, as detailed elsewhere in this book, black holes can generate jets in a wide variety of circumstances, the fundamental physics of black holes can be treated in a unified manner. The reason for this simplicity is that the mass of the black hole changes the physical length scale for events in its neighborhood, but little else. Once distance is measured in gravitational radii $- r_g \equiv GM/c^2 \simeq 1.5(M/M_\odot)\,\mathrm{km}$ – the mass becomes very nearly irrelevant. Thus, jets from black holes of roughly stellar mass (whether in mass-transfer binaries or in the sources of γ-ray bursts) can be expected to behave in a way very similar to those ejected from quasars, whose black hole mass can be 10^8 times larger.

Another unifying theme to the physics of jets from black holes is the central role of magnetohydrodynamics (MHD). In almost any conceivable circumstance near black holes, the matter must be sufficiently ionized to make it an excellent conductor. When that is the case, the fluid and the magnetic field cannot move across each other. Because magnetic fields are able to link widely separated locations, they can provide the large-scale structural backbones for coherent structures like jets.

10.2 General Relativity Review

To properly understand the mechanics of matter and electromagnetic fields near black holes, it is, of course, essential to describe them in the language of relativity. This is, therefore, an appropriate point at which to insert a brief review of the portions of this subject most necessary to understand how black holes can drive jets.

10.2.1 Kinematics

One of the fundamental tenets of general relativity is that mass–energy induces intrinsic curvature in nearby space–time; what we call gravity is the result. Thus, in order to describe the motion of anything under gravity, it is necessary to find the space–time metric, the relationship that determines differential proper time ds, the time as it is perceived by a particle in its own rest frame, in terms of a differential separation in four-dimensional space–time dx^μ:

$$ds^2 = g_{\mu\nu}dx^\mu dx^\nu. \tag{10.1}$$

By standard convention, all Greek indices vary over the integers 0, 1, 2, 3, and the zero-th index is the one most closely associated with time, with the other three associated (more or less) with the usual three spatial dimensions. We will also adopt the (very common, but not universal) convention that the metric signature is $-+++$. Proper time is a physically well-defined quantity that is invariant to frame transformations; in contrast, motion in space–time can be parameterized in terms of all sorts of different coordinate systems that, on their own, do not necessarily have to possess any sort of physical significance.

Of the many possible ways to erect coordinate systems around black holes, two are most commonly used: Boyer–Lindquist coordinates and Kerr–Schild coordinates.

Both are built on conventional spherical coordinates for the spatial dimensions. The former has the advantage that its coordinate time is identical to the proper time of an observer at large distances from the point mass, but the disadvantage that some of the elements in the metric diverge at the black hole's event horizon. The latter has the advantage of no divergences, but the disadvantage that its time coordinate bears a more complicated relation to time as seen at infinity.

In Boyer–Lindquist coordinates, the metric around a rotating black hole is

$$ds^2 = -\left(1 - \frac{2Mr}{\Sigma}\right)dt^2 - \frac{4aMr\sin^2\theta}{\Sigma}dtd\phi + \frac{\Sigma}{\Delta}dr^2 \tag{10.2}$$
$$+ \Sigma d\theta^2 + \left(r^2 + a^2 + \frac{2Mra^2\sin^2\theta}{\Sigma}\right)\sin^2\theta d\phi^2,$$

where

$$\Sigma \equiv r^2 + a^2\cos^2\theta,$$
$$\Delta \equiv r^2 - 2Mr + a^2, \tag{10.3}$$

and the polar direction of the coordinates coincides with the direction of the angular momentum. The "spin parameter" a is defined such that the black hole's angular momentum $J = aM$. Here, and in all subsequent relativistic expressions, we take $G = c = 1$.

When $r/M \gg 1$, the Boyer–Lindquist coordinate system clearly reduces to that of spherical spatial coordinates in flat, empty space. On the other hand, when $r/M \sim O(1)$, there are odd-looking complications, most notably a non-zero coupling, proportional to a, between motion in the azimuthal direction ϕ and the passage of time. That there is such a coupling suggests that particles must rotate around the black hole (i.e., change ϕ over time) because of the properties of the space–time itself, rather than because they have intrinsic angular momentum. The truth of this hint is seen clearly after a transformation to a closely related system of coordinates. In this new system, r, θ, and t are the same as in Boyer–Lindquist coordinates, but the azimuthal position is described in terms of what would be seen by an observer following a circular orbit in the plane of the black hole's rotation with frequency $\omega_c = -g_{t\phi}/g_{\phi\phi}$. The relation between the new azimuthal coordinate and the old is $d\phi' = d\phi - \omega_c dt$, and the resulting metric is

$$ds^2 = -\frac{\Sigma\Delta}{A}dt^2 + \frac{A}{\Sigma}\sin^2\theta\,(d\phi')^2 + \frac{\Sigma}{\Delta}dr^2 + \Sigma d\theta^2, \tag{10.4}$$

where the new function A is

$$A = (r^2 + a^2)^2 - a^2\Delta\sin^2\theta. \tag{10.5}$$

This metric is diagonal, so relative to its coordinates there is no required rotation. For this reason, it is sometimes called the "LNRF" (for "Locally Non-Rotating

Frame") coordinate system. It is also sometimes called the "ZAMO" (for "Zero Angular Momentum") system because (as one can easily show) particles moving in the equatorial plane with zero angular momentum follow trajectories with $d\phi' = 0$. In other words, near a black hole, even zero angular momentum orbits must rotate relative to an azimuthal coordinate system fixed with respect to distant observers. Another consequence of the metric's diagonality is that it is easy to read off the lapse function, the gravitational redshift factor: it is $\alpha = \sqrt{\Sigma\Delta/A}$.

Further physical implications of this metric can be found by studying the behavior of the four-velocity $u^\mu \equiv dx^\mu/ds$ for any physical particle with non-zero rest mass. As a consequence of the Equivalence Principle, particles travel along geodesics, paths that maximize accumulated proper time. The metric can then be viewed as akin to a Lagrangian and framed in terms of canonical coordinates and their conjugate momenta. Because $dx^\mu = u^\mu ds$, the covariant components of the four-velocity effectively become the momenta conjugate to the coordinates. In particular, when the metric is time-steady and axisymmetric about an axis, u_t is the conserved orbital energy and u_ϕ is the conserved angular momentum for motion about that axis. This conserved orbital energy is often called the energy-at-infinity (E_∞) – after all, if the orbit extends to infinity, that is still the energy it would have there. Consider, for example, a particle in a circular orbit with coordinate orbital frequency Ω; that is, $\Omega = u^\phi/u^t$. With the minus sign demanded by our sign convention, its conserved energy is

$$E_\infty = -u_t = -\left(g_{tt}u^t + g_{t\phi}u^\phi\right) = -u^t\left(g_{tt} + g_{t\phi}\Omega\right). \qquad (10.6)$$

Ordinarily, $E_\infty > 0$ because $g_{tt} < 0$ and $g_{t\phi}$ is small. However, close to a rapidly rotating black hole, these sign relations can change: g_{tt} becomes negative when $r < 2M$ (in the equatorial plane) and $g_{t\phi}$ can be $\sim O(1)$. This region is called the "ergosphere." Inside the ergosphere, rotations that are relatively large and retrograde can lead to $-u_t < 0$. That is, the orbital energy, even including rest mass, can become negative!

The event horizon is found at smaller radius than the edge of the ergosphere when the black hole possesses any spin. It forms the constant-r surface $r/M = 1 + \sqrt{1 - (a/M)^2}$. On that surface, $\Delta = 0$, so g_{rr} (in Boyer–Lindquist coordinates) diverges. As we shall see momentarily, this is only a formal divergence and indicates nothing odd physically (or at least nothing stranger than an event horizon!).

Alternatively, motions around a rotating point mass can be described in terms of Kerr–Schild coordinates. These differ from Boyer–Lindquist by the coordinate transformation

$$
\begin{aligned}
dt_{KS} &= dt_{BL} + \frac{2Mr}{\Delta}dr, \\
dr_{KS} &= dr_{BL}, \\
d\phi_{KS} &= d\phi_{BL} + \frac{a}{\Delta}dr, \\
d\theta_{KS} &= d\theta_{BL}.
\end{aligned}
\qquad (10.7)
$$

The corresponding metric is

$$ds^2 = -(1 - \beta_r)dt^2 + 2\beta_r dt dr + (1 + \beta_r)dr^2 + \Sigma d\theta^2 - 2a\beta_r \sin^2\theta dt d\phi$$
$$+ \left(r^2 + a^2 + \frac{2Mra^2\sin^2\theta}{\Sigma}\right)\sin^2\theta d\phi^2 - 2a(1 + \beta_r)\sin^2\theta dr d\phi.$$

$$(10.8)$$

Here $\beta_r = 2Mr/\Sigma$. Note that there is no divergence in any of the metric elements at the event horizon or anywhere else (except the origin). Because physical results cannot depend on our choice of coordinates, this fact demonstrates that the Boyer–Lindquist divergence at the horizon is purely an artifact of that coordinate system. On the other hand, the same divergence occurs in the relation between these two coordinate systems. The closer to the event horizon one probes, the faster Kerr–Schild time advances relative to Boyer–Lindquist time. Similarly, close to the event horizon, the Kerr–Schild azimuthal angle advances rapidly relative to the Boyer–Lindquist azimuthal angle. Thus, the convenience the Kerr–Schild system provides in avoiding divergences is partially offset by a price paid in ease of physical interpretation.

The last comment worth making here is that often the most interesting frame in which to evaluate quantities is that of the local fluid motion itself. To do so most conveniently, one erects a local system of orthonormal unit vectors called a "tetrad." By standard convention, the tetrad element pointing in the local time direction is $\hat{e}^\mu_{(t)} \equiv -u^\mu$ (the minus sign is the result of our choice of metric signature). In the fluid frame, the only non-zero component of the four-velocity is the rate of advance of proper time ($=1$, of course), so this makes a natural definition of the local sense of time. There is considerable freedom in how the spatial tetrads are oriented, but it is always possible to construct a complete set via a Gram–Schmidt procedure. When such a tetrad is available, any tensor quantity may be evaluated as it would be measured in the fluid frame. All that is required is to project onto the appropriate tetrads: for example, a four-vector X^μ as seen in the fluid frame is $X^\nu \hat{e}^{(\mu)}_\nu$.

10.2.2 Electromagnetic Fields

One way to develop an appropriately covariant formulation of electromagnetism is to begin with a four-vector potential A_μ. Its elements have the usual interpretation: the time component is related to the electrostatic potential, while its spatial components determine the three-vector whose curl is the magnetic field. Thus, we can construct the Maxwell tensor

$$F_{\mu\nu} = \nabla_\mu A_\nu - \nabla_\nu A_\mu, \qquad (10.9)$$

so that what we might call the electric field $\mathcal{E}^i = F_{ti}$ and the magnetic field is $\mathcal{B}^i = [ijk]F_{jk}$. Here $[ijk]$ is the antisymmetric permutation operator and the symbol

∇_μ denotes a covariant derivative, i.e., a derivative that accounts for changes in the direction of local unit vectors as well as changes in vector components.

To see the electric and magnetic field components as elements of a tensor rather than as vector quantities may seem in conflict with the usual way to think about these fields. However, the Maxwell tensor can be easily related to a vector version of the fields through the construction

$$E^\mu = u_\nu {}^* F^{\mu\nu},$$
$$B^\mu = u_\nu {}^* F^{\mu\nu}, \tag{10.10}$$

where $^* F$ is the dual of F, i.e.,

$$^* F^{\mu\nu} = (1/2)[\alpha\beta\mu\nu] F_{\alpha\beta}. \tag{10.11}$$

In the rest frame of the fluid, only the time component of u_μ is non-zero; thus, the spatial parts of E^μ and B^μ may be interpreted as the electric and magnetic fields as they appear in the fluid rest frame. Moreover, because the field tensor is antisymmetric, the time component of both E^μ and B^μ is always zero in the fluid frame.

In this language, Maxwell's equations are simply

$$\nabla_\mu F^{\mu\nu} = 4\pi J^\nu,$$
$$\nabla_\mu {}^* F^{\mu\nu} = 0. \tag{10.12}$$

Here the four-current density is

$$J^\mu = qnu^\mu, \tag{10.13}$$

for particle charge q, particle proper number density n, and four-velocity u^μ.

10.2.3 Dynamics: The Stress–Energy Tensor

Mechanics can be thought of as the conservation laws at work in the world. To apply the laws of conservation of energy and momentum in an electromagnetically active relativistic setting, we first define the stress–energy tensor

$$T_\nu^\mu = \rho h u^\mu u_\nu + p g_\nu^\mu + \frac{1}{4\pi} \left[F^{\mu\alpha} F_{\nu\alpha} - \frac{1}{4} g_\nu^\mu F^{\alpha\beta} F_{\alpha\beta} \right], \tag{10.14}$$

where $h = 1 + (\epsilon + p)/\rho$ is the relativistic enthalpy, ρ is the proper rest-mass density, ϵ is the proper internal energy density, and p is the proper pressure. The first two terms in this expression are manifestly the relativistic generalizations of the flux of momentum. The second two terms give the electromagnetic contribution, which can also be written in a more familiar-appearing way:

$$F^{\mu\alpha} F_{\nu\alpha} - \frac{1}{4} g^\mu_\nu F^{\alpha\beta} F_{\alpha\beta} = \left(||E||^2 + ||B||^2 \right) u^\mu u_\nu - E^\mu E_\nu - B^\mu B_\nu \qquad (10.15)$$

$$+ \frac{1}{2} \left(||E||^2 + ||B||^2 \right) g^\mu_\nu.$$

In other words, the electric and magnetic energy densities contribute to the total energy density conveyed with the fluid (the term $\propto u^\mu u_\nu$); the pressure (the final term); and the momentum flux (the $E^\mu E_\nu$ and $B^\mu B_\nu$ terms).

When the characteristic rate at which the fields are seen to change in the fluid rest frame is slow compared to the electron plasma frequency $(4\pi_e n e^2/m_e)^{1/2}$, the electrons can quickly respond to imposed electric fields and cancel them. Because the most rapid conceivable rate at which anything can change is $\sim c/r_g$, this criterion translates to $n_e \gg 14(M/M_\odot)^{-2}$ cm^{-3}. If $E^\mu = 0$ in the fluid frame, it must be zero in all frames. In other words, the MHD condition, the limit in which charges can flow freely and quickly in response to imposed fields, places the constraint

$$E^\mu = u_\nu F^{\mu\nu} = 0. \qquad (10.16)$$

In the MHD limit, then, which ordinarily is very well justified in the vicinity of a black hole, energy–momentum conservation is expressed by

$$\nabla_\mu T^\mu_\nu = \nabla_\mu \left\{ \rho h u^\mu u_\nu + \left(p + \frac{||B||^2}{8\pi} \right) g^\mu_\nu + \frac{1}{4\pi} \left[||B||^2 u^\mu u_\nu - B^\mu B_\nu \right] \right\} = 0.$$
$$(10.17)$$

10.3 Candidate Energy Sources for Jets

Fundamentally, there are only two possible sources to tap for the energy necessary to drive a jet: the potential energy released by accreting matter and the rotational kinetic energy (more formally, the reducible energy) of a rotating black hole.

Consider the accretion energy first. The net rate at which energy is deposited in a ring of an accretion disk is the difference between the divergences of two energy fluxes: Inter-ring stresses do work, thereby transferring energy outward, while accreting matter brings its energy inward. However, there is an inherent mismatch between these two divergences, which in a steady-state disk is always positive. That is, the energy deposited by the divergence of the work done by stress always outweighs the diminution in energy due to the net inflow of binding energy brought by accretion. If disk dynamics alone are considered, there are only two possible ways to achieve balance: by radiation losses or by inward advection of the heat dissipated along with the accretion flow. However, in principle the heat could also be used to drive an outflow, e.g., a jet.

These qualitative statements are readily translated to the quantitative stress-tensor language just formulated. In the simple case of a time-steady and axisymmetric disk (viewed in the orbiting frame),

$$\int dz \left\{ \partial_r \left[\rho h u^r u_t + \frac{1}{4\pi} \left(||B||^2 u^r u_t - B^r B_t \right) \right] + \partial_z T_t^z \right\} = 0 \quad (10.18)$$

$$\partial_r \int dz \left[T_t^r(\text{matter}) + T_t^r(\text{EM}) \right] + 2F = 0.$$

Here T_t^z is nothing other than the vertical flux of energy, perhaps in radiation, hence the natural identification of T_t^z at the top and bottom surfaces as the outward flux F. Because that outward flux cannot be negative in an isolated accretion flow, the net energy deposited by work must always exceed the loss due to inflow.

How this dissipation takes place is, of course, unspecified by arguments based on conservation laws. Given the preeminence of magnetic forces in driving accretion, we can reasonably expect most of the heat to be generated by dissipating magnetic field. Reconnection events and other examples of anomalous resistivity such as ion transit-time damping are good candidate mechanisms, but definite little is known about this subject (see, e.g., [17] and references therein). Nonetheless, given the nature of all these mechanisms, which are characteristically triggered by sharp gradients, it is very likely that the heating is highly localized and intermittent. For this reason, it is quite plausible that the dissipation rate is sufficiently concentrated as to drive small amounts of matter to temperatures comparable to or greater than the local virial temperature. If this is the case, disk dissipation could substantially contribute to driving outflows.

Disk heating may also help expel outflows through a different mechanism: radiation forces. As mentioned before, the energy of net disk heating can be carried off by radiation. Because the acceleration due to radiation is $\kappa \mathcal{F}/c$, wherever the opacity per unit mass κ is high enough, radiation-driven outward acceleration may surpass the inward acceleration of gravity. Although most other elements of accretion dynamics onto black holes are insensitive to the central mass, the opacity, through its dependence on temperature, can depend strongly on it. In particular, the opacity of disk matter in the inner rings of AGN disks, where the temperature is only $\sim 10^5$ K, may be several orders of magnitude greater than in the inner rings of Galactic black hole disks, where the temperature is so high ($\sim 10^7$ K) that almost all elements are thoroughly ionized. It may consequently be rather easier for AGN disks to expel radiation-driven winds than for galactic black holes, even though the accretion rate is still well below Eddington.

Black hole rotation may also power outflows. The Second Law of Black Hole Thermodynamics decrees that the area of a black hole cannot decrease, but diminishing spin (at constant mass) *increases* the area. Consequently, a black hole of mass M and spin parameter a has a reducible mass, that is, an amount of mass–energy that can be yielded to the outside world, of $1 - [(1 + \sqrt{1 - (a/M)^2})/2]^{1/2}$.

Given the intuitive picture of a black hole as an object that only accepts mass and energy, one might reasonably ask how one can give up energy. One way to recognize that this may be possible is to observe that the rotational frame-dragging a spinning black hole imposes on its vicinity permits it to do work on external matter and in that way lose energy. Another way to see the same point is to recall that it is possible for particles inside the ergosphere to find themselves on negative energy orbits. If they cross the event horizon on such a trajectory, their negative contribution to the black hole's mass–energy results in a net decrease of its mass–energy, or a release of energy to the outside world, as originally pointed out by [23]. These negative energy orbits in general involve retrograde rotation, and so likewise bring negative angular momentum to the black hole. Collisions between a pair of positive energy particles that result in one of the two being put onto a negative energy orbit and then captured by the black hole are called the "Penrose process" and are the archetypal mechanism for deriving energy from a rotating black hole. Perhaps regrettably, the kinematic constraints for such collisions are so severe as to make them extremely rare [2].

It is also possible for electromagnetic fields to bring negative energy and angular momentum through the event horizon. Consider, for example, an accretion flow that is in the MHD limit, so that the magnetic field and the matter are tied together. The Poynting flux at the event horizon can then be outward even while the flow moves inward if the electromagnetic energy-at-infinity is negative. As shown by [16], the density of this quantity can be written (in Boyer–Lindquist coordinates) as

$$
\begin{aligned}
e_{\infty,EM}/\alpha &= -T_t^t(EM) \\
&= -(1/2)g^{tt}\left[g^{rr}(\mathcal{E}^r)^2 + g^{\theta\theta}(\mathcal{E}^\theta)^2 + g^{\phi\phi}(\mathcal{E}^\phi)^2\right] + (1/2)\left(g^{t\phi}\mathcal{E}^\phi\right)^2 \\
&\quad + \left[g^{\theta\theta}g^{\phi\phi}(\mathcal{B}^r)^2 + g^{rr}g^{\phi\phi}(\mathcal{B}^\theta)^2 + g^{rr}g^{\theta\theta}(\mathcal{B}^\phi)^2\right].
\end{aligned}
\tag{10.19}
$$

Here $T_\nu^\mu(EM)$ is the electromagnetic part of the stress–energy tensor and $\mathcal{E}^i$, $\mathcal{B}^i$ are the elements of the Maxwell tensor. If all the spatial metric elements were positive-definite, $e_{\infty,EM}$ would be likewise ($g^{tt} < 0$). However, inside the ergosphere $g^{\phi\phi} < 0$. Thus, wherever inside the ergosphere the poloidal components of the magnetic field and the toroidal component of the electric field are large, the electromagnetic energy-at-infinity can become negative.

There is a close analogy between negative electromagnetic energy-at-infinity and negative mechanical energy-at-infinity. In the case of particle orbits, the energy goes negative when the motion is in the ergosphere and is sufficiently rapidly retrograde with respect to the black hole spin. As [15] has shown, electromagnetic waves have negative energy-at-infinity when their normalized wave-vector as viewed in the ZAMO frame has an azimuthal component $< -\alpha/\omega_c$. This becomes possible only inside the ergosphere because it is only there that $\alpha < \omega_c$. In addition, this constraint demonstrates that negative electromagnetic energy-at-infinity is also associated with motion that is sufficiently rapidly retrograde.

10.4 The Blandford–Znajek Mechanism

As we have already seen, it is possible for black holes to give up energy to the outside world both by swallowing material particles of negative energy and by accepting negative energy electromagnetic fields. The latter mechanism can be much more effective. That this is so is largely due to the large-scale connections that magnetic fields can provide, as first noticed by [6].

In that extremely influential paper, Blandford and Znajek pointed out that a magnetic field line stretching from infinity to deep inside a rotating black hole's ergosphere can readily transport energy from the black hole outward. For its original formulation, the Blandford–Znajek mechanism was envisioned in an extremely simple way: a set of axisymmetric purely poloidal field lines stretch from infinity to the edge of the event horizon and back out to infinity and are in a stationary state of force-free equilibrium. There is just enough plasma to support the currents associated with this magnetic field and to cancel the electric field in every local fluid frame, but its inertia is entirely negligible. These assumptions were relaxed slightly in the work of [24], who showed that inserting just enough inertia of matter to distinguish the magnetosonic speed from c does not materially alter the result.

Given these assumptions, the Poynting flux on the black hole horizon can be found very simply in terms of the magnitude of the radial component of the magnetic field there and the rotation rate of the field lines. Because $E_\mu = 0$, the electromagnetic invariant $\boldsymbol{E} \cdot \boldsymbol{B} = E_\mu B^\mu = 0$. This, in turn, implies that $^*F^{\mu\nu}F_{\mu\nu} = 0$. When the fields are both axisymmetric and time-steady, the product of the Maxwell tensors reduces to a single identity

$$\left(\partial_\theta A_\phi\right)\left(\partial_r A_t\right) = \left(\partial_\theta A_t\right)\left(\partial_r A_\phi\right). \tag{10.20}$$

Also because of the assumed stationarity and axisymmetry,

$$\begin{aligned}
\partial_r A_t &= \mathcal{E}^r, \\
\partial_\theta A_\phi &= \mathcal{B}^r, \\
\partial_\theta A_t &= \mathcal{E}^\theta, \\
\partial_r A_\phi &= \mathcal{B}^\theta.
\end{aligned} \tag{10.21}$$

Consequently, the identity of Eq. (10.20) may be rewritten as

$$\frac{\mathcal{E}^\theta}{\mathcal{B}^r} = \frac{\mathcal{E}^r}{\mathcal{B}^\theta}. \tag{10.22}$$

Because rotation through a magnetic field creates an electric field, both ratios in the previous equation can be interpreted as a rotation rate Ω_F associated with the poloidal field lines.

With all these relations in hand, it is straightforward to evaluate the electromagnetic energy flux, i.e., the electromagnetic part of $-T_t^r$. As [21] pointed out, the algebra to do this is more concise in Kerr–Schild coordinates than in Boyer–Lindquist, and the result is

$$\mathcal{F} = -\Omega_F \sin^2 \theta \left[2(\mathcal{B}^r)^2 r \left(\Omega_F - \frac{a/M}{2r}\right) + \mathcal{B}^r \mathcal{B}^\phi \left(r^2 - 2r + (a/M)^2\right) \right]. \tag{10.23}$$

At the horizon, $r^2 - 2r + (a/M)^2 = 0$, and the rotation rate of the black hole itself $\Omega_H = (a/M)/(2r)$, so

$$\mathcal{F}_H = -2\Omega_F \sin^2 \theta (\mathcal{B}^r)^2 r \, (\Omega_F - \Omega_H). \qquad (10.24)$$

Note that outgoing flux depends critically on the field lines rotating more slowly than the black hole. Because space–time itself immediately outside the horizon rotates at Ω_H (as viewed by a distant observer), if there is some load that always keeps the field lines moving more slowly, there is a consistent stress through which the black hole does work on the field.

This formula neatly describes the Poynting flux in terms of the strength of the magnetic field and its rotation rate, but on its own it cannot tell us the luminosity of the system because both the strength of the field itself and the field line rotation rate are left entirely undetermined. In [6], $\mathcal{B}^r$ and $\mathcal{B}^\phi$ are found in terms of a field strength at infinity by assuming a specific field configuration (split monopole or paraboloidal) and performing an expansion in small a/M. Separately, Mac-Donald and Thorne [20] argued that the power generated was maximized when $\Omega_F = (1/2)\Omega_H$, where Ω_H is the rotation rate of the black hole itself. However, the Blandford–Znajek model per se has no ability to determine either the magnitude of the field intensity or the field line rotation rate.

Raising a specific version of the general question about how rotating black holes can give up energy, Punsly and Coroniti [25] questioned whether this mechanism can operate on the ground that no causal signal can travel outward from the event horizon. A summary answer to their question is that the actual conditions determining outward energy flow are determined well outside the event horizon and that accretion of negative energy is possible when the black hole rotates. In addition, only rotating black holes have reducible mass that can be lost. This summary can be elaborated from several points of view, all of which are equivalent.

One, which we have already mentioned, but was not explored in the original Blandford–Znajek paper, is that EM fields deep in the ergosphere can be driven to negative energy by radiating Alfven waves outward. Accretion of those negative EM energy regions then amounts to an outward Poynting flux on the event horizon. In the special case of stationary flow, a critical surface outside the event horizon can be found within which information travels only inward. The inward flux of negative electromagnetic energy may then be considered to be determined at this critical point [30, 5].

Another, which can be found in [32], is to note that the enforced rotation of field lines within the ergosphere creates an electric field. This electric field can in turn drive currents that carry usable energy off to infinity. Indeed, [15] argues that such a poloidal current is part and parcel of the plasma's electric field screening.

A third way to look at this same process is that frame-dragging forces field lines to rotate that would otherwise be purely radial. As a result, toroidal field components are created – and transverse field is the prerequisite for Poynting flux.

Although the Punsly–Coroniti question has now been put to rest, there are a number of other questions left unanswered by the classical form of the Blandford–Znajek model. Because it is a time-steady solution, by definition it does not consider how the field got to its equilibrium configuration. One might then ask, in the context of trying to understand why certain black holes support strong jets and others do not, whether *intrinsic* large-scale field is a prerequisite for jet formation, or whether field structures contained initially within the accretion flow can expand to provide this large-scale field framework. Another question is whether the force-free (or nearly force-free) assumption, while surely valid in much of the jet, yields a complete solution: for example, in the equatorial plane of the accretion flow, one would generally expect a breakdown in this condition; could that affect the global character of the jet generated? Still another question would be how we can extend this model quantitatively to more general field shapes and higher spin parameters. Lastly, if field lines threading the event horizon can carry Poynting flux to infinity, perhaps they can also carry Poynting flux to the much nearer accretion disk: is there an interaction between Blandford–Znajek-like behavior and accretion?

As we shall see, explicit MHD simulations make these approximations and limitations unnecessary and allow us to answer (or at least begin to answer) many of the questions left open by the original form of the Blandford–Znajek idea. These numerical calculations explicitly find the magnetic field at the horizon on the basis of the field brought to the black hole by the accretion flow, as well as the coefficients that replace Ω_F when, because of time variability and a breakdown of axisymmetry, it is no longer possible to give that quantity a clear definition. They can also determine the shape of the field, its connection to the disk, etc., and work just as well when a/M approaches unity as when it approaches zero.

10.5 What Simulations Can and Cannot Do

Before presenting the results of jet-launching simulations, it is worthwhile first to put them in proper context by explaining which questions they can answer well, and which not so well. Some of the considerations governing these distinctions are built into the very nature of numerical simulations, but others merely reflect the limitations of the current state of the art.

First and foremost, simulation codes are devices for solving algebraically complicated, nonlinear, coupled partial differential equations. After discretization, these equations can be solved by a variety of numerical algorithms designed so that, when applied properly, the solutions that are found converge to the correct continuous solution as the discretization is made finer and finer. Analytic methods are far weaker at solving problems of this kind, particularly those with strong nonlinearities. The ability to cope with nonlinearity makes numerical methods especially advantageous for studying problems involving strong turbulence (an essential ingredient of accretion disks [1]), as turbulence is fundamentally nonlinear. In addition, algorithms can be devised that maintain important constraints (e.g., $\nabla \cdot \boldsymbol{B} = 0$, energy and momentum conservation) to machine accuracy. Employing these built-in constraints, most jet/accretion codes are very good, for example, at conserving angular momentum and using the induction equation to follow the time dependence of the magnetic field.

A brief discussion of the two principal varieties of code that have been employed to date will serve to illustrate how their methods achieve these ends. One such family (exemplified by the Hawley–De Villiers general relativistic MHD code *GRMHD* [7]) derives from the *Zeus* code [29]. In these codes, hyperbolic partial differential equations that are first order in time but of arbitrary order in space are written as finite difference equations of the form

$$U_i^{k+1}(\mathbf{x}) = U_i^k(\mathbf{x}) + \left[S_i^k(\mathbf{x}) + T_i^k(\mathbf{x}) \right] \Delta t, \tag{10.25}$$

where the quantity U_i^k is one of the dependent variables (velocity, density, magnetic field, etc.) at the kth time step at spatial grid point $\mathbf{x}$ and Δt is the length of the time step. The various terms that may enter into defining the time derivative are divided into "source" terms S_i, terms that are local in some sense (e.g., the pressure gradient), and "transport" terms T_i, terms that describe advection (e.g., the terms describing the passive transport of momentum or energy with the flow). The source and transport terms are generally handled separately. When conserved quantities are carried in the transport terms, time-centered differencing can improve the fidelity with which they are conserved. Organizing the grid so that the dominant velocity component is along a grid axis also improves the quality of conservation for the momentum component in that direction. Although not strictly speaking a require- ment of this method, it is most often implemented with an energy equation that follows only the thermal energy of the gas, ignoring any interchange between that energy reservoir and the orbital and magnetic energy, except as required by shocks. Coherent motions automatically conserve orbital energy through the conservation of momentum, but this approximation ignores losses of kinetic and magnetic energy that occur as a result of gridscale numerical dissipation. The reason for this choice is that in many cases the thermal energy is so small compared to the orbital energy that it would be ill-defined numerically if a total energy equation were solved and the thermal energy only found later by subtracting the other, much larger, contributions to the total.

The other family organizes the equations differently. In this approach (e.g., the general relativistic MHD codes *HARM* [11] and *HARM3D* [22]), conservation laws are automatically obeyed precisely because the equations of motion are written in conservation form, i.e.,

$$\frac{\partial U_i}{\partial t} = -\nabla \cdot \mathbf{F_i} + S_i, \tag{10.26}$$

with U_i a density, $\mathbf{F_i}$ the corresponding flux, and S_i again the source term. Individ- ual Riemann problems are solved across each cell boundary in order to guarantee conservation of all the quantities that should be conserved. However, because the conserved densities and fluxes are often defined in terms of underlying "primitive variables" (e.g., momentum density is ρv), after the time advance one must solve a set of nonlinear algebraic equations to find the new primitive variables implied by the new conserved densities. Clearly, in this method, it is hoped that the advantages of conserving total energy outweigh the disadvantages of local thermal energies that may be subject to large numerical error.

In both styles, an initially divergence-free magnetic field can be maintained in that condition by an artful solution of the induction equation called the "constrained transport" or CT method [9, 33]. The essential idea behind this scheme is to rewrite the differenced form of the induction equation as an integral equation over each cell and then use Stokes' theorem to transform cell-face integrals of $\nabla \times (\mathbf{v} \times \mathbf{B})$ into cell-edge integrals of $\mathbf{v} \times \mathbf{B}$. Because the latter can be done exactly, the induction equation itself can be solved exactly and the divergence-free condition preserved.

Unfortunately, however, this list of the strengths of numerical methods does not encompass all possible problems that arise in black hole jet studies. Contemporary algorithms are particularly weak in those aspects involving thermodynamics. Two large gaps in our knowledge, one about algorithms, the other about physics, make this a very difficult subject: First, as in most areas of astrophysics, temperature regulation is the result of photon emission, but there are as yet no methods for solving 3-D time-dependent radiation transfer problems quickly enough that they would not drastically slow down a dynamical simulation code. Second, although the dense and optically thick conditions inside most accretion disks make it very plausible that all particle distribution functions are very close to thermodynamic equilibrium, outside disks, whether in their coronae or, even more so, in their associated jets, this assumption is, to put it mildly, highly questionable. We would need a far better understanding of plasma microphysics to improve upon this situation. The combination of these two gaps makes it very hard to determine reliably the local pressure (whether due to thermalized atoms, non-thermal particles, or radiation) and associated hydrodynamic forces. As a result, the most reliable results of these calculations have to do with dynamics for which gravity and magnetic forces dominate pressure gradients.

Limitations in computing power create the next set of stumbling blocks. The nonlinearities of turbulence may be thought of as transferring energy from motions on one length scale to motions on another. Generally speaking, turbulence is stirred on comparatively long scales, these nonlinear interactions move the energy to much finer scale motions, and a variety of kinetic mechanisms, generically increasing in power as the scale of variation diminishes, dissipate the energy into heat. Unfortunately, the computer time required in order to run a given 3-D simulation with spatial resolution better by a factor of R scales as R^4 (three powers from the increased number of spatial cells, one power from the tighter numerical stability limit on the size of the timestep). Consequently, they are generally severely limited in the dynamic range they may describe between the long stirring length scale and the shortest scale describable, the gridscale. The gridscale is therefore almost always many many orders of magnitude larger than the physical dissipation scale.

If MHD turbulence in accretion behaves like hydrodynamic Kolmogorov turbulence in the sense that it develops an "inertial range" in which energy flux from large scales to small is conserved, the fact that the gridscale is much larger than the true dissipation scale would not matter: All the energy injected on the large scales is ultimately dissipated by dissipation operating on fluctuations of some length scale, and we do not much care whether that happens on scales smaller by a factor of 10^{-2} or 10^{-10}. However, it is possible that MHD is different because it is subject to (at least) two dissipation mechanisms, resistivity and viscosity. If the ratio between these two rates (the Prandtl number) is far from unity, the nature of the turbulence could be qualitatively altered, potentially in a way that influences even behavior on the largest scales [10]. These questions are particularly troubling in regard to jet launching because (as we shall see later) the efficiency with which magnetic fields generated by MHD turbulence in accretion disks can be used to power jets may depend on the field topology, and magnetic reconnection, which depends strongly on poorly understood or modeled dissipation mechanisms, alters topology.

Another problem whose origin lies in finite computational power is the difficulty in using one simulation to predict behavior of the system under different parameters. In contrast to analytic solutions, a single numerical simulation only rarely points clearly to how the result would change if its parameters were altered. However, the scale of the effort required to run simulations makes it nearly impossible to do large-scale parameter studies: Typical computer allocations allow any one person to do at most 5–10 simulations per year; each one may take a month or so to run to completion; and analysis of a single simulation often takes several months of human time. Thus, scaling the results to other circumstances is in general very challenging.

Lastly, there is a fundamental limitation to numerical methods: the solution they find depends on the initial and boundary conditions chosen as well as on the equations and their dimensionless parameters. Our goal is generally to find what Nature does, but the initial and boundary conditions for a simulation are usually chosen on the basis of human convenience. This means, for example, that there may be equilibrium solutions that are never encountered in a simulation because the initial condition was, in some sense, "too far away"; put in other language, the radius of convergence for the iterative solution represented by the time advance of the simulation may not be large enough for the simulation to find the equilibrium. Similarly, when there are several stable equilibria available, any one simulation can settle into at most one of them. On other occasions, boundary conditions can subtly prevent a simulation from evolving into a configuration that Nature actually permits. In the context of jet maintenance, the most important of these imponderable issues (or at least, the most important ones of which we are currently aware) have to do with the magnetic field structure. We know neither the topology of the field supplied in the accretion flow nor to what degree it has a fixed large-scale structure imposed by conditions at very large distances from the black hole. It is also very hard to imagine a way in which we might learn more about either of these two questions. Thus, the best we can do is to explore the consequences of a variety of choices; if we are fortunate, we may find that some of the options do not make much difference to the astrophysical questions of greatest interest.

10.6 Results

As we have seen, what determines the strength and structure of the magnetic field near the black hole is the central question for studies of relativistic jet launching. Any attempt to answer this question must therefore be carefully structured so as to avoid embedding the answer in the assumptions. Because it is not always easy to predict which assumptions are truly innocuous, here we will report what has been found to date and then discuss potential dependences upon parameters, boundary conditions, and the set of physical processes considered.

10.6.1 The Simplest Case

The configuration that has been studied most is arguably the simplest: A finite amount of mass in an axisymmetric hydrostatic equilibrium is placed in orbit a few tens of gravitational radii from the black hole, its equatorial plane identical to the equatorial plane of the rotating space–time. The initial magnetic field is entirely contained within the matter, so that there is no net magnetic flux and no magnetic field on either the outer boundary or the event horizon. Because the gas density declines monotonically away from a central peak, one can identify the initial magnetic field lines with the density contours, so that they form large concentric dipolar loops [21, 8, 13].

Starting from this configuration, the field line segments on the inner side of the loops are rapidly pushed toward the black hole, arriving there well before much accretion of matter has taken place. Unburdened by any significant inertia, they expand rapidly into the near-vacuum above the plunging region, where a centrifugal barrier prevents any matter with even a small amount of angular momentum from ever entering. Because the vertical component of the magnetic field on the inner field lines has a consistent sign in both hemispheres, within a short time individual field lines run from far up along the rotation axis in the upper hemisphere to equally far down along the axis in the lower hemisphere, passing close outside the horizon when they cross the equatorial plane. Just as predicted by the Blandford–Znajek picture, a rotating black hole forces an otherwise radial field to develop a transverse component, as shown in Fig. 10.1. Note how the winding of the field lines is tightest close to the event horizon, where frame-dragging is strongest.

We stressed earlier that the classical Blandford–Znajek model predicts the luminosity in terms of the radial component of the magnetic field on the horizon and the rotation rate of the field lines, but gives no guidance about how to estimate the former and only a guess about the latter. In the Newtonian limit far from the innermost stable circular orbit, the field strength is tightly coupled to the accretion rate because angular momentum conservation requires

$$ -\int dz \, \langle B_r B_\phi \rangle \simeq \frac{\dot{M}\Omega}{2\pi}. \tag{10.27} $$

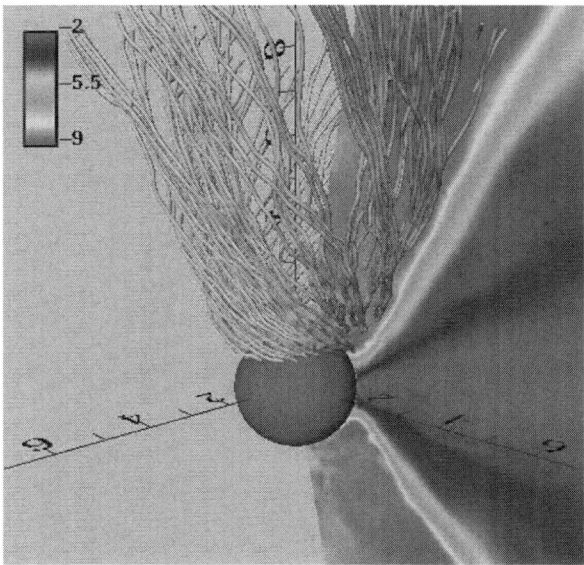

Fig. 10.1 Field lines near a black hole rotating with $a/M = 0.9$ [14]. The background colors illustrate matter density on a logarithmic scale, calibrated by a color bar found in the *upper left-hand* corner. The axes show Boyer–Lindquist radial coordinate

Close to the black hole, this relation remains correct as an order of magnitude estimator. Consequently, when the field is directly associated with accretion, its intensity near the black hole should be proportional to $\dot{M}$. Using simulation data, [13] confirmed this expected proportionality, but found that spin matters, too:

$$\left\langle \frac{B^2}{8\pi} \right\rangle \simeq \frac{0.01}{1 - a/M} \frac{\dot{M}c^5}{(GM)^2},$$ (10.28)

where the magnetic energy is measured in the fluid frame at the event horizon.

Despite the fact that the accretion flow sets the scale of magnetic intensity, the inertia of matter has little to do with field dynamics near the rotation axis. As already mentioned, any matter with even the angular momentum of the last stable orbit is excluded from a cone surrounding the axis. Within that cone, the field is force-free in the sense that $||B||^2/(4\pi\rho c^2) \gg 1$. However, it should also be recognized that simulations like those done to date are not able to define quantitatively just how large this ratio is. Precisely because the matter's angular momentum makes the interior of the cone forbidden territory, any matter in the jet cone got there by some numerical artifact, either through exercise of a code density floor or through numerical error associated with insufficient resolution of the extremely sharp density gradient at the centrifugal barrier. The enthalpy of the gas is equally poorly known, but that is because the treatment of thermodynamics in global simulations thus far is so primitive. Thus, the most one can say at this point is that, in a purely qualitative sense, the interior of the jet cone should be magnetically dominated. For exactly these reasons, the Lorentz factor of the outflow is equally ill determined.

Although even defining a field line rotation rate for time-dependent non-axisymmetric fields is a bit dicey, it is possible to do so in an approximate way by monitoring the azimuthal velocity of the matter attached to the field lines and subtracting off the portion attributable to sliding along the field lines. For example, if one defines the "transport velocity" by $V^i \equiv u^i/u^t$, in Boyer–Lindquist coordinates the local field line rotation rate ω can be written as

$$\omega = V^\phi - B^\phi \frac{V^r B^r g_{rr} + V^\theta B^\theta g_{\theta\theta}}{(B^r)^2 g_{rr} + (B^\theta)^2 g_{\theta\theta}}. \tag{10.29}$$

As Fig. 10.2 shows, after averaging radially, the rotation rate is close to the MacDonald–Thorne guess, perhaps 10–20% less, with surprisingly little variation in polar angle through the jet [21, 13].

Strikingly, even when the accreting matter orbits *opposite* the sense of rotation of the black hole, the field lines in the jet rotate with the black hole, not the matter [13]. This fact makes it clear that the primary motive power for the jet is drawn from the black hole, *not* from matter circulating deep inside the ergosphere to which the field lines are attached.

Normalized to the rest-mass accretion rate in traditional fashion, the jet luminosity can be sizable when the black hole spins rapidly. Both [21] and [13] made rough analytic fits to the dependence on spin:

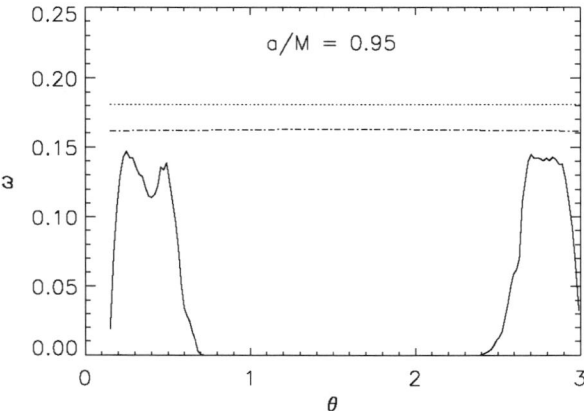

Fig. 10.2 Radially and time-averaged local field line rotation rate (*solid curve*) as a function of polar angle for a simulation in which the black hole has spin parameter $a/M = 0.95$ [13]. The *dotted curve* is half the black hole's rotation rate, the *dash-dot curve* is half the rotation rate of the inner boundary of the simulation, which is slightly outside the event horizon

$$\eta_{EM} \simeq \begin{cases} 0.068 \left\{ 1 + [1 - (a/M)^2] - 2\sqrt{1 - (a/M)^2} \right\} & [21] \\ 0.002/(1 - |a/M|) & [13] \end{cases} \qquad (10.30)$$

Although these expressions do not agree precisely, they agree at a qualitative level: the efficiency can be $\sim$0.1 at the highest spins, but is much smaller for $|a/M| < 0.9$. That the efficiency can be so high when the black hole spins rapidly is of some interest, given qualitative arguments (e.g., [18]) that the luminosities of jets driven directly by a black hole could never be significant.

Because the field lines in the jet rotate, it carries angular momentum as well as energy away from the black hole. The electromagnetic angular momentum delivered to infinity by the jet can be comparable to the conventional angular momentum brought to the black hole by accretion [12, 13, 4], particularly when the black hole spins rapidly. In fact, [12] argued that the electromagnetic angular momentum lost in the jet rises so steeply with black hole spin that it may limit a/M to $\simeq$0.93, a level considerably below the limiting spin proposed by [31], who suggested that preferential capture of photons emitted by the disk on retrograde orbits would cap a/M at $\simeq$0.998.

10.6.2 A Slightly More Complicated Case: The Effect of Field Geometry

Beckwith et al. [4] explored other options for the magnetic field's initial geometry beyond that of the simplest model, including a pair of quadrupolar loops, one above and one below the plane, with their field directions coinciding on adjacent edges; a purely toroidal field; and a sequence of four dipolar loops, each rather narrow in radial extent. All of these were far less effective in terms of time-averaged jet luminosity than the large dipolar loop initial state: the quadrupolar case by two orders of magnitude, the toroidal by three!

The explanation for these strong contrasts changes with the geometric symmetry in question. Quadrupolar loops residing in a single hemisphere can rise buoyantly as a single unit and collapse, reconnecting with themselves. Jets arising from that kind of field therefore tend to be highly episodic, with only brief moments of high luminosity. Toroidal fields cannot produce poloidal fields on scales larger than roughly the disk thickness, so their jets are always weak. A train of dipolar loops leads to a succession of jet-launching and jet-destruction events, separated in time by the difference in inflow times between their inner and outer edges.

10.6.3 Open Questions

Before directly applying the results of this simplest model to black holes in Nature, we must first answer several questions about its generality.

10.6.3.1 Do Zero Net-Flux Simulations Describe a Steady-State Jet?

Although these jets are long lasting, they behave in certain ways as if the flux had been placed on the black hole as an initial condition (it arrives well before the accretion flow and then stays there, with little change). Moreover, although it may take a long time, eventually the far end of the large dipole loops must reach the black hole, and at that point reconnection will eliminate the field driving the jet. To gain a sense of the timescales involved, in the $\sim 10^4 GM/c^3$ duration of these simulations, $\sim 10\%$ of the disk mass was accreted. One might guess, then, that the ultimate field annihilation would occur after $\sim 10^5 GM/c^3$. For galactic black hole binaries, this is only $\sim 5\,\mathrm{s}$; for AGN, it might be $\sim 5 \times 10^5$–$5 \times 10^8\,\mathrm{s}$, or no more than a few decades.

On the other hand, we do not know the true extent of such loops. It is at least conceivable that they might be much larger in radial extent, and the rapid increase of inflow time with radius might lead to much longer intervals between jet-field destruction events. If so, though temporary, these jets (especially in AGN) might be sufficiently long-lived as to be observationally interesting. Alternatively, one of the salient empirical facts about the jets we observe is that they are generally very unsteady. If another field loop follows close on the heels of the one that just closed, the jet might be restored equally quickly. In this sense, the picture just described could give a better sense of the typical state of the jet.

10.6.3.2 What Is the Generic Zero Net-Flux Magnetic Configuration?

Different zero net-flux magnetic geometries can lead to jets of drastically different character – how can we know which geometry (or mix of geometries) is present in any particular real object? It might seem "natural" to suppose that Nature is messy and serves up all possibilities at once, but when the different geometries differ in their results by orders of magnitude, the specific proportions matter a great deal. On the other hand, perhaps when the source of the accretion flow is a relatively ordered structure (e.g., a companion star in a binary system), the magnetic field delivered to the black hole, even while possessing no net flux, might be predominantly of a single topological character. One might speculate (as did [4]) that changes in the predominant topology might be related to observable changes in jet strength, for example, as seen in galactic black hole binaries.

10.6.3.3 What Would Be the Effect of Net Magnetic Flux?

Major open questions also remain in regard to a different sort of ill-understood magnetic geometry: the possible presence of large-scale magnetic field threading the accretion disk and possibly the immediate environs of the black hole. This is a very controversial issue: advocates exist for almost the entire range of possible answers, from very large to nil. On the one hand, if the accreting gas is truly infinitely conductive, it should hold onto any large-scale field threading it and bring that flux to the event horizon of the black hole. Even if successive parcels carry oppositely directed flux, the magnitude of the flux will have a non-zero expectation value $\sim\sqrt{N}\Phi$, where N is the number of accumulated flux ropes and Φ is their typical individual magnitude [32]. On the other hand, when gas is accreted by a black hole, the characteristic length scale of variation for the magnetic field contained within it must shrink by many orders of magnitude. Severe bends must then be created in any field lines stretching to large distance, and extremely thorough reconnection may therefore be expected. In principle, *all* the field within the inner part of the accretion flow may lose contact with large-scale fields by this mechanism. Presumably, the correct story lies somewhere between these two extremes.

A sense of the range of views brought to bear on this problem may be gained by mentioning a few of the contending approaches. Reference [19] attempted to define the problem in terms of the ratio between the accretion flow's magnetic diffusivity and effective viscosity (in modern language, $\langle -B_r B_\phi \rangle / (6\pi\rho\Omega)$). They argued that this ratio, an effective magnetic Prandtl number, determines whether fields are locked to the accretion flow (and must therefore bend sharply as they leave the disk) or can slip backward relative to the flow (and therefore stay pretty much where they started, without being forced to bend substantially). Unfortunately, the distance between this sort of "lumped parameter" approach and the actual microphysics is great enough that this approach hardly suffices to decide the issue. More recently, some (e.g., [28]) have argued that clumps of flux could self-induce inflow by losing angular momentum through a magnetic wind, while others (e.g., [26]) have suggested that little net magnetic flux inflow would occur because turbulent resistivity inside the disk effectively disconnects large-scale field lines from the accretion flow in the interior of the disk. As this sampling of the literature indicates, the matter remains highly unsettled.

It is, however, potentially an important question because large-scale flux running through the disk could have important effects both on its accretion dynamics and on its ability to support a jet. Simulational studies of MHD turbulence in shearing boxes suggest that net vertical flux can strongly enhance the saturation level of the turbulence [27]. How much of this effect is automatically embedded in global simulations by field loops that connect different radii remains unclear. In regard to jets, a consistent sense of vertical field would certainly serve to stabilize a base luminosity against the disruption that quadrupolar loops, etc. can create.

This issue could have significant observable consequences because the specific character of the large-scale field may depend strongly on black hole environment. One could well imagine that the field brought to the black hole by accretion from a stellar companion has more organized large-scale structure than the field threading turbulent interstellar gas accreting onto a black hole in a galactic nucleus. It is possible that long-term magnetic cycles in the companion star of a binary black hole system are connected to long-term changes in the state of the accretion flow and jet. There could even be cases intermediate between field entirely contained within the accretion flow and field with large-scale constraints: this might be the situation in a collapsing star in which extremely rapid accretion onto a nascent black hole drives a jet that creates a γ-ray burst [3].

10.6.3.4 Does the Condition of the Matter Matter?

As remarked earlier, neither the density nor the enthalpy of the matter in the jet can be determined quantitatively by simulations done to date. Although part of the difficulty is numerical, there are also serious unsolved physics (and astrophysical contextual) problems standing in the way. If there were, for example, a supply of matter with substantially smaller angular momentum than the matter in the accretion flow proper, it would see only a low centrifugal barrier and could enter the jet cone. Its thermal state would surely depend on whether there are numerous low-energy photons passing through the jet for the gas's electrons to upscatter (as in the case of AGN) or very few (as in the case of a microquasar in a low-hard state). In the sort of collapsing star that might be the central engine for a γ-ray burst, the thermal state of gas in the vicinity of the jet would depend on its nuclear composition and the neutrino intensity. Although the launching of relativistic jets by black holes may be only weakly dependent on the state of the matter it carries so long as $B^2/(4\pi\rho h) \gg 1$, the subsequent dynamics of the jet – its ultimate Lorentz factor, for example – are likely sensitive to these considerations.

10.7 Conclusions

Because work in this field is moving forward rapidly, any conclusions pronounced at this stage must be limited and preliminary. Nonetheless, results to date are certainly strong enough to give us some confidence that jet-launching mechanisms within the Blandford–Znajek conceptual family play a major role in this process. Magnetic fields that link distant regions with regions deep in the ergosphere have now been shown by explicit example to have the power, at least in principle, to tap the rotational energy of spinning black holes.

The progress that has been achieved so far rests on computational solutions of equations that, at least within the terms of the MHD approximation, express essentially ab initio physics. It is the direct connection to bedrock physics (momentum–energy conservation, Maxwell's equations, etc.) that gives us confidence that their results are meaningful and robust. Greater contact with observations will become possible when the physics contained in these calculations is expanded to include better descriptions both of how matter enters the jet and of how it couples to radiation. Although it will be difficult to build foundations for these parts of the problem as securely based as those of the dynamics, it may yet be possible to do so at a level permitting some reasonable testing by comparison with real data.

References

1. S.A. Balbus, J.F. Hawley: RMP, **70**, 1 (1998)
2. J.M. Bardeen, W.H Press, S.A. Teukolsky: ApJ, **178**, 347 (1972)
3. M.V. Barkov, S.S. Komissarov: MNRAS, **385**, L28 (2008)
4. K.R.C. Beckwith, J.F. Hawley, J.H. Krolik: ApJ, **678**, 1180 (2008)

5. V.S. Beskin, I.V. Kuznetsova: Il Nuovo Cimento B, **115**, 795 (2000)
6. R.D. Blandford, R.L. Znajek: MNRAS, **179**, 433 (1977)
7. J.-P. De Villiers, J.F. Hawley: ApJ, **589**, 458 (2003)
8. J.-P. De Villiers, J.F. Hawley, J.H. Krolik, S. Hirose: ApJ, **620**, 878 (2005)
9. C.R. Evans, J.F. Hawley: ApJ, **332**, 659 (1988)
10. S. Fromang, J. Papaloizou, G. Lesur, T. Heinemann: A&A, **476**, 1123 (2007)
11. C.F. Gammie, J.C. McKinney, G. Tóth: ApJ, **589**, 444 (2003)
12. C.F. Gammie, S.L. Shapiro, J.C. McKinney: ApJ, **602**, 312 (2004)
13. J.F. Hawley, J.H. Krolik: ApJ, **641**, 103 (2006)
14. S. Hirose, J.H. Krolik, J.F. Hawley: ApJ, **606**, 1083 (2004)
15. Komissarov, S.: arXiv:0804.1912 (2008)
16. J.H. Krolik, J.F. Hawley, S. Hirose: ApJ, **622**, 1008 (2005)
17. M.M. Kuznetsova, M. Hesse, L. Rastätter, A. Taktakishvili, G. Tóth, D.L. De Zeeuw, A. Ridley, T.I. Gombosi: JGRA, **112**, 10210 (2007)
18. M. Livio, G.I. Ogilvie, J.E. Pringle: ApJ, **512**, 100 (1999)
19. S.H. Lubow, J.C.B. Papaloizou, J.E. Pringle: MNRAS, **267**, 235 (1994)
20. D. MacDonald, K.S Thorne: MNRAS, **198**, 345 (1982)
21. J.C. McKinney, C.F. Gammie: ApJ, **611**, 977 (2004)
22. S.C. Noble, J.H. Krolik, J.F. Hawley: ApJ, **692**, 411 (2009)
23. R. Penrose: Riv. Nuovo Cimento, **1**, 252 (1969)
24. E.S. Phinney: A Theory of Radio Sources, PhD thesis, Cambridge University, Cambridge (1983)
25. B. Punsly, F.V. Coroniti: ApJ, **350**, 518 (1990)
26. D.M. Rothstein, R.V.E. Lovelace: ApJ, **677**, 1221 (2008)
27. T. Sano, S. Inutsuka, N.J. Turner, J.M. Stone: ApJ, **605**, 321 (2004)
28. H.C. Spruit, D.A. Uzdensky: ApJ, **629**, 960 (2005)
29. J.M. Stone, M.L. Norman: ApJ, Suppl., **80**, 753 (1992)
30. M. Takahashi, S. Nitta, Y. Tatematsu, A. Tomimatsu: ApJ, **363**, 206 (1990)
31. K.S. Thorne: ApJ, **191**, 507 (1974)
32. K.S. Thorne, R.H. Price, D.A. MacDonald: Black Holes, The Membrane Paradigm, Yale University Press, New Haven (1986)
33. G. Tóth: J. Comp. Phys. **161**, 605 (2000)

Index

Belloni, T.: *Index*. Lect. Notes Phys. **794**, 289–291 (2010)
DOI 10.1007/978-3-540-76937-8   © Springer-Verlag Berlin Heidelberg 2010